A
History of
Microtechnique

A
History of
Microtechnique

Brian Bracegirdle

CORNELL
UNIVERSITY PRESS
Ithaca, New York

First published 1978 by Cornell University Press.

International Standard Book Number 0-8014-1117-3
Library of Congress Catalog Card Number 77-78658

Printed in Great Britain by Morrison & Gibb Ltd,
Edinburgh and London

Preface

This book sets down the main facts of the evolution of the microtome and of the development of histological methods. It deals only in passing with the history of slide-making for botany and mineralogy. It is not concerned merely with the first introduction of particular stains, for such dates may or may not be meaningful; such details are usually noted, but so also are the first effective introduction of such stains, and of mixtures, reagents, and apparatus which were to become of established importance.

The basis of the work was a comprehensive survey of the large literature, and full documentation has been provided in over a thousand references. In addition, more than 40 000 microscopical preparations have been inspected and evaluated as a check on written accounts, as have some 55 microtomes: some of these were used to cut sections as they would have been used when first introduced.

The work is based on that accepted for a doctorate of the University of London, which was itself based on a long-standing interest in the subject. It could not have been carried through without the active interest and support of a number of people. Savile Bradbury first drew my attention to the books, instruments, and preparations in store in Oxford: to these collections of the Royal Microscopical Society Gerard Turner allowed full access, and gave me frequent benefit of his authoritative knowledge of the history of scientific instruments in general and of the microscope in particular. It is a pleasure also to acknowledge how much I have learned from Edwin Clarke, who supervised my doctoral work: his total encouragement and frequent masterly advice have been invaluable.

The Wellcome Trust gave generous support to enable me to visit important collections in Europe, and a number of private individuals have allowed access to their collections.

All the photographs illustrating this intensely visual topic have been made by myself specially for the work; in particular, the text figures have been copied direct from the originals to maintain their quality.

Facilities for recording instruments and preparations were freely given in this country by the Museum of the History of Science, Oxford; the Quekett Microscopical Club; the Hunterian Museum of the Royal College of Surgeons of England; the Science Museum; the Wellcome Institute for the History of Medicine; and the Whipple Science Museum, Cambridge. Overseas the Goethehaus, Weimar; Rijksmuseum voor de Geschiedenis der Natuurwetenschappen Leiden; Teyler's Museum, Haarlem; Universiteitsmuseum, Utrecht; VEB Carl Zeiss Jena-Archiv; and the Optiches Museum der Carl-Zeiss-Stiftung, Jena, similarly provided full access. The libraries of the Wellcome Institute for the History of Medicine; University College, London; the British Museum (Natural History); the Royal Society of Medicine; the University of London; the Museum of the History of Science, Oxford; the Quekett Microscopical Club; and the Science Museum, have been unfailing in their kindness and help in the search for many of the books and periodicals which are nowadays obscure but which played important parts in the history of preparative techniques.

Finally, it is a pleasure to acknowledge the encouragement of my publishers in undertaking the work, and to offer them my congratulations in making so attractive an end product.

London, March 1978   B.B.

Contents

Preface   v

List of Text Figures   ix

List of Plates   xiii

1 Introduction   1

The approaches used in the work
Review articles and other basic literature sources
The main periodical literature
References

2 A Brief Survey of Microtechnique before 1830   8

The seventeenth century
Microscopic injections
Other results from the eighteenth century
Hill and the microscopic anatomy of timber
Custance's microtome for timber sections
Books and preparations for the amateur
Ypelaar and his preparations
Hewson's preparations
The first thirty years of the nineteenth century
Test objects
The development of the slide from the slider
References

3 Works on Microtechnique 1830–1910: a select descriptive bibliography   31

Works published in England
Works in German
French books
American books
Other works
References

4 A Survey of Substances used in Microtechnique 1830–1910   57

Introductory remarks
Important survey literature
Killing and preserving methods
Hardening and fixing agents
Dehydration techniques and reagents
Staining techniques : general remarks
Natural staining substances
Aniline colours and their use as stains
Metallic colouring
Colouring living tissues

Embedding, infiltration and sectioning : introductory remarks
Infiltration with paraffin
The use of celloidin
Clearing agents
Injection techniques
Microdissection and maceration
Mounting media : non-aqueous
Cell mounts
Histochemical tests
References

5   A Survey of Instruments used in Microtechnique 1830–1910   111
Slides and covers
Small apparatus used in microtechnique 1830–1910
Freehand blades for sectioning
The development of the microtome from 1830 to 1870
The freezing method
The hand microtome
Microtomes 1870–1910 (a) : movable knife with vertical feed
Microtomes 1870–1910 (b) : movable knife with inclined feed
Microtomes 1870–1910 (c) : with fixed vertical knife
Microtomes 1870–1910 (d) : with fixed horizontal knife
Ancillary embedding equipment
References

6   Notes on Commercial Mounters 1800–1910   289
Introductory notes and the earliest mounters
Mounters at work in and from the middle years of the century
Specimens for amateur mounting
Mounters working from the beginning of the twentieth century
References

7   Microscopy, Microtomy and Histology in the Nineteenth
Century   308
Secondary sources
The earliest years of histology
The rise of histomorphology
Histophysiology and medical education
The work and influence of Virchow
Nerve and sense-organ histology
Histological errors
Histology and medical education in Europe
The situation in the United States
Histology in Europe at the turn of the century
References

Appendix   First use of synthetic dyes in histology :
a chronological summary   343

Index of Subjects   345
Index of Names   353

List of Text Figures

1. Page 3 of Hill, *The construction of timber*, 2nd edn, 1774.   14
2. Hill's microtome of about 1770.   15
3. Custance's microtome of about 1775.   17
4. The first slider, 1691.   19
5. The Adams microtome, 1787.   21
6. The Schmidt micro-dissector, 1859.   86
7. Quekett's microtome, from before 1848.   119
8. Oschatz microtome, 1843.   120
9. Fisher's illustration, 1846.   120
10. Capanema microtome, 1848.   121
11. Currey microtome, 1853.   122
12. Hand microtome, 1855.   122
13. Welcker instrument, 1856.   123
14. Welcker microtome illustrated, 1882.   124
15. Gibbon's section-cutter, 1856.   125
16. Smith's cutter, 1859.   125
17. Notcutt's cutter, 1859.   126
18. Side view of Schmidt's microtome, 1859.   127
19. Plan view of Schmidt's microtome, 1859.   128
20. Beck's section machine, 1865.   128
21. Follin's microtome, 1865.   129
22. Hensen's microtome, 1866.   130
23. Microtome by His, 1866.   131
24. Luys microtome, 1868.   132
25. Rivet's microtome, 1868.   132
26. Hawksley's microtome, 1870.   133
27. Stirling section-cutter, 1861.   134
28. Rutherford's freezing microtome, 1871.   136
29. Rutherford's freezing microtome of 1873.   137
30. Tait freezing microtome, 1873.   138
31. Hughes ether freezing microtome, 1876.   139
32. Lewis ether freezing microtome, 1877.   139
33. Rutherford microtome adapted for ether freezing, 1887.   140
34. Williams freezing microtome, 1881.   141
35. Groves' modification of the Williams design, 1881.   142
36. Fearnley's modification of the Groves-Williams design, 1883.   142

37. Taylor's microtome, 1882.   201
38. Microtome by Satterthwaite & Hunt, 1882.   201
39. Original Cathcart freezing microtome, 1883.   202
40. Improved Cathcart microtome, 1888.   203
41. Hayes freezing microtome, 1887.   204
42. Jung freezing microtome, 1887.   204
43. Delepine instrument, 1900.   205
44. Carbon dioxide microtome by Bardeen, 1901.   207
45. Osterhout instrument, 1904.   207
46. Krause solid carbon dioxide specimen platform, 1908.   208
47. A hand microtome of 1909.   210
48. Reichert's hand microtome of 1885.   210
49. A hand microtome in use, 1892.   211
50. Leitz hand microtome, 1907.   211
51. Stirling-type bench microtome, 1890.   212
52. Large Stirling-type bench microtome, 1890.   213
53. Roy's microtome, 1879.   213
54. Polymicrotome by Hailes, 1880.   214
55. Pearson-Teesdale instrument, 1880.   215
56. Katsch's large microtome, 1882.   216
57. Flatter's section-instruments, 1905.   217
58. Flatters & Garnett microtome, 1955.   217
59. The Curtis microtome, 1871.   218
60. Hoggan's section-cutter, 1874.   219
61. Seiler's microtome, 1879.   220
62. Swift's microtome, 1882.   220
63. Roy's microtome, 1881.   221
64. Boecker's microtome, 1882.   222
65. The Providence microtome, 1885.   222
66. King's microtome, 1889.   223
67. Francotte's microtome, 1889.   224
68. Beck microtome, 1892.   224
69. Fromme's instrument, 1891.   225
70. The Yankawer microtome, 1897.   226
71. Thate's instrument, 1900.   226
72. Korting microtome, 1881.   227
73. Bausch & Lomb student microtome, 1885.   228
74. Leitz 'support' microtome, 1889.   229
75. Reichert automatic microtome, 1884.   229
76. Beck universal microtome, 1885.   230
77. Malassez-Roy microtome, 1884.   231
78. Schiefferdecker's microtome, 1886.   231
79. Schiefferdecker's microtome in section, 1886.   232
80. Miehe microtome, 1889.   233
81. Strasser's paper-ribbon microtome, 1890.   234
82. Jung student instrument, 1892.   234
83. Bruce's brain microtome, 1895.   235
84. Beck-Becker instrument, 1897.   236

85. Minot sliding microtome, 1901. 236
86. Fiori's microtome, 1900. 237
87. Krefft's rotating-blade microtome, 1903. 238
88. Broek microtome, 1907. 239
89. Jung lever-operated sledge microtome, 1910. 239
90. Cambridge lever-operated sledge microtome, 1910. 240
91. Leyser-Brandt microtome, 1870. 241
92. Rivet-Leiser modified carrier, 1880. 242
93. Windler's microtome, 1880. 242
94. Lelong's instrument, 1883. 243
95. Thoma's microtome, 1881. 243
96. Becker's microtome, 1885. 244
97. Vinassa microtome for pharmacognosy, 1885. 245
98. Hildebrand's microtome, 1885. 246
99. Nachet's design, 1886. 246
100. Jung's microtome, 1886. 247
101. Naples laboratory pattern microtome, by Jung, 1886. 248
102. Walmsley's microtome, 1887. 249
103. Reichert instrument for cutting under water, 1900. 249
104. Cambridge rocking microtome, 1885. 251
105. Cambridge rocking microtome, 1900 model. 251
106. Van der Stad rocking microtome, 1909. 252
107. Pfeifer's automatic microtome, 1886. 253
108. Minot's microtome, 1887. 254
109. Minot design of 1892, from behind. 254
110. Zimmermann-Minot microtome, 1908. 255
111. Minot-Blake design, 1899. 256
112. Large brain microtome, Minot pattern, 1898. 257
113. Reinhold-Giltay microtome, 1892. 257
114. Fromme's microtome, 1896. 258
115. Reichert microtome, 1897. 259
116. Block trimmer from the Reichert instrument. 260
117. Leake microtome, 1901. 260
118. Cambridge flat-cutting microtome, 1899. 261
119. Ryder's microtome, 1887. 261
120. Improved Ryder design, 1895. 262
121. The Radais design, 1904. 263
122. Triepel's microtome, 1905. 264
123. Caldwell's microtome, 1885. 264
124. De Groot microtome, 1887. 266
125. Hart's microscope microtome, 1885. 266
126. Lathe as microtome, 1905. 267
127. Original Leuckart embedding moulds, 1881. 268
128. Dimmock embedding moulds, 1886. 268
129. Coplin jar, 1897. 269
130. Schulze section-stretcher, 1883. 269
131. Kornauth section-stretcher, 1896. 270
132. Bernhard's dropper, 1891. 271

133. Schaffer block trimmer, 1899.   271
134. Ryder's embedding apparatus, 1887.   272
135. Sherwald's embedding bath, 1888.   272
136. Kollosow embedding equipment, 1894.   273
137. Kollosow ancillary equipment, 1894.   274
138. Cambridge embedding bath, 1901.   274
139. Hoffman vacuum embedding apparatus, 1884.   275
140. Fuhrmann vacuum embedding oven, 1904.   275
141. Materna vacuum embedding apparatus, 1908.   276
142. Borrmann's rack for multiple staining, 1894.   277
143. Neumayer circular multiple rack, 1905.   278
144. Sanzo's embryo fixer, 1904.   278
145. Arendt's automatic tissue processor, 1909.   279
146. Arendt's processor basket, 1909.   280
147. Retina as illustrated by Hassall, 1849.   314
148. Retina as illustrated by Kölliker, 1854.   317
149. Retina as illustrated by Kölliker, 1860.   322
150. Retina as illustrated by Stricker, 1873.   324

List of Plates

Colour

I. A group of sliders illustrating typical types and the transition to glass.   145
II. Early balsam mounts from the 1830s.   147
III. A sequence of dated mounts, 1824–1869.   149
IV. Preparations by John Quekett.   151

Monochrome

1. An original Leeuwenhoek microscope.   153
2. Original pattern of the Hill microtome.   153
3. Later pattern of the Hill microtome.   153
4. Hill microtome in use.   153
5. An outfit of sliders of about 1770.   155
6. An Adams microtome.   157
7. A typical Ypelaar cabinet of medium size.   157
8. Examples of Hewson's hermetically-sealed preparations.   159
9. Photomacrograph of specimen 7, magnified × 50.   159
10. Photomacrograph of specimen 2, magnified × 50.   159
11. Examples of Hewson's dried preparations.   161
12. Photomacrograph of specimen I.9, magnified × 50.   161
13. Photomacrograph of specimen G.3, magnified × 50.   161
14. Photomicrograph made with the 1826 Lister microscope.   163
15. Photomicrograph of 1849 preparation with modern microscope.   163
16. Photomicrograph of fresh compressed liver with a microscope of 1780.   163
17. An uncleared timber section.   165
18. The timber section cleared.   165
19. Ypelaar resin mounts of about 1795.   165
20. Injected preparations by Hyrtl.   167
21. Examples of transparent and opaque commercial injected mounts.   169
22. A selection of fluid mounts.   171
23. A selection of dry mounts.   173
24. A group of live cells and related apparatus.   175
25. Compressoria and live boxes.   177
26. Small equipment for specimen handling.   179
27. Section knives.   181
28. Pritchard microtome, circa 1835.   183
29. Topping microtome in use.   183

30. Beck microtome, 1860s.   183
31. Rutherford freezing microtome, 1876.   183
32. Fearnley freezing microtome knife in use.   185
33. Hand microtome of the 1880s.   185
34. Cathcart microtome.   185
35. Cathcart microtome.   185
36. Zeiss bench microtome, 1877.   187
37. Korting microtome, 1880.   187
38. Korting instrument, 1883.   187
39. Cambridge rocking microtome, original pattern.   189
40. Minot microtomes.   189
41. Cambridge automatic microtome.   189
42. 'Caldwell' automatic microtome, 1883.   189
43. Examples of commercial preparations (1).   191
44. Examples of commercial preparations (2).   193
45. Examples of commercial preparations (3).   195
46. Examples of commercial preparations (4).   197
47. Examples of commercial preparations (5).   197
48. Examples of commercial preparations (6).   199
49. Examples of commercial preparations (7).   199

... Dr. Ogle, applying his eye to the microscope, screwed a quarter-inch right through the object; and Dr. Kidd, after examining some delicate morphological preparation, while his young colleague explained its meaning, made answer first, that he did not believe in it, and, secondly, that if it was true he did not think God meant us to know it ...

The above quotation is taken from a first-hand account of practical demonstrations in histology given by Henry Acland in Oxford from 1845; see page 47 of W Tuckwell—*Reminiscences of Oxford*. (London: Smith Elder, 2nd edn, 1907.)

1. Introduction

The approaches used in the work

This book traces the origins and development of the manipulations needed to prepare specimens for examination by the light microscope. It is restricted in the main to animal, especially mammalian, tissues: botanical or mineral specimens are considered only incidentally. The techniques of the seventeenth and eighteenth centuries are included, but special attention has been given to the progress seen in the last twenty years of the nineteenth century; developments after about 1910 are included only if they further earlier processes destined to become important.

A survey of the abundant literature was the basis of the work, but errors of omission, commission, and interpretation are no less abundant in microscopy and histology than they are in any other field of human endeavour. Fortunately, some objective checking is possible, as many instruments and preparations have survived against which it is reasonable to judge much of the literature. This approach, using surviving artifacts to recreate earlier observations, is not original, but is still surprisingly uncommon. Clarke has shown the importance of this kind of work, and provided excellent examples of its fruitfulness.[1]

More than 40000 microscopical preparations have been inspected for this book, some being set aside for further study. The basis of selection was, first, to choose all slides bearing a credible date, whatever their subject. Second, a group of histological preparations was selected, with special attention to nervous tissues. A third group included a variety of unusual mounts, while a fourth was of commercial mounters of the nineteenth century. The final group was representative eighteenth-century sliders.

These five groups totalled about 3500 specimens, all of them inspected carefully to give a bird's eye view of microscopical preparations spanning one hundred and fifty years. A modern set of high-quality histological slides served as a reference standard. Many microscopes were inspected concurrently with the slides, and used with contemporary preparations. This is a very valuable experience, as difficulties in manipulation can affect details of observation; further, if the historian does not immerse himself in the practices of the period with which he deals, a true picture of the history of the times cannot be obtained. A range of stands from about 1690 to 1920 was used, as far as possible in their original condition, and with appropriate illumination sources.

More important still was the opportunity to work with a wide range of microtomes. This is a curiously neglected instrument, but some diligence allowed more than twenty to be used to cut sections as they would have been cut at dates ranging from 1790 to 1910. Thirty-five others were inspected, compared with their original descriptions, and evaluted.

Review articles and other basic literature sources

In addition to invaluable personal tuition from those thanked in the preface to this book, much had to be learned of the general history of microscopy if serious error was to be avoided and technical details appreciated. Students of microscopical technique will look in vain for a modern detailed survey of the history of the instrument, although much has been written in the past. Petri[2] gave an excellent survey, and it is not necessary to consult any earlier works. Disney[3] provides some useful material, on instruments in the collection of the Royal Microscopical Society, now in care of the Museum of the History of Science in Oxford. Some of his findings must be treated with caution, as must those of Clay & Court.[4] A short but modern survey is that by Bradbury,[5] while Frison has dealt with the neglected optical parts of the microscope.[6] Turner[7] has given a good picture of the instrument as a unit in action, as also has Bradbury in his accompanying paper.[8] This work has been extended by Bracegirdle.[9,10] A wide perspective is offered in the survey by Turner.[11]

Turning now to the preparative side, a book by Gray & Gray[12] is useful for certain literature references, but is disappointing in other respects. The authors tried to provide a full list of books devoted to microtechnique, and from this alone to deduce the history of specimen preparation. Apart from serious omissions in their lists, they omit all original papers and also books which do not have specimen preparation as their primary content. Clearly, these limitations make it quite impossible for a proper history of specimen preparation to be deduced. A paper by Smith,[13] on the other hand, gives a good account of the introduction of botanical specimen preparations, but includes some errors and misinterpretations, and the literature references cannot be relied on for accuracy. A few other works include useful sections: Hughes[14] has given an excellent account of one aspect, while for a large number of references the book by Henneguy[15] is still difficult to better. For excellent translated passages, Clarke & O'Malley[16] are most helpful in the neurological field, and in the same area Farrar's paper,[17] written at the beginning of this century, reviews highlights as they appeared shortly after they had occurred.

Two earlier bibliographies deserve mention. That by Roper[18] was possibly the first microscopical bibliography in any language, and although it lists only books in the possession of its author, it is very well done. A similar list published twenty-five years later by Deby[19] is still useful for its listing of otherwise obscure serials and unusual books. In similar vein, the last library catalogue of the Royal Microscopical Society[20] remains a handy guide to obscure works, although the library itself is now dispersed.

Finally, two major catalogues require acknowledgement as invaluable sources. The catalogue of nineteenth-century scientific papers issued by the Royal Society provides accurate confirmation of often vague references from elsewhere, and is almost comprehensive.[21] It is also very useful for its lists of other papers by an author: these may have a bearing on a problem while being unquoted elsewhere. Since so many earlier serial publications no longer continue, they do not have short titles in the current *World List of Scientific Periodicals* and similar works. For this reason their titles are generally given in this book in the form in which they appear in the Royal Society catalogue. The fine series of catalogues issued by the Surgeon-General in Washington, with their cross-indexing and excellent subdivision, have become the single most important source for histological and microscopical publications.[22] These catalogues are almost, but not completely, comprehensive, representing the holdings of the largest collection of medical literature in the world.

The main periodical literature

By far the most important is that produced by or in association with the Royal Microscopical Society: this alone spans all of the critical period from 1840 onwards. There have been several different titles and series, and a summary of these may be helpful.

The Society was originally founded at the end of 1839. A meeting of those interested in forming a society was held on 3rd September of that year,[23] and on 20th December a public meeting founded the

Microscopical Society of London, with Richard Owen as first President. In his first annual report[24] Owen said

> ... Accurate reports of the papers communicated at each of our ordinary meetings, have appeared in a widely circulated journal, noted for the promptness with which it diffuses each new addition to science.

He then described the journal started by the young Daniel Cooper, which the Council agreed should contain full accounts of the papers read to the Society, with the permission of their authors. He said that Council had intended to publish its own *Transactions*, and had not lost sight of this intention. Cooper's journal[25] regularly appeared, but he died suddenly two years later, thus ending its short life. This prompted the Society to issue its own *Transactions*,[26] which appeared at irregular intervals and with considerable delays to papers, which must have proved daunting to potential contributors. As a result, a new series of *Transactions*[27] was published, in conjunction with the *Quarterly Journal of Microscopical Science*.[28] This last became increasingly devoted to a wide range of work in the life sciences, not all of it obviously microscopical: it was to have a very long life, but only the first thirty or so volumes are of much interest in the history of microscopy.

This joint publishing arrangement (whereby the two journals were actually published bound together) continued until 1868. The Society had received its Royal Charter in 1866, and had decided that it required more control over its publication; financial matters also played a part in the decision.[29] Thus between 1869 and 1877 the Society put out a journal devoted to matters microscopical every month.[30] This journal marks the beginning of a wider scientific interest by the Society, for it included abstracts of foreign papers as well as those from other British publications.

When the editor died another publication was started,[31] but lasted for only three years, as the first series of the *Journal of the Royal Microscopical Society*. The current series commenced in 1881, but volumes were not always numbered, the year alone being used as a reference for long periods. Nonetheless, the volumes were issued annually until 1958 (volume 78), since when volumes have appeared more frequently, volume 100 being reached in 1974.[32] It may be of interest to record that the volume for 1888 was 90 mm thick unbound, and weighed almost 3 kg. The importance of these journals in the history of microscopy is considerable. The number of original papers was not high in the earlier years, as specialist articles were even then sent to specialist journals, but some significant ones were included, while the abstracts of other papers and books are excellent. Unfortunately for our present purposes, they include only the minimum of histology.

Starting at the very pinnacle of microscopic involvement with technique, in 1884, a German journal[33] rapidly became highly important. Well-arranged references and abstracts in all branches of microscopy are given in detail, with some significant original papers. This journal started much later than the publications of the Royal Microscopical Society, but it was in time to cater for the deluge of German work which gave such pre-eminence to their histologists and microscopists in the last twenty years of the nineteenth century. German-speaking workers tended to neglect work from other countries, but this publication abstracts them all.

Another journal of some interest is that of the Quekett Microscopical Club: this was started in 1868.[34] The Club still exists, catering essentially for the amateur, and its journal has a practical bias with occasional papers by well-known authors. In similar vein, a journal started in the 1930s proves useful on occasion, in its earlier numbers, largely for its interest in commercial mounters.[35]

Many other journals will be referred to, but the above form the mainstay. Some other European journals carried occasional important papers, but are less important than the British and German serials mentioned. Similarly, the United States, as they became during the period covered in this book, offered very little that was original until the end of the nineteenth century.

Those interested in tracing the references to books and journals given in this book will, on occasion, have to be diligent in seeking them out. Some of the serials and many of the books are quite rare. For those with access to the libraries of Washington, only occasional difficulty will be met, and those working in London (with the facilities of the Wellcome Institute Library, the Royal Society of Medicine, the British Library, and the rest) will need only much patience. Elsewhere, it is quite unlikely that all the references listed herein will be found.

Two other matters which have a bearing on the progress of microscopy should be kept in mind. The first is the role of the various societies founded to encourage microscopical work: in this context the Royal Microscopical Society has made quite outstanding international contributions, but many others have been very active from time to time. Springing from this is the second consideration: the means whereby microscopical discoveries were communicated. A discovery affects progress hardly at all until it is widely known and acted upon, and in so intensely visual a discipline as microscopy wide dissemination of illustrations is of critical importance. A survey of microscopical communication has been given by Turner,[36] who has shown that the critical barrier to the provision of continuous-tone pictures in large numbers was early lack of means. Early illustrations in books touching microscopical subjects are often startlingly beautiful, but were not always accurate when they had passed through the hands of author, artist, engraver, and printer!

References for Chapter 1

1. Clarke describes the approach, which he calls 'practical history', in a number of papers. A clear statement is in: E Clarke & J G Bearn (1968)—The brain 'glands' of Malpighi elucidated by practical history, *J. Hist. Med.* 23:309–330. See pp. 309–311.
2. R J Petri—*Das Mikroskop. Von seinen Anfängen bis zur jetzigen Vervollkommnung für all Freunde dieses Instruments.* (Berlin: R Schoetz, 1896.)
3. A N Disney (with C F Hill & W E Watson-Baker)—*Origin and development of the microscope, as illustrated by catalogues of the instruments and accessories in the collections of the Royal Microscopical Society, together with bibliographies of original authorities.* (London: Royal Microscopical Society, 1928.)
4. R S Clay & T H Court—*The history of the microscope compiled from original instruments and documents, up to the introduction of the achromatic microscope.* (London: C Griffin, 1932.)
5. S Bradbury—*The evolution of the microscope.* (Oxford: Pergamon, 1967.)
6. E Frison—*L'évolution de la partie optique du microscope au cours du dix-neuvième siècle.* (Leiden: Rijksmuseum voor de Geschiedenis der Natuurwetenschappen, 1954.)
7. G L'E Turner (1966)—The microscope as a technical frontier in science, *Proc. R.M.S.* 2:175–199.
8. S Bradbury (1966)—The quality of the image produced by the compound microscope: 1700–1840, *Proc. R.M.S.* 2:151–173.
9. B Bracegirdle (1977)—J J Lister and the establishment of histology, *Medical History* 21:187–191.
10. B Bracegirdle (1978)—The performance of seventeenth and eighteenth century microscopes, *Medical History* 22:187–195.
11. G L'E Turner (1969)—The history of optical instruments—a brief survey of sources and modern studies, *History of Science* 8:53–93.
12. F Gray & P Gray—*Annotated bibliography of works in Latin alphabet languages on biological microtechnique.* (Dubuque: W C Brown, 1956.)
13. G M Smith (1915)—The development of botanical microtechnique, *Trans. Amer. M.S.* 34:71–129.
14. A Hughes—*A history of cytology.* (London: Abelard-Schuman, 1959.)
15. L F Henneguy—*Leçons sur la cellule.* (Paris: Carve, 1896.)
16. E Clarke & C D O'Malley—*The human brain and spinal cord.* (Berkeley: Univ. California Press, 1968.)
17. C B Farrar (1905)—The growth of histologic technique during the nineteenth century, *Rev. Neurology & Psychiatry* 3:501–515, 573–594.
18. F C S Roper—*Catalogue of works on the microscope and of those microscopical subjects in the library of Freeman C S Roper.* (London: printed for private circulation by J E Adlard, 1865.)

19. J Deby—*Bibliotheca Debyana being a catalogue of books & abstracts relating to natural science, with special reference to microscopy, in the library of Julien Deby.* (London: published for Julien Deby, 1889.)

20. Anon—*Catalogue of the printed books and pamphlets in the library of the Royal Microscopical Society.* (London: Royal Mic. Soc., 1929.)

21. Royal Society of London—*Catalogue of scientific papers.* (London, 19 vols published between 1867 and 1925.)

22. U.S. War Department—*Index-catalogue of the library of the Surgeon-General's office.* (Washington.) The first two series, of 18 and 21 volumes, take the listings into this century.

23. The details of this meeting were written apparently in retrospect, in the manuscript minute book of the meetings of the Council, vol. 1, in the archives of the Society.

24. Anon—*Report of the first anniversary of the Microscopical Society of London.* (London: Luxford, 1841.) See p. 17.

25. D Cooper (Ed.)—*The microscopic journal, and structural record.* (London: van Voorst, 1841, 1842.)

26. Anon—*The transactions of the Microscopical Society of London.* (London: van Voorst, 1, 1844; 2, 1849; 3, 1852.)

27. Anon—*Transactions of the Microscopical Society of London.* (London: S Highley, 1853–1868.)

28. E Lankester & G Busk (Eds)—*Quarterly journal of microscopical science.* (London: S Highley, vols 1–8, 1853–1860; vols 1–8 n.s., 1861 on.)

29. J B Reade (1870)—The President's address, *M.M.J. 3:113–133.* See pp. 130–133.

30. H Lawson (Ed.)—*The monthly microscopical journal: transactions of the Royal Microscopical Society, and record of histological research at home and abroad.* (London: R Hardwicke, 1869–1877.)

31. Anon—*Journal of the Royal Microscopical Society; containing its transactions and proceedings, with other microscopical information.* (London: Williams & Norgate, 1878–1880.)

32. F Crisp—*Journal of the Royal Microscopical Society; containing its transactions and proceedings, and a summary of current researches relating to zoology and botany (principally invertebrata and cryptogamia), microscopy, &c.* (London: Williams & Norgate, 1881 on.)

33. W J Behrens (Ed.)—*Zeitschrift für wissenschaftliche Mikroskopie und für mikroskopische Technik.* (Braunschweig: Schwetschke, 1884 on.)

34. Anon—*The journal of the Quekett Microscopical Club.* (London: Hardwicke, 1868 on.)

35. A L E Barron (Ed.)—*The microscope.* (London: Barron, 1937 on.)

36. G L'E Turner (1974)—Microscopical communication, *J.R.M.S. 100:3–19.*

2. A Brief Survey of Microtechnique Before 1830

The seventeenth century

The first microscopists were preoccupied with making their micro-scopes, and paid less attention to specimens, for anything visible was impressive by its sheer novelty once attention was directed towards it by a lens. Relatively little was written at this time: Nelson's survey[1] is short, in spite of including all that even remotely touches on microscopy in the seventeenth century. The work reported by Leeuwenhoek[2] was thoroughly scientific in approach, but much of the other microscopical literature before 1830 was directed to the amateur dilettante.

There was, of course, some scientifically-important work: some of this must be evaluated from the literature alone, but some of it can also be re-created quite closely.[3] No preparations from the seventeenth century have survived, for it is almost certain that all were only of a temporary kind, for viewing on one occasion only. From the eighteenth century, the only preparations to have come down to us are those sliders made for the amateur: these can be compared with their descriptions.

Hooke[4] provided the first picture of what we now call cells,[5] and in preparing the specimen deliberately sectioned his material in two planes:

> I took a good clear piece of Cork, and with a Pen-knife sharpen'd as keen as a Razor, I cut a piece of it off, and thereby left the surface of it exceeding smooth, then examining it very diligently with a Microscope, me thought I could perceive it to appear a little porous; but I could not so plainly distinguish them, as to be sure that they were pores, . . . I with the same sharp Pen-knife, cut off from the former smooth surface an exceeding thin piece of it, and placing it on a black object Plate, because it was itself a white body, and casting the light on it with a deep plano-convex Glass, I could exceedingly plainly perceive it to be all perforated and porous, . . .[6]

Hooke's microscope at that time was for reflected light only, but the advantages of transmitted light were soon apparent, for, as he stated a few years later,[7] by fixing a piece of Muscovy glass (mica) to the instrument, he could transilluminate liquid specimens, and even achieve dark-ground illumination.[8] More important still, he also

8

described the technique of mounting some specimens in olive oil:

> But there are other substances which none of these ways I have yet
> mentioned will examine, and these are such parts of animal or vegetable
> bodies as have a peculiar form, figure, or shape ... such are the Nerves,
> Muscles, Tendons, Ligaments, Membranes, Glandules, Parenchymas, &c.
> of the body of Animals ... but if the same be put into a liquor, as water, or
> very clear oyl, you may clearly see such a fabrick as is truly very admirable,
> and such as none hitherto hath discovered ...

This was indeed a new and, we now see, very important advance: the
failure of other microscopists to use the method of clearing specimens
must have held back histology for a century. The detail visible in the
usual dry mounts is minimal, and they are prepared with little finesse, so
that although they held sway until as late as the 1830s, they have so little
resemblance to life that scientific interpretation is almost impossible.
Hooke's microscope magnified about × 30, so that fine detail would not
have been apparent to him in any case: he did not pursue microscopy
after about 1680 and bemoaned the fact that few were interested in the
subject.[9] This was true so far as the number of workers was concerned,
but in terms of output and scientific worth we have to recall that
Leeuwenhoek was active for fifty years, communicating his results to
the Royal Society to leave a lasting monument to his industry and
imagination.[10]

Unfortunately, Leeuwenhoek was secretive about his methods, and
in all his writings there is very little about technique. He did, however,
bequeath 26 instruments to the Royal Society, and as it was his usual
practice to make a microscope specially to view one specimen, 26
specimens were also bequeathed. Folkes examined the collection soon
after it arrived, and his description[11] shows that the specimens included
blood corpuscles, hair, teeth, spermatozoa, muscle, and other
histological specimens, among more general objects. If the specimen
was solid, it was simply glued to the point of the needle which was the
only specimen support on the usual Leeuwenhoek microscope (see plate
1). If the specimen was liquid, it was spread on thin glass or mica, which
was then glued to the needle. This is probably the first use of a smear of a
liquid tissue, a technique still widely employed.

Sixteen years later, Baker wrote a further description of the
bequest,[12] and said that most of the objects had fallen from their
needles. This is, of course, not unexpected, for they would have been
handled quite frequently in the intervening period and I have
established at first hand how difficult it is to fix a specimen to the needle-
point of a Leeuwenhoek microscope, both to achieve a suitable

orientation and to secure firmness. Baker's paper is, though, of much interest, as it gives details of the optical characteristics of the instruments, showing that the most powerful of them magnified × 200 by modern standards.[13] It may be, as Dobell suggests,[14] that Leeuwenhoek could achieve dark-ground illumination (Hooke knew the technique,[15] and may have communicated it); in any case, with either a Hooke-type instrument or a Leeuwenhoek, it is easy accidentally to move the microscope off-centre relative to the light source, thus getting oblique lighting which approximates a dark-ground condition visually. It has been shown[16] that the Leeuwenhoek microscope at Utrecht magnifies × 275, and has a numerical aperture of about 0·4: this would resolve blood corpuscles easily, and perhaps objects of only 1 μ diameter when in original state. Leeuwenhoek also mounted liquid specimens in capillary tubes:[17] this mount suited his instruments quite well, of course, but was not widely adopted.

Leeuwenhoek's instruments were sold off at auction in 1747, and a full list of the specimens fixed to them at that time is given by Harting.[18] 247 instruments were sold, most of them with specimens, but neither Protozoa nor bacteria were represented. The lists of specimens, and the published work describing them, show clearly that Leeuwenhoek worked in a truly scientific spirit, preparing specimens specifically to pursue a line of enquiry sometimes over many years. He did not always draw what we now know to be the correct conclusions, but he did investigate both sectioned and fresh material to a far greater extent than any other worker for over a century.

So far as other seventeenth century workers are concerned, little is known of their methods. Malpighi says nothing at all of his techniques, while in all Grew's published works he says only:

> And to do all this by several Ways of section, Oblique, Perpendicular, and Transverse; all three being requisite, if not to Observe, yet the better to Comprehend, some Things. And it will be convenient sometimes to Break, Tear, or otherwise Divide, without a Section. Together with the Knife it will be necessary to joyn the Microscope; and to examine all the Parts.[19]

This advice is a model of clarity and good sense, being still applicable: unfortunately, the techniques of sectioning were not to be perfected for a further two centuries.

Microscopic injections

The technique of vascular injection gives results which are immediate and striking, and as Cole[20] has shown, it passed rapidly into use. Even today, histologists unfamiliar with the achievements of earlier times are fascinated by a good injected preparation, which can still have useful teaching potential. It soon becomes obvious, however, that the amount of histological information to be derived from injected preparations is very limited. Once the microscope revealed the amazing spectacle of blood capillaries, science tended to be abandoned for display. Even in the seventeenth century the technique was widely applied, as by Malpighi. His work rarely makes clear if he was speaking of the gross or the microscopic, but his major work on the kidney certainly demonstrates the Malpighian bodies of that organ.[21] In the next century Swammerdam worked at the technique, achieving excellent results, as typified by his investigations on the honey-bee.[22] He used a wax mass for the finest injections, clearing the surrounding tissues with oil of turpentine. This is, of course, a well-known clearing agent in microscopy, but he was not using it for such a purpose. In spite of this, he could have given a valuable lead to those microscopists who read his works, had they followed it.

A significant, but largely disregarded, early use of a microscopic stain was that by Muys.[23] In 1714 an account of his use of a coloured liquid injection which also stained the surrounding muscle fibres was published,[24] and noted in the *Philosophical Transactions*.[25] This staining was an accidental accompaniment of the injection process, but was also part of the work of Muys on muscle structure, and may be noted as a true microscopical stain. At about the same time some of the inadequacies of the injection process were commented on by Boerhaave,[26] his main criticism being that too great a pressure during the injection forces the medium into spaces not continuous with the vessel. This was the more likely when using mercury metal as the injection medium. Ruysch, however, the acknowledged master of injection, achieved remarkable success in the art[27] at the macro level, and cleared surrounding tissues with oil of lavender as well as oil of turpentine.

The first to prepare injections at the specifically microscopic level was Lieberkühn. A number of his preparations are still extant,[28] and reports indicate that they stand high magnifications without loss of detail: certainly the plates in his book on the intestine[29] are quite remarkable in the detail of the capillaries figured.

A further result of injection work, this time of cavities rather than vessels, was that of le Cat, who used a medium of wax and resin to show the relationship of organs.[30] He prepared sections (gross rather than micro) of his embedded organs, in a remarkable prevision of the paraffin section technique of 240 years later.

By the end of the eighteenth century the superiority of gelatin-based media had been established for work at the microscopic level. The method claimed adherents during much of the nineteenth century, largely for commercial reasons, as is discussed below (see page ooo).

Other results from the eighteenth century

In addition to a number of popular works on the microscope, and some good original research on timber, some striking scientific results were achieved: a bare summary of some of these must suffice here. In 1718 Joblot discovered the contractile vacuole of Protozoa,[31] and in 1744 Trembley published his fundamental work on *Hydra*.[32] *Amoeba* was first described by von Rosenhof in 1755,[33] and in 1773 Müller first wrote about diatoms.[34] Fontana was another prominent worker, describing the nucleolus in 1773,[35] and making other notable observations.[36] Other examples could be quoted, but the underlying significance is that people worked in isolation, and the coincidence of their interest being in things small seems to produce a coherence which is in fact lacking. There was no organisation of research until later in the nineteenth century. It is likely that the instruments and techniques available were not the deciding factors in choice of work pursued: the underlying philosophy of the eighteenth century taught men to deduce all phenomena from the idea of the absolute—making it easy to explain things without troubling with actual observations. Apart from a few investigators such as those mentioned above, the amateur was the microscope's only user, apart from a notable investigator, John Hill, who requires more detailed attention.

Hill and the microscopic anatomy of timber

Hill's books on timber[37] are of much importance in the history of microtechnique, for his approach provided probably the first micro-tome, advances in technique of major worth, and accurate information on plant anatomy. Fortunately several of his microtomes have survived for actual use: the technique of using them is described in the caption to plate 4. Hill stated that Cummings actually invented the microtome for him, with Ramsden subsequently undertaking their manufacture. Curiously, Adams later stated that his father was the originator, in a passage which seems to have been misinterpreted subsequently (see page 20 below). In addition to the microtome which

Hill described in detail, he made another,[38] which could cut sections even thinner than the 1/2000 in. claimed for the former. Even more important in the history of the microtome, this other instrument was made to

> ... shift its own screw at every revolution of the handle, so that very little time was left for the pith to shrink; as a hundred slices could easily be cut in a minute.

This can only mean the provision of an automatic advance mechanism, a very remarkable innovation and one which was not to become established for over another hundred years. There were, however, difficulties, for Hill says:[39]

> It performed extremely well, but was judged less fit for general use than that which has already been described, it being more complex, and liable to disorder, as well as more difficult to manage.

This criticism would also apply in due time to its successors a century later! No other reference to this automatic microtome has been traced, and no example seems to have survived.

Apart from making the first microtome and the first automatic microtome, Hill may have been the first to section material other than timbers. His microtome uses a length of ivory to raise the specimen, and he specified that a full length of this, with roughened end, be supplied, to allow sealing wax to be used to cement pith, cork, agaric, etc., to it for sectioning.[40] Freehand techniques aside, this is the only reference to the sectioning of such specimens to occur before the nineteenth century. In addition, it is possible that Hill may have used a relatively large volume of wax, thus partially surrounding the specimen. The author has repeated this use with a Hill microtome, and unless a large amount of wax is used, the specimen of a mushroom cannot be kept cemented in position when the blade reaches it. This would constitute the first example of what was later called embedding—that is, surrounding the specimen with a supporting medium without actually infiltrating it with the medium in solution (see Figures 1 and 2, and plates 2, 3 and 4).

Besides his remarkable instruments, Hill introduced several important methods, for he investigated the structure of his timbers in as many ways as possible. One such method is that of maceration, to loosen fibres from each other.[41] Reintroduced by Moldenhawer in 1812,[42] the method was later hailed as one of the great steps in the history of phytotomy.[43] Hill kept his material for further study in a solution of alum before transfer to spirit:[44] this is the direct equivalent of our modern fixing and hardening. He was also the first to use deliberate

The Cutting Engine is an invention of the ingenious Mr. Cummings. The two or three firſt were perfected under his own hand; and they are now made for general uſe by Mr. Ramſden.

DESCRIPTION of an INSTRUMENT

For cutting Tranſverſe Slices of Wood, for MICROSCOPICAL OBJECTS.

A A. Plate I. Fig. 1. repreſents *a cylinder of ivory*, three inches and a half long, and two inches in diameter; to the one end of which is fitted PLATE I.

B B. *A plate of bell-metal*; the ſection of which, with the manner of fitting it to the ivory, may be ſeen in Fig. 2. in which the ſeveral parts are marked with the ſame letters as in Fig. 1.

C. is *a plate of braſs*, fitted to the other end of the cylinder; through which and the ivory there paſs two long ſcrews, which take into the thick part of the bell-metal B B, ſo as to fix both plates ſtrongly to the ivory; into which they are alſo indented, ſo as to prevent ſuch ſhaking as might otherwiſe happen after ſwelling or ſhrinking.

D D. *The Cutter*; whoſe edge is a ſpiral, and the difference of whoſe longeſt and ſhorteſt radii is equal to the thickneſs of the largeſt piece of wood that the inſtrument would take in. The loweſt ſide of this cutter muſt be ground extremely flat and true, in order that all the parts of its edge may be exactly in the ſame plane; and that the middle part of it may be applied cloſely to the flat circular plane left at the center of the plate B B, to preſerve it in the proper direction when carried round by the handle.

All that part of the bell-metal, which *the edge of the cutter traverſes*, is turned ſo low as not to touch it, (ſee the Section:) the middle of the cutter is about ⅟₄ of an inch thick, and has in it a ſquare hole that fits on the end of a ſteel axis P P, one end of which turns on a pivot in the Plate C, the other end in the plate B B. This end has a conical ſhoulder which fits into a hole of the ſame ſhape in the under ſide of the plate, as repreſented in the Section.

e e. *A piece of braſs ſomewhat in the form of an index*, which is alſo put on the axis P P: this piece has a round hole in its center ſo large as to admit of its being turned into any poſition with regard to the cutter; and in order to keep it concentric thereto there is left on it a circular projection, which fits into a cavity made in the lower ſide of the handle, where it fits on the axis. (See the Section.)

B F. *The*

Figure 1. Part of page 3 of Hill's The construction of timber, *2nd edition, 1774.*
The attribution to Cummings is included, as is part of the description of the instrument.

staining as an aid in the investigation of the microscopic structure of plants. He employed an alcoholic tincture of cochineal in which to stand the cut stem, but he did not use the part which had been submerged: this is true staining, and he also used a mordant of lead acetate in conjunction with a solution of quicklime and orpiment to demonstrate the finest vessels of the phloem.[45] Hill also used circles of glass, instead of the usual mica, in the sliders of his time,[46] although it is not possible to claim that he was the first to do so.

Possibly his greatest advance was to use spirits of turpentine to clear tissues for microscopy:[47] this seems the first deliberate use at the micro

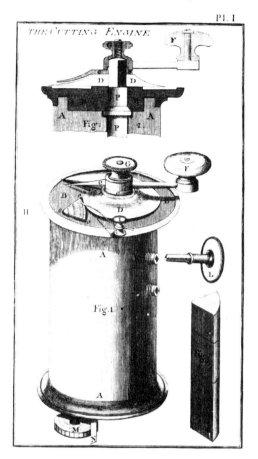

Figure 2. Hill's microtome of about 1770 (from plate I from Hill's *The construction of timber*, 2nd edition, 1774).

The plate is identical with that in the first edition of 1770, and shows the appearance of the instrument (compare with my plate 2). The sectional drawing at the top of the plate is interesting, as is the representation of the length of ivory specimen support. The screw M has 40 threads per inch, and its head carries 25 divisions, thus advancing the specimen by one division is to raise it by 1/1000th on an inch—in theory. In practice the nature of the specimen, with its considerable elasticity, does not allow for such precise lifting. It is necessary to clamp the twig (using the key L), once it has been raised, to prevent it being pushed down again when the knife contacts it.

level. Hill's methods, taken together, could have revolutionized microscopy had they been applied widely. In fact, they were ignored by both scientist and amateur until each was separately rediscovered up to sixty years later. Workers still relied on simple small-scale dissections, teasing, and compressing, much as Leeuwenhoek had done a century before. When transmitted illumination became the norm in the middle of the eighteenth century there was obvious need to make specimens thin enough to see through, but much would be falsified by using water as a medium for dissection and compression. However, the optics were poor and visible osmosis in fine structures does not seem to have attracted attention, for there are no reports describing it.

Custance's microtome for timber sections

Custance was a carpenter of Ipswich, famed for the superb quality of his sections of timber: he had the monopoly of supply of these for about thirty years, but always refused to divulge his methods. Thornton (physician to the General Dispensary, and lecturer in medical botany at Guy's Hospital) was an able botanist who required specimens for illustrating a book, and offered £50 to Custance for his secret.[48] This was refused, £105 being named as the price, to be refused in turn. When Custance died, his effects were auctioned off, and Thornton bought both of Custance's instruments, which he then described in print, so that others might benefit. Neither instrument seems to have survived, but the illustration from Thornton's paper (see Figure 3) shows why such good sections were produced. Although there is no evidence to support the possibility of Custance sectioning any materials besides timber, the microtome would certainly have made excellent sections of other specimens; as Hill had already written about sticking different objects in position in his own microtome the idea would not have been novel.

Books and preparations for the amateur

The eighteenth century produced a number of books on the microscope, some more scientifically-based than others. Fairly good scientific content is to be seen in Gleichen's book,[49] which contains a description of the instrument and some details of specimen manipulation. A more frankly amateur appeal is typified by Ledermueller,[50] which is a considerable work of rather scant content but appealing presentation. It was in England, however, that the most important books for the amateur were written, doubtless on account of the highly-developed scientific instrument trade in that country: books of instruction were clearly required to support sales of instruments and preparations.

Many, even of those who have purchas'd Microscopes, are so little

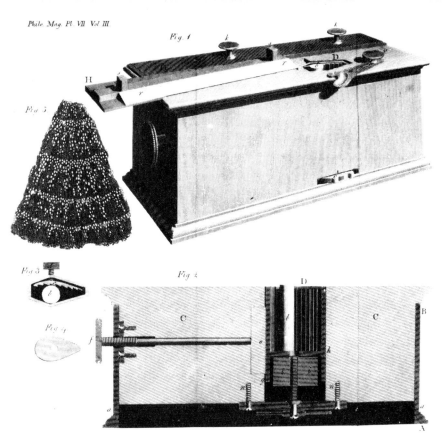

Figure 3. Custance's microtome of about 1775 (from plate VII volume III of *The Philosophical Magazine*, 1799).

Fig. 1 is a general view of the microtome; Fig. 2 is a longitudinal section; Fig. 3 is a section through the well, showing the serrated clamp; Fig. 4 shows the captive base of the advancement; Fig. 5 shows a section of charcoal (sic) made with the section cutter—a very remarkable achievement.

Although no examples of the instrument have survived, the drawings show why Custance was able to make such excellent timber sections, and why they were not surpassed for fifty years or more after his death: the design is by one who thoroughly understood the nature and behaviour of timber.

The main requirement in a microtome is that of rigidity, difficult to achieve when hard materials such as timber are to be cut. The design is in fact a rigid block of hard mahogany C, boxed in by rigid brass plates B. The block is pierced by a well of modified oval shape (Fig. 3), to contain the advance mechanism and prevent it turning as it is raised by the captive nut (i) driven by the micrometer screw turned by the wheel (o). For even greater rigidity, the knob (f) tightens the block against the side of the moving well and can be regulated to take up backlash without interfering with the smoothness of specimen advance. Inside the well the timber is held by a grooved holdfast, and the brass spur (u) is engaged behind the specimen to give total protection against movement when the knife first contacts the timber.

The rigidity might have been reduced by an inadequate knife, but this was of large section secured to a massive rule (H), carrying the actual edge just clear of the top plate and at a very narrow angle to the specimen. Not only did this maintain rigidity, but it eliminated frictional lack of even advance.

The microtome was possibly designed a little after Hill's instrument, say in the early 1770s, but in use it would certainly have surpassed both this and the Adams design with ease.

acquainted with their general and extensive Usefulness, and so much at a Loss for Objects to examine by them; that after diverting themselves and their Friends, some few times, with what they find in the Sliders bought with them, or two or three more common Things, the Microscopes are laid aside as of little farther value: and, a Supposition that this must be the Case, prevents many others from buying them: ...[51]

Some very comprehensive sets of objects were supplied, in the universal form of ivory or hardwood sliders having several apertures each containing a pair of mica discs retained against a shoulder by a brass circlip. Such a set is shown in plate 5, with spare talcs (mica discs) and circlips, and a typical manuscript list of objects. The only feature in common to objects on any one slider was that one power of objective would be needed to look at them! The first illustration of a slider is in Bonanni's book of 1691,[52] which even shows the chamfered end to allow easy insertion into the sprung carrier (see Figure 4). The objects included were rapidly stereotyped, and so uniform are the survivors that it is usually impossible to date an example. Baker[53] gives a chapter of six pages on preparing specimens, starting his advice with the following:

Most Objects require some Management, in order to bring them properly before the Glasses. — If they are flat and transparent, and such as will not be injured by Pressure, the best Method is to inclose them in Sliders, between two Moscovy Talcs or Isinglass. . . .

Sliders were to be filled according to the power of glass required to view them, and a slider could also be used to hold living fleas by pressing the talcs together just so far as was needed to hold them without crushing. Leeuwenhoek's method of capillary tubes for holding small amounts of liquid was recommended, as was the use of wider tubes to show the circulation of the blood in frog and fish. Fine dissecting was to be done under water, on a piece of glass the size of a slider, differently coloured glass being chosen to contrast with the object. Opaque objects were to be mounted on slips of card using thin gum, and his plate 6 shows a case made to contain a collection of these—surely the first cabinet for microscopic preparations? The boxes were sold by Cuff of Fleet Street to those disinclined to make them. Throughout the various editions of Baker, from the first in 1743 to the last in 1785, there is no difference in the advice on mounting.[54]

Two other important authors from the eighteenth century are the older and younger George Adams. Again, the older wrote his book[55] largely to help sales of his instruments, but it contains some useful pointers to microscopic technique, which again do not differ between

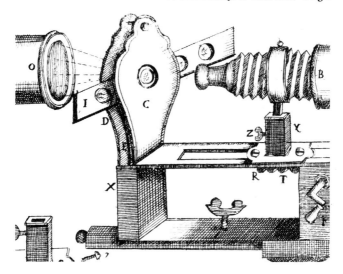

Figure 4. The first slider, 1691 (from Plate II from Bonanni's *Observationes*, 1691).

Quite apart from the very interesting technical details of the microscope illustrated, the design shown for the slider was to be the norm for well over a century.

editions. His advice deserves close attention even today:

> The greatest care imaginable should be taken in preparing objects for examination; otherwise the best skill'd in magnifying glasses may be misled, if they give too sudden a judgement on what they see, without assuring themselves of the truth by repeated experiments. ...[56]

The plates include the usual accessory apparatus: plate 2 of his variable microscope shows a cavity slide for supporting a frog's web, a very early representation of this still widely-used slide. Stage forceps are figured, as are watchglasses, and tubes for acquatic creatures. These apart, the slider is the only object support.

The younger Adams wrote a successor to his father's book.[57] He states that his father's book and that of Baker were very imperfect, and since much had been done since their time a much fuller work was required: he obviously acknowledged that the later editions of these books were mere reprints. There is a lot of difference between his book and its predecessors: the topics treated were wider in range, including details of the Linnean classification, of Lyonet's work on the willow caterpillar, of work on *Hydra*, as well as on smaller animalcules after Muller, the structure of timber, and a lot on minerals and salts.

A range of instruments is described at length, his chapter IV being concerned with using the microscopes and preparing the objects. He follows Boerhaave in recommending Swammerdam's technique of minute dissections with fine scissors, as knives and lancets are apt to disorder delicate substances. Swammerdam's use of oil of turpentine to clarify his dissections is mentioned, as is Lyonet's method of fixing small insects for dissection by partial immersion in melted wax. Each possible specimen is considered in turn, with details for preparing it. Hooke's technique of immersing specimens in water or oil is mentioned, specifically for nerves, tendons and muscular fibres.

His advice for preparing blood is interesting. A piece of glass is used on which to spread the film as thin as possible, the blood possibly being diluted with a little warm water. Hewson's method of using serum as a diluent is finally recommended, without stating its origin. There is no doubt that diluting with isotonic serum is the secret of making good smears, and the method is still in use in essence.

The structure of timber is discussed fully, and Adams says that as only Hill had handled the subject, his advice will be repeated, together with Adams' improvements in the cutting engine. Hill's advice is indeed repeated, and plate 9 shows an improved microtome; curiously, there is no mention of this anywhere in the actual text. Adams does say that his father invented an instrument for cutting transverse sections of wood in about 1770, implying that he was the actual inventor of Hill's microtome, Cummings only later improving it. This would mean that Hill was unaware of the real inventor of the instrument, which is difficult to accept; further, there is obvious confusion on the same page of Adams, where he couples the name of Hooke with that of Custance. Custance is credited with improving the optics of the microscope, which is very unlikely indeed!

Adams refers to another instrument for cutting sections of timber, more easily managed than Hill's, shown in plate 9, and to be described in a later chapter: as has been said, the description is not in fact given. The price list at the back of the book makes no mention of it either, and so little is made of this that one can hardly believe that the device was actually to hand at that time. For so novel a cutting engine, with obvious scientific and commercial value, much more would have been made of it had it been past the drawing board stage. On the same page as this discussion of microtomes (page 21), some recent workers have imagined that the younger Adams suggests that his father invented the later, non-Hill, microtome; on careful reading this is found not the case. It may be that the older man had some kind of hand in designing Hill's instrument in about 1770; the sliding-knife machine could have been thought up only a little while before the book was published in 1786, and could owe nothing to him. Figure 5 shows the original drawing of the instrument.

The device was eventually made, as shown by the fact that one was bought in September 1789, for £6, for Teyler's museum in Holland.[58] This machine survives, and the author has examined it and used it (see plate 6). The instrument is in very good order, smooth in action and with a good edge to the blade. It is much easier to use than the Hill design, and although quite rigid, is less so than Custance's would probably have been. It is very likely that this is the production version of a machine only partly conceived by Adams when his book was being written; there are some differences of detail between the illustration in the book and the actual instrument.

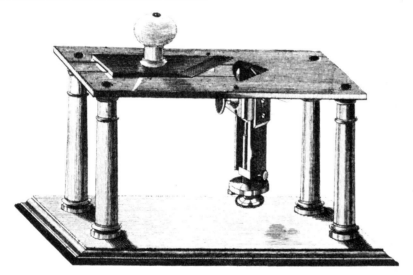

Figure 5. The Adams microtome, 1787 (from Plate IX from Adams' *Essays on the microscope*, 1787).
This represents, probably, what was at that time only the drawing-board stage in the manufacture of this instrument. Nonetheless, a lot of thought and practical understanding had gone into it, and the design was subsequently put into production.

A second edition of the book was published in 1798,[59] being largely a reprint with a few additions by the editor. The text of the first edition concerning the two microtomes is exactly repeated, and there is still no description of the Adams instrument! In addition, when describing plates of timber sections, it is stated that the specimens were provided by Custance.[60] The price list of microscopes and apparatus sold by Jones of Holborn does now include a 'cutting engine for slices of vegetable objects' at £3 3s. This is much less than the £6 paid for the Teyler's example bought nine years before, which might strengthen the supposition that this was a pilot version. Towards the end of the text there is a list of microscopic objects classified by type;[61] as might be expected, most of them are of very general kind, with specific reference to Custance timber sections. These are priced at £1 10s. per dozen, and are mentioned as being generally supplied with Jones microscopes. Others contained insect parts and whole mounts, a few fossils, a few histological specimens, and a variety of chemical salts.

The book did not advance the state of microtechnique, but was a competent compilation from the literature, showing us how little progress had been made in mounting techniques. It is a useful summary of the state of the art of microscopy at the close of the eighteenth century.

Ypelaar and his preparations

Although most surviving eighteenth-century preparations seem to be of British origin, one Continental maker deserves note for the high quality and unusual form of his mounts. Abraham Ypelaar was originally a diamond setter and merchant of Amsterdam, but after the Anglo–Dutch naval war, in 1780 he turned his hobby of mounting objects for the microscope into a profession. Rooseboom's paper[62] gives helpful details of his life and circumstances, and shows that he had a high opinion of his own worth and that of his preparations; nonetheless, he bought in sections of timber from England whenever he had to supply them, and this probably meant that they were made by Custance. This is supported by Turner,[63] who records that Van Marum bought 42 sliders of timber sections in London in July 1790, for £5 5s., which is also in accordance with the price list at the end of the 1798 edition of Adams. Some details of Ypelaar mounts are included in the caption to plate 7. A surprising number have survived, in England as well as abroad,[64] which testifies to the industry of Ypelaar.

Hewson's preparations

Of very different character are the few surviving preparations made by Hewson before 1774, the year of his death. So important are these from the scientific viewpoint, that I have included two pages of plates to illustrate the two types. Briefly, these are the preparations preserved in liquid in tubes (shown in plates 8, 9 and 10), and those spread on glass, dried, and varnished (shown on plates 11, 12 and 13). Both are of high quality, and both were made for scientific purposes, being unavailable commercially. A useful reference for these preparations, kept in the Hunterian Museum of the Royal College of Surgeons of England, is by Dobson.[65] She records that there were originally 70 preparations in tubes, and 147 dried. The collection as it existed in February 1975 consisted of 43 tubes (some of them empty of spirit but with tissue *in situ*), and 47 dried preparations. The list of them as written in volume 1 of the manuscript catalogue entitled *Royal College of Surgeons Histological Collection* is on pages 365–371 (only the left-hand pages, numbered odd, are used).[66] William Hewson was born in 1739, and came to London at the age of 20. He went to Edinburgh to study for a year, then became William Hunter's partner about 1762, taking an active part in the Windmill Street residence's school of anatomy which opened in 1768. It is likely that the preparations were made between that date and 1771, when the partnership with Hunter ended. The mounts were left to Magnus Falconar, and it is known[67] that on his death in 1778 no fewer than 300 microscopical objects originating from Hewson were auctioned, complete in a mahogany cabinet with 16 drawers. It is remarkable that of these, about one third still survive. Those selected for photography are captioned on pages 158 and 160, where technical details are recorded, but it should be said here that it has not been possible to check that the liquid used is ethanol, or at what concentration. It is likely to be so, and certainly there are no preservative salts in solution, for, where slight weeping has taken place, there is no trace of solids.

The first thirty years of the nineteenth century

As outlined above, the eighteenth century had provided sporadic advances of varying worth, not assimilated into an armoury of techniques generally in use. Another significant advance should be mentioned, as an even better example of overlooked possibilities: this is the first use of a cover-slip. Hoff[68] has shown that Ingen-Housz used a cover first of mica and then of thin glass, to prevent the evaporation of fluid mounts. These were used with both ordinary and solar microscopes (being doubtless highly necessary for the latter), and were described in the book published in 1789.[69] There is no equivocation about his description, and this is certainly the first recorded use of such a cover. However, this cover was clearly not intended for permanent preparations, and the topic is taken up again below (see page 113).

A quickening of interest in the microscope was evident in the early years of the nineteenth century. In Germany botanical work was carried out, in an organized manner, but in England there were few original observations at this time: important innovation would not follow here for some years. For the greater part of the century the English would regard a preparation as an end in itself, while the Germans would regard it merely as a means to an end. Be that as it may, before about 1830 most preparations were mounted dry if intended for sale for amateur use: those for scientific purposes were made by the investigator concerned as temporary wet mounts, by teasing, compressing, and minute dissection and maceration. It was usually the case, so far as can be made out from the scant technical details in the literature, that preparations were left uncovered, giving them a short life only.

Attempts were therefore made to render fluid mounts more permanent. Goring, a well-known worker of the time, recommended sealing in tubes:

> I have neglected to describe a kind of slider which I use in my microscope; it is composed of a glass tube, flattened, and drawn out to the size of a common slider, and polished on one side: its use is to hold microscopical objects which will not keep in a dry state, such as pieces of finely injected membrane, petals of flowers, and the like; these little preparations are introduced into the slider, which is then filled with spirits, and covered at the end with a bit of bladder secured by a wax thread.[70]

Such tubes have a limited life, as the seal cannot be made perfectly tight, as evidenced by inspection of the Hewson preparations. They are also difficult to support properly under the microscope, in addition to being optically poor. A better answer to the problem is to use a flat slider with a container for the liquid attached. Holland[71] described a deep cell which he had used satisfactorily for some time, and Cooper stated that a Dr Cook had used the method since about 1820, using white lead as the medium for building up the cell wall.[72] Such deep-cell preparations were to be very popular over the next forty years, for lack of adequate sectioning techniques resulted in much popularity for thick sections, injected and arranged for viewing by reflected light. Another, more rational use of a deep cell was to contain minute dissections: these are discussed in more detail below (see page 85).

A development from the simple wet mount was to use various media to observe how the tissue reacted to it, rather than as a preservative only: this is the beginning of histochemical techniques proper. Sporadic reports of such tests appear, often from the Continent, in the early 1800s. An example is Girod-Chantrans,[73] who used a variety of chemicals such as spirit, nitric acid, potassium carbonate, water at a variety of temperatures, and other reagents, before microscopic examination; the resulting colour reactions were noted and used as the basis of work on various mosses, lichens, and fungi. Raspail used the starch–iodine reaction (already known on the macro scale) under the microscope in 1825,[74] and in 1829 adopted methods for testing for albumen and sugar.[75] The later developments are touched on below (see page 95), since the techniques assumed much importance in microscopy and the life sciences.

Test objects

Pritchard gives an account of these special preparations,[76] showing how Goring had arrived at the idea from Leeuwenhoek's account of the dust on a silkworm's wing. The intense development of microscopic optics from the 1820s required means to check improvement in resolution, and Goring provided biological objects which needed definite degrees of perfection in lenses used to resolve them. A revealing first chapter in the important book by Goring & Pritchard[77] shows that Goring realised the increasing importance of the microscope: no book had been published in England since the second edition of the younger Adams in 1798,[78] and there was need for specific instruction to improve the quality of microscopy. The book appeared in quarterly parts, the first on 1st February 1829, and broke new ground by showing coloured engravings of living objects seen through the microscope—the first to be published in England, and a considerable technical achievement. The original descriptions of test objects are mentioned in this book, but are much more fully documented in a later book:[79] in this there is a full list of different degrees of difficulty in resolution. The explosion of interest in test objects, and especially in diatoms, which was to occupy English microscopists for so large a part of the rest of the century, is quite unheralded at this time. Goring does not go unchallenged as the originator of test objects. Chevalier[80] claimed that le Baillif in 1823 examined butterfly scales and other objects for such a purpose, and that he, Chevalier, told Goring of this. This is impossible to prove now, one way or the other. Chevalier further states that A. C. M. le Baillif, who worked in Paris, used venice turpentine from 1825 for mounting, and that Goring was also told of this by Chevalier, in 1826. Chevalier also makes a further claim, that he himself used cover glasses for the highest magnifications in 1825:

J'attache une grande importance aux petits carrés ou disces de verre mince. Le

mica dont on fesait usage autrefois, se brisait trop facilement, et ne jouissait pas
d'une transparence parfaite. Ces lamelles étaient traversées presque en tout sens,
par des stries que d'amplification rendait fort apparentes. Le moindre frottement
en augmentait encore le nombre. En 1825 je préparai des verres assez minces pour
être employés avec les plus grandes grossissemens et depuis cette époque, je puis à
peine satisfaire aux nombreuses demandes qui me sont addressées par les
observateurs.[81]

It is clear that this use was not for permanent preparations.

The development of the slide from the slider

Colour plate I shows a group of sliders which bridge the gap between
the ivory multi-place slider and the glass slide. This development is well
documented by Gould,[82] who first says that an object being examined
in a drop of water should be covered with a piece of talc, . . . *and you will*
have a perfect level surface.[83] He then goes on to say

> You will find clear slips of glass preferable to talc in forming the slides for
> transparent objects; they may be prepared in the following manner: take two
> slips of glass about the size of the ivory slides; then get a piece of writing
> paper, with holes, of the same size; wet one side of the paper with gum-water,
> and lay the glass upon it, suffering it to dry; then place your objects in the
> holes, wet the other side in the same manner, and lay on the other glass.
> Several objects may be placed on each slip for viewing the animalcule.[84]

This method became standard, and is quoted by, for example,
Griffith,[85] von Mohl,[86] Quekett,[87] and Harting.[88] Such sliders are
undoubtedly substantial, but as higher powers became available they
were cumbersome. By 1830 some glass sliders had mica covers, held in
place (in the absence of a mounting medium) by a paper cover: these
would continue to be provided for the next 75 years, as a means of
decoration and identification. This method is described by, for
example, Solly,[89] Varley,[90] and Pritchard.[91]

There can be no doubt that the microscope in 1830 was capable of
showing sufficient detail to demonstrate a number of fundamental
truths. That it was not often so used was responsible for this cry from
the heart by Goring:

> Great disgrace has been brought on microscopic science by the manner in
> which it has been perverted to the support of preconceived opinions and
> hypothetical views, as well as to a spirit of wondermaking. I hope that a new
> and golden age of observation will now Commence.[92]

Perhaps a good example of a man seeing what he thought he should
see—compounded by gross optical deficiencies in his instrument—was
the work published by Milne-Edwards in 1823.[93] All tissues were

described as being made up of globules about 1/300 mm diameter, regardless of the organ or even the species concerned. Pickstone[94] has shown that Milne-Edwards himself modified his views in the next three years, and that they were not shared by all his contemporaries in any case;[95] but Goring's hope for a new golden age was only answered by Lister.

The paper published by Hodgkin & Lister[96] was made possible by Lister's work on improving the microscope objective, details of which were later published,[97] albeit in somewhat attenuated form, by a Royal Society which failed to recognize their value in removing both achromatic and spherical aberrations. The difference in results achieved by Hodgkin & Lister as compared with Milne-Edwards can be described only as a breakthrough. Their paper must be regarded as the foundation of modern histology, for the true nature of tissues is made clear. Plate 14 shows a modern preparation photographed with the Lister-inspired microscope of 1826, with notable clarity.

Plate 15 is of a preparation of reasonably contemporary date photographed through a modern instrument. It is obvious from this that there remained a great deal to accomplish in preparative techniques!

References for Chapter 2

1. E M Nelson (1902)—A bibliography of works (dated not later than 1700) dealing with the microscope and other optical subjects, *J.R.M.S.* *22 :20–23.*
2. The most convenient source of the voluminous correspondence is: Anon—*The collected letters of Antoni van Leeuwenhoek*. (Amsterdam: Swets & Zeitlinger.) Publication continues, volume 1 being dated 1939, volume 8 (taking the correspondence to 1692), dated 1967.
   A very helpful summary of the various letters as they apply to zoology and histology, with notes on his techniques so far as they can be deduced, is given by F J Cole (1937)—Leeuwenhoek's zoological researches I, *Annals of Science* *2 :1–46,* and II, *Annals of Science* *2 :185–235.* See especially pp. 43–45.
3. See Clarke, note 1 in chapter 1, for good examples.
4. R Hooke—*Micrographia : or some physiological descriptions of minute bodies made by magnifying glasses*. (London: J Martin & J Allestry, 1665.)
5. ibid., plate XI, facing p. 115.
6. ibid., pp. 112–113.
7. R Hooke—*Microscopium*. (London: J Martyn, 1679.) This is part of a collection of lectures by Hooke, and has been reprinted in volume 8 of R Gunter—*Early science in Oxford*. (Oxford: Clarendon Press, 1931.) See pp. 307–308 of this reprint.
8. ibid., p. 310.
9. For Hooke's complaint see W Derham—*Philosophical experiments and observations of the late eminent Dr Robert Hooke*. (London: W & J Innys, 1726.) See p. 261.

10. loc. cit., note 2 above. A good account of the man and his work is in C Dobell—
Antony van Leeuwenhoek and his 'little animals'. (London: J Bale, 1932.)

11. M Folkes (1724)—Some account of Mr Leeuwenhoek's curious microscopes, lately
presented to the Royal Society, *Phil. Trans. R.S.* *32 :446–453.*

12. H Baker (1740)—An account of Mr Leeuwenhoek's microscopes, *Phil. Trans.
R.S.* *41 :503–519.*

13. loc. cit., note 10 (Dobell), p. 319.

14. loc. cit., note 10 (Dobell), p. 331.

15. loc. cit., note 8 above.

16. P H van Cittert—*Descriptive catalogue of the collection of microscopes in charge of the
Utrecht university museum with an introductory historical survey of the resolving
power of the microscope.* (Groningen: P Noordhof, 1934.) See pp. 13–16.

17. A v Leeuwenhoek (1674)—Microscopical observations from Mr Leeuwenhoek ...,
Phil. Trans. R.S. *9 :121–128.*

18. P Harting—*Het mikroskop, deszelfs gebruik, geschiedenis en tegenwoordige toestand.*
(Utrecht: Van Paddenburg, 3 vols, 1848–1854.) Vol. 3, p. 41 footnote describes
the auction catalogue, and p. 465 lists the objects.

19. N Grew—*The anatomy of plants.* (London: W Rawlins, 1682.) See p. 9.

20. F J Cole (1921)—The history of anatomical injections, pp. 285–343 in C Singer—
Studies in the history and method of science, vol. 2. (Oxford: Clarendon Press,
1921.)

21. M Malpighi—*De viscerum structura exercitatio anatomica.* (Bologna: J Montij,
1666.)

22. J Swammerdam—*Biblia naturae ; sive historia insectorum.* (Leyden: I Severinus &
B V der Aa & P V der Aa, 1737–1738.)

23. P W van der Pas (1970)—A note on the origin of staining techniques for
microscopical preparations, *Science History* *12 :63–72.*

24. Anon (1714–)—(a series of untitled paragraphs) *Journal Literaire.* *3 :238–241.*

25. Anon (1717)—An extract from the *Journal Literaire,* published at the Hague, for
the months of January and February 1714, p. 238, being, An account of several
observations concerning the frame and texture of the muscles, by Mr Muijs of
Franeker, *Phil. Trans. R.S.* *29 :59–61.*

26. H Boerhaave—*A method of studying physic.* (London: C Rivington, B Creake & J
Sackfield, 1719.)

27. F Ruysch—*Opera omnia anatomico-medico-chirurgica.* (Amsterdam: Jansson-
Waesberge, 6 vols, 1696–1729.)

28. V I Kosobutskiï (1967)—A brief survey of the exhibits of the museum of the history
of microscopy (Institute of the History of Science and Technology of the USSR
Academy of Sciences), *Proc. R.M.S.* *2 :356–361.*

29. J N Lieberkühn—*De fabrica et actione villorum intestinorum tenuium hominis.*
(Lugduni Batavorum: C & G J Wishof, 1745.)

30. C N le Cat (1741)—On the figure of the canal of the urethra, *Phil. Trans.
R.S.* *41 :681–686.*

31. M Joblot—*Observations d'histoire naturelle, faites avec le microscope.* (Paris:
Briasson, 1754.)

32. A Trembley—*Mémoires, pour servir à l'histoire d'un genre de polypes d'eau douce à
bras en forme de cornes.* (Leiden: J & H Verbeek, 1744.)

33. R von Rosenhof—*Der monatlich Herausgegebenen Insecten belustigung.*
(Nuremberg: Geben, 1755.)

34. O Müller—*Vermium terrestrium et fluviatilium seu animalium infusoriorum
helminthicorum et testaceorum non marinorum, succincta historia.* (Copenhagen:
Heineck & Faber, 1773–1774.)

35. F Fontana—*Traité sur le vénin de la vipere sur les poisons Americains.* (Florence: [no
publisher quoted], 1781.)

36. For a modern study of part of Fontana's work see E Clarke & J G Bearn (1972)—The spiral nerve bands of Fontana, *Brain* 95 :1–20.

37. J Hill—*The construction of timber, from its early growth ; explained by the microscope, and proved from experiments, in a great variety of kinds.* (London: printed for the author, 1770.)

38. ibid., p. 5, of second edition of 1774.

39. ibid., p. 4.

40. ibid., p. 4. An interesting modern appraisal of Hill's work is in G B Thomas (1968)—John Hill and 'The construction of timber', *Proc. R.M.S.* 3 :186–209.

41. op. cit. note 38 above, see p. 10.

42. J J P Moldenhawer—*Beitrage zur anatomie der pflanzen.* (Kiel: 1812.)

43. H E F Garnsey & I B Balfour—*Sachs' history of botany 1530–1860.* (Oxford: Clarendon Press, 1890.) See p. 258.

44. Hill, note 38 above, p. 11.

45. ibid., p. 13.

46. ibid., p. 14.

47. ibid., p. 29.

48. R J Thornton (1799)—Account of the new machine invented by the late Mr Custance, for making vegetable cuttings for the microscope, *Philosophical Magazine* 3 :302–309.

49. W F F von Gleichen—*Les plus nouvelles decouvertes dans le regne vegetal ou observations microscopique.* (Nuremberg: widow of C de Launoy, 1770.)

50. M F Ledermueller—*Amusement microscopique tant pur l'esprit que pour les yeux.* (Nuremberg: A W Winterschmidt, 3 vols, 1764, 1766, 1768.)

51. H Baker—*The microscope made easy.* (London: R Dodsley, 1743.) See p. 51.

52. F Bonanni—*Observationes circa viventia, quae in rebus non viventibus reperiuntur. Cum micrographia curiosa.* (Rome: D A Hercules, 1691.) See plate II, p. 28.

53. op. cit. note 51, pp. 56–62.

54. The background to Baker's work is discussed by G L'E Turner (1974)—Henry Baker, FRS: founder of the Bakerian lecture, *Notes & Records R.S.* 29 :53–79.

55. G Adams—*Micrographia illustrata : or the microscope explained.* (London: for the author, 1746; 2nd edn, 1747; 4th edn, 1771.)

56. ibid., p. xliii.

57. G Adams—*Essays on the microscope ; ...* (London: R Hindmarsh, 1787.)

58. G L'E Turner & T H Levere—*Van Marum's scientific instruments in Teyler's museum.* (Leyden: Noordhoff International, 1973.) See p. 304.

59. G Adams—*Essays on the microscope ; ... Second edition, with considerable additions and improvements by F Kanmacher.* (London: Dillon & Keating, 1798.)

60. ibid., p. 599.

61. ibid., pp. 698–712.

62. M Rooseboom (1940)—Some notes upon the life and work of certain Netherlands artificers of microscopic preparations at the end of the XVIIIth century and the beginning of the XIXth, *Janus* 44 :24–44.

63. op. cit. note 58, see p. 306.

64. Good examples are to be seen in the Zeiss collection in Jena, Tyler's museum in Haarlem, and the Whipple museum in Cambridge.

65. J Dobson (1960)—John Hunter's microscope slides, *Annals R. Coll. Surgeons Eng.* 28 :175–188.

66. The manuscript volume is one of two which record the 10000 slides originally in the Collection of Surgeons collection, most of them placed there by John Quekett. Further discussion of these sources is given on p. 69 below, for they are important in the discussion of Quekett's own preparations, although he did not personally write these volumes.

67. Dobson (note 65 above, see p. 186) describes the finding of an annotated copy of the auction catalogue of the Falconar sale. Many of the eminent medical men of the time attended and made purchases.

68. H E Hoff (1962)—Jan Ingen-Housz and the cover-slip, *Bull. Hist. Med.* *36 :365–368.*

69. J Ingen-Housz—*Nouvelles expériences et observations sur divers objets de physique.* (Paris: Barrois, vol. 2, 1789.)

70. C R Goring (1819)—Description of an improved microscope, *Annals of Philosophy* *13 :52–59.* See p. 58.

71. J Holland (1833)—Triplet for a microscope, *Trans. Soc. Arts* *49 :120–126.*

72. D Cooper (1841)—Microscopical memoranda, *Mic. J. & Str. Rec.* *1 :32.*

73. J Girod-Chantrans—*Recherches chimiques et microscopiques.* (Paris: Bernard, An X.) (An X = 1802.)

74. F V Raspail (1825)—Développement de la fécule dans les organes de la fructification des céréales, et analyse microscopique de la fécule, suivie d'éxperiences propres à en expliquier la conversion en gomme, *Ann. Sci. Nat.* *6 :224–239, 384–427.* See p. 229 *et seq.*

75. F V Raspail (1829)—Neues Reagens zur Unterschiedung des Zuckers, Oeles, Eiweisstoffes und Harzes, *Froriep Notizen* *24 :129–136.*

76. A Pritchard—*The microscopic cabinet of select animated objects.* (London: Whittaker, 1832.) See p. 135 ff.

77. C R Goring & A Pritchard—*The natural history of several new, popular, and diverting living objects for the microscope.* (London: A Pritchard, 1829.)

78. op. cit. note 59.

79. op. cit. note 76.

80. C Chevalier—*Des microscopes et de leur usage.* (Paris: chez l'auteur, 1839.) See p. 169.

81. ibid., p. 164.

82. C Gould—*The companion to the microscope.* (London: W Cary, 1827.)

83. ibid., p. 8.

84. ibid., p. 9.

85. J W Griffith (1843)—On the different modes of preserving microscopic objects, *Ann. & Mag. Nat. Hist.* *12 :115–117.*

86. H von Mohl—*Mikrographie.* (Tubingen: Fues, 1846.) See p. 329.

87. J Quekett—*A practical treatise on the use of the microscope.* (London: H Bailliere, 1848.) See p. 285.

88. P Harting—*Das Mikroskop.* (Braunschweig: F Viewig, 1866.) See vol. 2, p. 65.

89. R H Solly (1831)—Letter from R H Solly on certain parts of vegetable structures, *Trans. Soc. Arts* *48 :400–412.*

90. C Varley (1831)—Improvement in the microscope, *Trans. Soc. Arts* *48 :332–400.* See p. 398.

91. op. cit. note 76. See p. 230.

92. op. cit. note 77. See p. 12.

93. H Milne-Edwards (1823)—Mémoire sur la structure élémentaire des principaux tissus organiques des animaux, *Archiv. Gen. de Med.* *3 :165–184.*

94. J V Pickstone (1973)—Globules and coagula: concepts of tissue formation in the early nineteenth century, *J. Hist. Med.* *28 :336–356.*

95. F V Raspail (1827)—Premier memoire sur la structure intime des tissus de nature animale, *Repert. Anat. Physiol.* *4 :148–161.*

96. T Hodgkin & J J Lister (1827)—Notice of some microscopic observations of the blood and animal tissues, *Phil. Mag.n.s.* *2 :130–138.*

97. J J Lister (1830)—On some properties in achromatic object-glasses applicable to the improvement of the microscope, *Phil. Trans. R.S.* *130 :187–200.*

3. Works on Microtechnique 1830–1910: a Select Descriptive Bibliography

Although many of the important advances were reported in the periodical literature, books on microscopical specimens and their preparation also formed a valuable means of communicating such discoveries, as well as codifying them in handy form. Many books were intended only for beginners, and although the English amateur was an important original worker in some ways, reference to all such books in all languages would be both tedious and superfluous. Therefore, certain works of merit in their geographical and historical contexts, have been chosen to present a discursive survey of their content and importance, in approximately chronological order.

Works published in England

In the 1830s there were a few current books. Gould[1] reached its 7th 'edition' in 1831, and its 15th in 1848 (issues between these two are not recorded). This last version is significant in its inclusion of details of mounting in canada balsam, which the 1831 edition omits. The book by Goring & Pritchard would then still be in print, but it is not easy to determine how many of its parts were ever issued.[2] In 1832, however, Pritchard published a work valuable for its excellent directions for mounting.[3] In its 246 pages are many woodcuts, plus 13 coloured plates. Various mountants are described, including the mica discs in vogue for many years, but on page 230 is an original mountant, thick gum. This was a very important advance, and is discussed in more detail on page 88 below. In 1834 Pritchard also published what seems to be the first textbook of protozoological technique in any language.[4] The advice on mounting is rudimentary, dry preparations being suggested, although it is stated that specimens can be 'relaxed' in a moist chamber later if need arise. A second edition followed in 1843, a third in 1851, and a fourth in 1861, each rather larger than its predecessor. The first edition includes two pages of price-list, dated 1834 from 18 Picket-street, Strand; it includes outfits ranging from a Pritchard pocket microscope at £1 18s., to a Goring Operative Aplanatic Engiscope at

£50. Forty microscope slides of transparent objects cost 7s. 6d.

This author appears to be very prolific, for in 1835 he published his small but very significant list of microscopic objects.[5] This somewhat rare work is of the first importance as containing the original published description of balsam mounting, discussed on page 88 below. Of its thirty-six pages, twelve are essentially a list of names to be cut up and stuck on prepared slides. The other pages extol the virtues of specimens for exhibition. Three pages of catalogue are included at the end (see page 290 below). A second version was issued in 1847, having 18 pages of labels and a much fuller text of 192 pages. This issue is anonymous, but does laud Mr Pritchard for being the first to describe balsam mounting! The book might have been intended for trade distribution by various opticians to supply with their microscopes and Pritchard's preparations.

Pritchard also published in 1840 a second edition of his illustrations,[6] and a third in 1845. Little had been altered in these, but their relatively frequent reappearance suggests that a demand for works on the microscope was sustained.

Another, very rare, work for the amateur is that by Fisher,[7] which has a section on mounting objects, and a remarkable figure of a microtome of relatively advanced design (page 48). This has a knife carried in a frame, but no reference is made to its origin. This is the first illustration of a microtome to appear after that in the 1798 edition of Adams, and is discussed on page 121 below.

By far the most important book to have appeared to its date was that of Quekett,[8] which was the first major work on object preparation in any language. The first section is a history of the microscope, the second deals with its use, and the third (more than half the book) with specimen preparation. This is a thoroughly practical work set out in a properly scientific way. A second edition was issued in 1852, and a third in 1855. An edition in German was published in 1850.

Although intended to cover a much wider field, the six volumes of Todd's cyclopaedia do contain a large amount of information on microscopical techniques, and are well worth looking through: sets are not uncommon.[9]

Perhaps surprisingly, the next major work to appear was Beale's manual on medical microscopy.[10] Based on a course of lectures given in 1853, the work is thoroughly practical, and deals with the value of the instrument in diagnosis, the apparatus needed, specimen preparation, reagents required, and details of the various organs seen in health and disease. The work is detailed, and provides a useful insight into the standards of its time. A second edition, much enlarged, was issued in 1858, a third in 1867, and a fourth in 1878.

The book by Hogg[11] was a different kind altogether, designed expressly for the amateur. This popular need was supplied in 440 pages, with about 500 woodcuts: the book is well suited to its purpose. Various editions followed, a second in 1856, a third in 1858, a sixth in 1867, and a tenth in 1883. None of these differed much from each other, all being modelled on Quekett, but with less attention to mounting and more to collecting. A fifteenth edition was issued as late as 1898: this was reset in larger format, and had new illustrations. It remains an attractive book for those interested in older instruments. Hogg was a medical man, and an active member of the Medical Microscopical Society, where he read entirely scientific papers, so one must suppose his writing of an amateur's book was in answer to a clear need.

Rather more scientifically inclined was Carpenter's work.[12] He was also a medical man, well-known for his work in physiology. His book requires rather more of the reader than Hogg's book, and it also went into many editions over a long time: a second in 1857, third in 1862, and so until a sixth in 1881 was followed by a much enlarged and reset edition of 1891—revised by Dallinger. The last edition was in 1901, and its 1136 pages remain valuable for reference. The treatment of section cutting is rudimentary, and this may be taken as representative of the relative lack of interest attached to histology in England.

Griffith's elementary book[13] contains only a little material of significance on mounting objects, while the book by Beck[14] is really an instruction manual for his firm's instruments; it does, however, have excellent illustrations, including a section on mounting equipment, and a rare illustration of a simple microtome in use (page 118). His plate XXI shows a range of compressoria and other instruments, and his page 128 carries actual specimens of covering papers pasted in.

Beale's other well-known book[15] was issued in 1857, as a very slight volume of only 124 pages. He stated in his preface that he wished to put out a book which would assist and stimulate work with the instrument, and that the book grew out of a series of lectures given in 1856. However, much of the content is histological, unlikely to appeal to most amateurs. The two serious competitors, Hogg and Carpenter, were obviously well established, and had earned good reviews, but Beale wished to separate technical instruction from clinical results, by offering separate books. A second edition followed in 1861, and a third, much enlarged, in 1865. This has a frontispiece of a photograph of 13 objects prepared by Maddox, but although more material is included little has been done to improve the section on specimen preparation; this may mirror Beale's preoccupation with a relatively limited range of techniques. A useful bibliography is included on pages 264–265, and the recently introduced 'anilin colours' are mentioned. The last edition was the fifth in 1880, and the volumes are still of much interest, in spite of Beale's highly individual views on the nature of tissues and their development.

A scarce work by Notcutt[16] is interesting, for in spite of its wholly amateur approach, it does include, in addition to full instructions for mounting, details of the Topping microtome and a very simple V-groove for sectioning timber. This detail is unusual in books of this kind as early as 1859, as is his small selective bibliography.

Much better known is the *Micrographic Dictionary* of Griffith & Henfrey.[17] It was first issued in 1856, with a second edition in 1860, a third in 1875, and a fourth and last in 1883. The arrangement is that of an encyclopaedia, with most of the entries concerned with specific organisms or organs. Issued in occasional parts for each edition, and being very expensive in sum, this book was still the only one likely to be possessed by most ordinary amateurs. Inspection reveals that the work was not adequately revised between editions, and this was certainly recognized in contemporary reviews.[18] The 696 pages, 41 plates, and 816 woodcuts were an adequate treatment for their time and purpose, but remarkably, on page xxi, they advise that slides should not exceed $2\frac{1}{2} \times 1$ in. in size. Their directions for specimen preparation are not full, and certainly not up-to-date, which is surprising in view of the eminence of the authors, who would have been well read in their subjects.

Much more down-to-earth is a small book by Davies,[19] who makes the point in his preface that although much has been published on mounting objects, most of it is in supplementary chapters only, requiring a search for anything like a complete knowledge of the subject. In his 156 pages, he gives unequivocal advice on virtually all possible procedures, even to the extent of being useful for some things to the present day. By 1880 the book had sold 11 000 copies, a good total and indicative of its merit.

Frankly amateur in scope was the work by Cooke,[20] a well-known microscopist and botanist of the time. This was reissued in many undated reprints for years after its first appearance, and in its 179 pages does condense much of value on mounting, although it lacks the total practicality of Davies. That both should sell so well is significant of the kind of work being carried out in microscopy in England at the time: it must be doubted if they would have sold so well in Germany, for example.

A further example of an intensely practical book is that by Suffolk,[21] who was very well known for his abilities as a mounter. His book, going into a second edition in 1875, is excellent in its treatment of what by that date were the amateur aspects of specimen preparation; it is much less satisfactory in the necessities for section cutting, which by then was important in scientific work.

The same would have to be said, but much more forcefully, of Martin's book.[22] He was a very amateur microscopist, although a professional analyst, and his book is mostly secondhand in its information. His illustrations are remarkably ill executed (as were those of his earlier work on objects seen through the microscope).[23] The reviews of this and his other book are among the most searing of any in Victorian journals, where feelings were not spared.[24]

Outlines of Practical Histology, by Rutherford, is a very different book.[25] Originally issued as notes on lectures, it is a full and scientific account of the tissues, with an especially useful guide to the manipulation of his freezing microtome. A second edition was issued in the following year, and the book is certainly equivalent to the scientific level of contemporary German works; it marks a significant stage in scientific microscopy in England.

Marsh wrote the first book in any language specifically devoted to section-cutting.[26] In its ninety pages useful and clear instruction is given: he favours the Rutherford freezer in its ice/salt form, and condemns all ether freezers as expensive and inefficient (page 25). An indication of his practical experience is that he had suffered from the cylinder of wax in an ordinary microtome becoming detached from the bottom of the well, causing uneven advance; he offers an effective remedy for this (page 24). A second edition followed in 1882, after which time most serious works included information on section-cutting as part of their general treatment. Nonetheless, the book is a landmark of some importance.

An example of this comparatively full treatment is seen in the book by Davis,[27] which is intermediate between the purely amateur and the exclusively scientific. Here, not only is Marsh's book referred to, but also four types of microtome are described, including those of Hailes and Rutherford. Generally, the treatment is at least as detailed as German books of the time. There was a second edition in 1889, and a third in 1895, by which time the Cathcart instrument was included at the expense of the Rutherford. In this last edition the standard had, however, fallen below that of German works.

A most important series of works by Cole must now be described in some detail.[28] These are of major significance on account of their expert authorship, their date of issue, their completeness of subjects covered, their very full illustrations and descriptions, and most of all because matching preparations were issued and have survived to be compared with their descriptions. Originally issued in parts, the volumes as bound are made up as follows.

Volume I contains 52 descriptions, of 26 histological preparations interspersed with both plant and mineral sections. The articles were written by Ady, and the preparations made by Martin J. Cole (Arthur C Cole's son). Taking the first specimen as an example, we have yellow fibrocartilage from cow pinna VLS, described first in general terms, then specifically, then as affected by various reagents including water, acetic acid, gold chloride, osmic acid, and iodine. Methods of preparation are discussed, and there is a good chromolithograph plus a bibliography of 37 entries. The whole forms a full and up-to-the-minute account, covering nine pages. This is a typical entry for this volume, and the five matching slides inspected by this author are all of the highest quality for their time.

The second volume starts with an introductory essay, *The Methods of Microscopical Research*, dealing with the microscope and the eye, the preparation of tissues, section cutting and staining, mounting and drawing specimens, in a total of 83 pages. In addition to the bound volumes, a few of the original parts sometimes occur. For example, Part I is dated Saturday, 16th June 1883, and contains some useful details of this volume, and of volume I. On the inside front cover the contents of the first volume are listed: it was stated to be *a Weekly Periodical for the use of Students, Professors and Teachers, the Medical Profession, and others interested in the progress of the Natural Sciences*. The bound volume was offered at £1 11s. 6d., or £4 10s. 0d. with the 52 matching preparations. The inner rear cover carries an impressive list of those who had thought well of the work, including Cardinal Manning, Richard Owen, Professors Balfour and Herdman, and Drs Dallinger and Hogg. Subscriptions were received by appointed agents not only throughout the United Kingdom, but also in the United States, France and Germany. The rear cover contains many review extracts, the general tone of which can be described only as eulogistic.

The cover also sets out details of the parts to be included in volume II. The three sections, of animal and of plant histology, and of popular microscopical studies, each cost 21s. for the annual subscription, for twelve monthly parts with preparation. Interestingly, the first two parts give the name of the author as John Ernest Ady, but the rest state only that the editor is Arthur C Cole. After the first part of this introductory essay, the succeeding covers carry advertisements for a wide range of microscopical items—mounting materials from Stanley, living specimens from Bolton, microscopes and books from Collins, instruments by Swift and Watson, etc. The subject-matter is all up-to-date (this is significant in English publications of this date), and clear in its practical directions.

Volume II proper, as mentioned, contains thirty-six preparations. One example must suffice here—human cerebellum stained with anilin blue-black. Another excellent chromolithograph accompanies notes on etymology, anatomy and histology, and special mention of Purkinje cells. The method of preparation used was detailed (either freezing or embedding being possible), but only two items (Quain and Klein) are mentioned in the bibliography. This is typical of the less detailed scientific treatment in this volume; in such a commercial situation the lengthy details in the first volume could hardly be sustained.

Volume III includes four sections, each containing 12 preparations. The first consists of botanical histology, the second animal histology, the third pathological histology, and finally popular microscopical studies. An example from the third section is typical of the whole; some of the preparations were illustrated not by chromolithographs, but by pasted-in albumen photomicrographs; and all the pathological subjects are illustrated in this way. The section on anthracosis, dated January 1886, covers four pages of fairly discursive description, including the gross appearance at post-mortem, and refers to an earlier section for references, which it is stated will certainly require modification in the near future. The photomicrograph is not as sharp as it might be, but does adequately convey the nature of the preparation.

Volume IV again includes four sections, identical with, and in the same order as, those of volume III. The third section was contributed by Sheridan Delepine, of St George's Hospital Medical School. Like the other sections in this volume, it is once more illustrated only by lithographs. Interestingly, all the twelve slides are of kidney, in normal state followed by a variety of pathological conditions. The normal kidney is described in some detail over thirteen pages, and has two plates. The specimens were prepared with Müller's fluid, cut on the freezing microtome, and stained with logwood and eosine.

These volumes are a valuable crystallization of the best English practice of the 1880s, and compare well with Continental standards; but Cole was an exceptional man, and the venture was not well supported.[29]

Nonetheless, the whole series is an outstanding landmark in the history of microscopy and histology.

At about this time interest in bacterial work was developing apace, and Crookshank[30] was early in the field with a substantial book of 249 pages, dealing with all aspects of the subject, including microscopical preparation. This had not progressed very far at that time, and it is not without significance that Cole fails to include such preparations, although they would have had considerable commercial value on account of their novelty.

Interest in histology in the medical schools was obviously growing in the 1880s, for several smaller practical textbooks appeared. That by Gibbes[31] has only 107 pages, but covers fixing, sectioning, staining, and mounting, with directions for individual organs. The book is clear and practical, and designed for use at the laboratory bench, unlike most others of previous date. Fearnley's book is smaller still, but follows the same pattern; neither is a textbook of histology, but only of histological technique.[32]

1885 saw the publication of what has become the most important of all books on microscopical specimen preparation: the work by Arthur Bolles Lee.[33] This is still an invaluable source, for Lee often merely added extra material in successive editions, with literature references, which though often vague, are traceable. The second edition was in 1890, the third in 1893, the fourth in 1896, the fifth in 1900, the sixth in 1905, and the seventh in 1913. The last (11th) edition was issued in 1950, still unchanged in some particulars. The first edition was obviously influenced by Frey (see page 43, below), and must count as the first scientific, as opposed to popular, account of micro-technique in English. In his preface, Lee states that it was his aim

> ... to put into his [his reader's] hands a concise but complete account of all the methods of preparation that have been recommended as useful for the purposes of Microscopic Anatomy. ... The Collection of Formulae here brought together is, I believe, practically exhaustive; no process having any claim to scientific status having been rejected ...

Even more significantly, on page 1 he writes:

> From a superficial inspection of the formulae set out in the following pages, an uninformed person might not unreasonably conclude that microscopical methods are as numerous as microscopical anatomists, or even that to every particular investigation there belongs a particular process; he would not suspect that there is a backbone of constentaneous practice running through modern researches. That conclusion, however, would be erroneous; such a backbone exists; the great majority of recent investigations have been mainly

carried out by carefully fixing the structures to be examined, staining them with a nuclear stain, dehydrating with alcohol, and mounting series of sections of the structures in balsam.

Only in the early 1880s, then, had our modern methods started to gain currency.

His first chapter, an introductory survey, covers ten valuable pages, showing that Lee regarded as normal the preparation of sections *by the microtome*. Lee, of course, though English, worked and was trained on the Continent! Subsequent chapters are on fixing (a word he has to define); on the theory of staining; and on carmine (in much detail); then others on haematoxylin; vegetable stains; gold; other metallic salts; and coal-tar colours, in general and in particular. Of these Lee is very critical. He admits they give brilliant colours quickly, but doubts their permanence. A chapter on staining bacteria includes the methods published between 1879 and 1884, amounting to only twelve references. Multiple stains are discussed: he approves them for teaching, but not for original research, and his passage on pages 138–139 might still be borne in mind:

> I know that in England of late years a great many sections of young rats' tails and children's larynges have been polystained in many different colour combinations; but I am not aware that any new fact of scientific importance has been brought to light by any of these methods.

His chapters on microtomes and embedding methods are also enlightening, for he clearly distinguishes between merely surrounding an object by a supporting mass, and actually infiltrating that object's interstices with the support. He has pertinent things to say of the design of microtomes, demonstrating familiarity with many of them: the usual thickness of section is stated to be 5 μ. Zeiss, Schanz, and Reichert are all mentioned, but the best he says is the Thoma pattern, by Jung. Various embedding masses are listed, and the mounting of serial sections covered.

Other reagents are described in detail, followed by a number of chapters concerned with specific organs. The same pattern is followed in later editions, with further methods included as they appeared in the literature. The first edition has 424 small pages, the fourth has 536 larger ones, while the eleventh edition has 753 pages. In addition to the eleven English editions in 65 years, the book was published in the United States, and had 3 editions in French and 4 in German. This is an impressive record, and indicative of the authority enjoyed by the work.

Given such a magnum opus, other authors might have been expected to have used at least some of the information in their own books, but three years later Colman's book on sectioning showed little evidence of this.[34] Although only 107 pages long, the subject is inadequately treated even for the beginners to whom it was directed, and Lee's book is not even mentioned!

Very much more satisfactory is Stirling's book.[35] This is well illustrated, includes exercises for the individual student, and is clearly set out. Descriptions of tissues are included, in addition to practical techniques, and in its 356 pages a good standard is maintained. A useful section on microtomes is offered, and up-to-date methods of infiltration are examined. The second edition, of 1893, had 419 pages, and earned good reviews.

The first book of its kind in English, the little work by Squire,[36] giving only a terse list of available reagents and formulae, was written by a well-known supplier of such substances. It is essentially a recipe compendium, giving no references whatsoever, a practice followed also by its German counterpart (see Behrens, page 44 below).

Published in the same year, 1892, was the book by Wethered,[37] offering in its 412 pages suitable guidance to medical man and student, but with much less satisfactory background material than Stirling. The book is mentioned here as an example of a poorly-documented work, one of the last to be published in England before the German trend to over-documentation started to influence authors in this country.

That there was still a place in England for a general textbook of microscopy is shown by the success of Cross & Cole.[38] ('MI Cross' was a pseudonym for W Watson-Baker, an eminent manufacturer, making the book a collaboration between two commercial writers, a combination unknown on the Continent.) In this general textbook, however, the section on specimen preparation occupied a greater portion than it did in the earlier texts which were still being reissued in new editions. The book went into a second edition in 1895, a third in 1903, a fourth in 1912, and a fifth and last in 1922, by which time it was much expanded and improved.

The older Cole issued a substantive edition of his methods book[39] in 1895, and says that the reception accorded to his first edition (sent out as part of the periodical/slide package ten years before) prompted him to do this. This second version contains the results of his thirty years of experience in microscopical work: all the formulae from his private notebook, and all his own special processes and methods of working are revealed, to produce a thoroughly practical book which can still be of use today. Of the 207 pages, 121 are devoted to the preparation of slides, and the rest to the instrument and methods of drawing and photographing from it. The survey of fixing agents is adequate, with many practical details; that of injection is very full, but Cole was a commercial man, and such slides were still in demand. A full range of aniline dyes is listed, and many clearing and mounting agents. The notes on embedding and sectioning include mention of carrot and pith, and also of paraffin and celloidin. Much attention is given to freezing microtomes, which by this time were little used for scientific work: Cole used a freezer frequently, for speed. In spite of its somewhat grandiose title, this book has a decidedly amateur flavour when compared with German books of its time.

Very different is the work by Mann.[40] This is impressive, giving for the first time in English, in accessible form, the detailed scientific background of the mechanisms underlying the use of the various techniques, as well as detailed notes on the processes themselves. For our present purposes, Mann's summary of the history of various reagents and their applications, and his summary of the development of the microtome, are very useful, although neither can (in the light of modern knowledge) be said to approach completeness or accuracy. It is difficult to know to what extent Mann's book influenced the course of scientific histology in England; as he says in his preface:

> There was a time when British histologists were in the forefront both as regards the nature and the extent of their researches, but now—there has been a tendency of recent years to alienate the physiologist from the study of Histology ... [and so] the book has been written, and it represents a first attempt to explain the principles underlying histological methods. ... I hope that this book, if it serve no other purpose, will bring out deficiencies in theories and in methods.

There is ample evidence that histological advances were in the hands of German workers in the last quarter of the nineteenth century, and none at all to show that this position changed before the first world war, in spite of this splendid book. Mann discussed various *in vitro* experiments, and he gives quite exhaustive treatment of the theoretical basis of the action of differential staining. This had been discussed in the literature, but was ignored in other English works on histological technique. For a possible reason, we might finally consider one other English textbook.

This is another work by yet another commercial preparer,[41] with another grandiose title. Flatters lectured at the Manchester School of Technology, and the book is a summary of his lectures there in the session 1902–3. This is another intensely practical work, with photographs showing the various operations, and fully-detailed instructions for all manner of plant preparations. Photomicrographs, in monochrome and colour, are printed as half-tones to show the results expected: these are overlaid by translucent sheets with line drawings labelling the various parts. The result is a superb didactic and practical manual, but it contains not one reference to the literature, and is most restricted in its range of instruments and reagents. No explanation is offered of the mode of action of any of the processes described: the work is a recipe book pure and simple.

England was concerned with the amateur user of the microscope, and had been for most of the century. Further, many of those who wrote had no scientific training, but were self-taught artisans of the instrument, proceeding on a rule-of-thumb basis. In Germany the worth of the microscope and its techniques was early appreciated for the light it could shed on histological and physiological details. In England the slide, like the microscope, was a means in itself: in Germany both were just a means to an end. This attitude is mirrored in the relative provision for technical education in the two countries (see page 335).

Works in German

The first significant book is that by Schacht,[42] which was also the first textbook of plant histological technique to be published anywhere. There was a second edition in 1855, and a third in 1862. It appeared in French in 1855, and in two English editions, of 1853 and 1855.[43] Of the English versions, it can be said that the first included details of an improved version of Schacht's very simple microtome, which is discussed on page 122 below; the second edition figured what would later be described as a Ránvier instrument. All versions include the same, rather limited, range of reagents, and the advice that the successful outcome of any enquiry rests on the methods of investigation adopted: if the method be accurate, the result will be valuable. This is far-sighted advice for its time. A very practical caution is given: not to mistake fragments of skin from the thumb (inadvertently cut off with the freehand section) for significant plant structures! At this time, botanists were more apt to use microtomy than animal histologists, and certainly more forward in their use of microchemical tests.

There were also some popular works on the microscope, such as that of Willkomm.[44] This was similar to the English pattern, giving a few instructions for specimen preparation, and much information on the objects which might be viewed. In the same year, however, a very different book, now of considerable scarcity, was published. Welcker's booklet[45] gives details of his microtome for animal tissues (based on the earlier instrument by Oschatz), but much more than this. It gives notes on the local microscopical society, laying down rules for almost all possible contingencies in its 44 pages, including the size of slides to be used, and a survey of other sizes then in use. Mounting agents are listed and their use described, and agreed abbreviations are actually given for the society's members to use! The importance of serial sections, obtainable with his microtome, is clearly stated (pages 33–39), and this publication has to be regarded as one of great significance, far surpassing anything which had been published in any country beforehand.

Naturally, other works dealing with histology had been published in Germany by this time, but none had contained more than rudimentary directions for very limited manipulations. Some advanced texts had been translated into English, a good example being that by Kölliker.[46] In a total of 932 pages, nothing at all is given of the methods of preparation of the tissues so well described. Such works were valuable surveys of results, but contributed nothing to histological technique as such.

In the next decade, however, this situation was remedied. Frey's book,[47] which was to pass through eight editions by 1886, is the earliest German work to deal comprehensively with microtechnique. Of the twenty-two chapters, five deal with the instrument, five with the principles of microtomy, and the rest with the application of these principles to the various tissues and organs. This was the plan adopted by most subsequent authors. Little is said of the microtome in any of the editions, the Valentin knife being regarded as sufficient where freehand sections could not be made: doubtless this advice would have been altered if the book had survived beyond 1886. At the end are price lists of 15 Continental and 6 British manufacturers, offering a convenient comparative summary of their wares. A slightly edited version of the fourth German edition was published in America in 1872[48] containing little additional material apart from a description of the Curtis microtome, otherwise difficult to obtain. The price lists at the end of this volume include American makers.

A small book of only 84 pages, which was to become the standard elementary text for German schools for the next sixty years, was issued in 1866, but contains only ten pages on slide production.[49] Another small text was that by Exner,[50] interesting for its early section on embryological techniques, but with only a few scattered references in its 93 pages.

A major work by Dippel[51] was published in 1882. The Valentin knife is described, as is Welcker's microtome, with a good illustration (page 671). This was still available, and would cut to 1/100 mm. The Ranvier hand-held instrument is also mentioned, as well as its bench equivalent by Loewe. Other more advanced instruments are described, and a total of 12 pages is given to this topic, more than in any earlier book published anywhere. Full instructions are included for embedding and staining, and the work is a landmark in practical histology.

A rather different book is that by Flemming,[52] for this contains only very scattered technical directions for fixing and staining throughout the 424 pages of text. Nonetheless, some of these are highly original, and the book is important.

Friedlander's book[53] carries a twelve-page section on the microscope, followed by detailed descriptions of equipment and methods. Other editions were issued in 1884, 1886, 1889, and 1894; particular attention is given to bacteriological methods and the clinical microscopy of body fluids and secretions. Versions of the book were published in New York in 1884 and 1885, and in Italy in 1885 and 1888; such rapid translations suggest that a need for such a book was widely felt at the time.

Behrens' book on plant histology[54] was published also in Boston in 1885, again a rapid translation. About one quarter of the 398 pages deal with slide preparation, and scattered references occur in the text. A rather simpler version was published in 1890, for by then much of the detail of the earlier work was inappropriate.

The smaller, zoological equivalent of Behrens was written by Kükenthal in 1885.[55] This has only 37 pages, but is a model of clarity, its three woodcuts showing specially-drawn figures of a Jung sliding microtome, and paraffin embedding-baths, unusual for its size and date.

A textbook of bacteriology which contained useful and up-to-date techniques for slide preparation is that of Hueppe,[56] which went into five editions in only seven years, increasing in number of pages from 174 to 495 in the process. An English translation was issued in New York in 1886, and has clear references to slide making.

Very condensed tables for reference were published by Behrens in 1887,[57] more or less the equivalent of Squire's book (see page 40 above). The first edition had only 76 pages, but offered a vast amount of information in tabular form. In addition to various equivalents and chemical constants, tables of the characteristics of dyestuffs and other reagents, and of the various Nobert rulings, are included. One cannot help but remark on the difference between this highly scientific work and the amateurish one by Squire: the one by a trained scientist and the other by a shopkeeper. Other editions of the book followed; that of 1892 contained 205 pages and included more formulae, including ones for bacteriology and photography. The fourth edition of 1908 had 245 pages and even more detail. These books are still valuable for checking formulae and individual reagents: the listed constants help to decide if present-day stains, for example, are likely to be identical with those of the same name used 80 years ago.

The same author took part in producing an exceptionally good handbook published in 1889.[58] In 315 pages and with 193 woodcuts, Behrens himself dealt with the microscope and its operation, Schiefferdecker covered preparative techniques, and Kossel described microchemical tests. Plenty of space is given to microtomes and their knives, but a surprisingly large part is still devoted to injection techniques. Literature references are scanty, but the flow of words is much easier to read than in the earlier German works.

One of the classic works on microscopical techniques, equivalent to the English book by Lee, is that of Bohm and Oppel.[59] Eight editions were published between 1890 and 1912, and under different authorship it has continued to the present day. The first edition appeared four years after the last edition of Frey, and it may be considered a continuation of that work. There is the usual initial section on general matters, then sections on specific tissues. The first edition had 155 pages; the fourth, of 1900 had 240; while the 1948 edition has no fewer than 695 pages, with a bibliography of about 2000 entries. Much of the earlier work is omitted in the later editions, and for modern work the references are unbeatable. It is interesting that no translations have ever appeared.

The standard work on plant micro technique appeared in 1892[60]. There is a general first section, of 42 pages, then one on microchemical methods occupying 93 pages, and a final section of 108 pages on specific applications. An appendix of 12 pages on bacteriological methods is added, with a bibliography of about 300 titles. A rather rare edition in English was published in 1893,[61] in America, and reissued in England in 1896. The translation closely follows the original.

An unusual approach is used by Apáthy.[62] His two volumes together contain an historical sequence devoted to the discussion of improvements, to some extent in technique but mainly in instruments, from 1660. There are no technical directions, but copious literature references. These little-known volumes are a potentially very useful source, due to their references to events which were relatively recent when the books were written.

A specialized text dealing with the preparation of the nervous system was issued in 1897,[63] the first of its kind. There was a second edition in 1898, and a third in 1905 : a French edition was published in 1900. The specific instructions cover all the neurological techniques known at the time, with especially good coverage of stains, and there is a useful bibliography of about 200 titles.

The classical work by Fischer[64] is of considerable importance, and in its 362 pages there are scattered references and directions of merit.

The most complete work ever to be written on microtechnique was first published in 1903. Paul Ehrlich and his co-authors produced two volumes totalling 1400 pages,[65] with alphabetical entries covering all reagents, processes, and tissues. Each entry has a bibliography, and major articles are signed. To take a few examples : the article on blood covers 10 pages and has 24 references; that on embryological techniques covers 48 pages and has over 300 references; that one the nervous system occupies 43 pages and has over 200 references. The treatment is authoritative and terse, and the work is still of value.

The second edition, of 1910,[66] had 1480 pages, and was fully revised. Blood now filled 27 pages and has over 190 references; embryological technique filled 47 pages with 300 references (plus a further section on experimental embryology); the nervous system needed 46 pages (with a number of entries for specific neurological components in addition). The treatment is still complete and authoritative.

The third edition (of 1926) must also be mentioned, although it is strictly outside the period covered by this chapter.[67] In three volumes and 2444 pages, each of them thoroughly revised, the treatment is, to say the least, full. Each entry is substantial; that on the microtome, for example, covers 30 pages, has 33 illustrations, and is a most useful summary and classification.

Only two other German works need be referred to. The first is a specialist text on embryological methods.[68] This is still a standard reference, being by far the most comprehensive treatment ever published. Each phylum is treated separately, with especially comprehensive details on vertebrates.

The last is a very interesting historical summary, with few technical instructions, but full references to the literature.[69] As a summary of progress in the previous 35 years, written very soon after some of the particularly important advances took place, it gives interesting viewpoints not only of the advances themselves, but of those who made them.

This brief survey of German books emphasizes the fundamental difference between the amateur English workers and those in Germany who were early organized into proper laboratories for research. Those few English scientists who were engaged in fundamental histological work had trained in Germany, and must certainly have kept up to date from German works!

French books

Chevalier was an important writer, not only for his claims on the very early use of coverslips (see page 25, above), but also for his general influence. His book[70] is largely concerned with the instrument alone, but does include some details of preparative methods. It has an important place in history in that the word 'microtome' is used in it for the first time (see page 117, below).

Another early work, of a fairly popular character, was that by Dujardin.[71] In its 330 pages it has a more scientific bias than the purely popular English works of its time, but it is not truly scientific in its approach to preparative methods. This is not, however, the case with the book by Donné.[72] This text is specifically for the medical man, and covers the results which are obtained by inspecting the various body fluids: nothing is said about tissues as such, which may be a reflection of French interests of the time. Similarly, the atlas which accompanied the book shows nothing other than crystals and blood cells, although it is of major interest in being based on the use of very early photomicrographs to secure accuracy in its representations.[73] Such few directions for slide preparations as are scattered in the text are lacking in detail.

The work by Robin is somewhat remarkable for its total preoccupation with injections as the only form of preparation mentioned. Injections were popular at this time, but not normally exclusively so: perhaps the author was indulging his idiosyncrasies in the text.[74] A work by Saurel, published only a little later, gives a more balanced view.[75] Again, directions are rudimentary, very little being said beyond advice to use fresh tissues in the compressorium.

Another, quite rare, work by Robin appeared in 1856.[76] This bears more than a slight resemblance to Pritchard's list (see page 32 above) in content, being little more than a simple listing of possible preparations which might be suitable for examination by the microscope. He suggests advantages for both opaque and transparent objects, and seems to suggest that they be purchased, actually recommending various suppliers for this purpose. This really is a most remarkable publication:

if it could be taken as typical of French microscopy at that time, then that science hardly existed at all. Donné was an influential senior member of the medical faculty in Paris, and well known throughout Europe.

An early French work on the clinical applications of the microscope is that by Michel.[77] There is neither an index nor a list of contents in its 200 pages, but there are scattered references, and the whole is much more satisfactory than the work by Donné. The work by Coulier which[78] followed two years later has an adequate section on the microscope, but is still unsatisfactory on preparations. He advocated drying the tissue on the slide, and considering that Quekett's book had been published nearly ten years earlier, this is curiously backward. He did mention (page 57) the use of such liquids as glycerine and canada balsam, but seemed to regard them more as reagents than as mountants, preferring to mount in olive oil for best results. He did, however, include details of some tissues, but nowhere was there mention of any sectioning technique.

A book by the younger Chevalier, published in 1864 and much enlarged for a new edition the following year, is a notable advance.[79] Good coverage is given of various preservative liquids of the day, and the art of section cutting is described. A Topping microtome is figured, and so is one designed by Follin: both are recommended only for wood sections. In general, though, this book marks the turning point in French works on microscopical preparations. Still behind the times compared with English books of the period, it is nonetheless a considerable advance over earlier native works. Another book by Robin appeared in 1871,[80] and although there is still much preoccupation with injections, the book is more up-to-date. It is a substantial tome of 1028 pages, with 317 text-figures and 3 plates. It does not claim to be another edition of any of Robin's earlier efforts. In spite of this, the first section, of no fewer than 87 pages, deals with injections. The microscope is then described in 133 pages, and preparative techniques are covered in the next 167 pages. The Follin, Luys, and Nachet microtome are all mentioned, and this section is adequate in terms of its date. This is the most important French work to appear until that time.

A most important book, by Ranvier,[81] was published in 1875. This textbook of histology had a section of 145 pages describing appropriate techniques. The microscope is described (quite remarkably, on page 12, is the description of a rack-and-pinion arranged to move only the eyepiece for focusing!), and a microtome is recommended for use only where sections are required with surety and without difficulty (page

49). This is strongly reminiscent of earlier German writers. The only microtome figured is the very simple hand-held model, after Ranvier's own design, although the Rivet type is briefly referred to. This book is a landmark for its treatment of descriptive histology; in the rest of its 1109 pages is a comprehensive account of organ and tissue structure, with good illustrations. Ranvier went on to write other works on histology, and one cannot help but suppose that he was an unusual worker for his country at the time.

In the following year, Pelletan published his tome of 772 pages.[82] With 278 text figures and 4 plates, an account on the usual plan was written, with special attention to histology. A brief description of microscopes is followed by sections on human, plant, and animal histology. This is a further worthwhile book for its time, irrespective of country of origin.

In 1887 appeared another noteworthy book,[83] with a good first section offering advice on slide-making, and containing many illustrations. The rest of its 511 pages, however, include applications of these principles to botany, zoology, and medicine. While the standard remains at a proper level, it would have been unusual in Germany to put out a book which included this assortment of specialisms at this date.

Three years later a book was published which gives a good idea of the state of French microtechnique towards the end of the century,[84] at a time when much was becoming standardized elsewhere. Each chapter deals with a specific organ or tissue, with very complete directions for fixing and staining. An appendix lists formulae, but the book differs from contemporary German works in that there are no references to the literature and no bibliography. These are serious criticisms of a scientific text, but, as we have seen, could also be applied to English works of the same date.

American books

Books published in the United States in the earlier years tended to be direct pirated copies of English texts, and this pirating continued for many years. The first book published in America on this subject is an especially blatant example of piracy. The author is Wythes[85] but the book is simply a shameless shortened version of Quekett. Wythes' Figure 16 is Quekett's Figure 66, Wythes' Figure 19 is Quekett's 75, Wythes' Figure 20 is Quekett's Figure 76, while Wythes' Figure 12 is Quekett's Figure 49 turned through ninety degrees! Apart from this, the book is quite undistinguished.

The first book to be written, as opposed to merely published, in America was that by King.[86] Even in this, the first part draws very heavily on Quekett, in both arrangement and detailed content. The second part is a new dictionary of terms met with in specimen preparation and general technique. Injections and mounting fluids are treated quite fully. Yet another American book was drawn very heavily from Quekett, that by Phin.[87] It was first published in 1875 (nearly 30 years after the book from which most of it was copied!), and went through six 'editions' by 1890. The second part of the work describes mounting techniques, but these are poorly covered, and the book is unremarkable except for its country of publication.

In 1881 Seiler's book[88] was issued. Its 130 pages do not lean very heavily on Quekett, but much of its content certainly relies on other European authors. The pages on microtomes are adequate for the student, and, refreshingly, he advocated use of the Valentin knife only in the post mortem room where hasty microscopic examination is required. He illustrates a Beck microtome, and had devised a knife carrier to support the blade at each end to increase rigidity and evenness of cut. This leads one to suppose that the author had had some experience of his subject-matter, which seems unlikely to have been the case with earlier writers from his country. The Beck microtome shown is also adapted for freezing, and sensible remarks are offered on this topic. Adequate advice is included on staining, and an interesting fold-out section shows the various characteristics of pathological tissues under the microscope.

Ten years later another useful book was issued, with a slightly misleading title.[89] Its 255 pages are a conventional textbook of microscopical technique, with the usual division into general and special descriptions. A very good account of microtomes is included, with the Thoma and Schanze models figured, together with a very rare figure of the Caldwell automatic instrument. The Rutherford and the Jacobs freezing models are also described and illustrated, as are a number of section stretchers. The detailed descriptions of embedding and cutting show that the author had done all these things for himself, and the literature references are satisfactory. This is the best book on the subject published in the United States in the whole of the nineteenth century.

A book by Gage[90] was published in 1886, and went into ten editions by 1908, after which it became less of a textbook of microscopy plus histology, and much more of a book on the instrument alone. In 1886 there were 32 pages, but by 1908 this had grown to 359. About half the book deals with microscopical preparations, and it was (and, in different form, is still) an important and widely-used college text; the level is roughly similar to those produced in England at the time.

The book by James[91] appears to be only the first part of a projected three-part work: the other two parts never appeared, and this explains the preoccupation of the whole 107 pages with mounting media, especially whole fluid mounts. For this specific purpose, the text is still useful on occasion today.

Clark's book,[92] which deals with slide making for over half its length, went into five editions between 1894 and 1925. The main feature is a large appendix of 'useful formulae and information', but there are no references and the level is only that of a beginning student.

A more specialized work appeared in 1895,[93] giving a complete picture of methods applicable to dental microscopy at the time. This is a difficult subject, on account of the great differences in hardness between adjacent tissues of which sections are required: this book was the only one in the world on the topic to be issued in the nineteenth century. It is very adequately done, and it is significant that this edition and the second one of 1899 were published simultaneously in Philadelphia and London. Not only was there a demand, but simultaneous publication ensured that the book was protected by copyright in both countries.

A major work on plant histology, by Chamberlain, was first issued in 1901,[94] and went into five editions by 1932, remaining the standard text for over forty years. In the same period, the number of pages rose from 159 to 416. Interestingly, this book also grew out of a series of lectures, in common with the book by Flatters,[95] but the standard is very different, although literature references are not plentiful. There are good photomicrographs, and full and clear instructions.

A further specialized book was issued in 1902.[96] The work is not very satisfactory, consisting as it does of 20 techniques in neurology, set out for laboratory purposes and including gross dissection procedures as well as microtechnical ones. Although literature references are included for the first few procedures, they are omitted from later ones, thus reducing it to another simple recipe book. It is much less satisfactory than the earlier German work by Pollack.[97]

Only one other book requires mention here, the college text by Guyer.[98] It went through five editions by 1953, and is still current, being one of the most widely-used books of its kind in the United States. A special feature is the description of the reconstruction of a whole animal from serial sections—a procedure which had attracted attention when the book was first published.

The American scene, then, started with blatant pirating but went on to produce some worthwhile books, especially at the college level. Nothing was produced which rivalled the best of the German texts, but neither was this the case anywhere else.

Other works

Noteworthy books written in Belgium come from two pens only. Van Heurck wrote his book on general microscopical technique as a very modest little text of only 108 pages in 1865,[99] although it did have a content of plant histology. By the time of its last edition the book had grown considerably, and the English translation[100] is a particularly handsome volume of 382 large pages. A part is devoted to the preparation of objects, and six types of microtome are included. This work remains one of the classics of microscopy.

The other author is Francotte, who published a book of 426 pages in 1886, after the usual plan of a general section followed by one on the preparation of specimens.[101] The author was a leading member of the Belgian microscopical society, and did much to introduce the techniques of paraffin embedding into the country. His book reflects his preoccupation with the subject, for the section on embedding and microtomy is long and well illustrated for its date.

In Italy Garbini in 1885 wrote the first native book.[102] In its 207 pages it gives an adequate account of the subject, although much is owed to contemporary German texts. In the next fourteen years it went into four editions. Much more original, however, is the work by Carazzi & Levi.[103] Between 1896 and 1916 this went into three editions, and is a good book of its kind. One other book of the period requires mention, the work by Maggi,[104] from Italy, for this remains one of the most complete accounts of protozoological technique ever written.

Spain produced only two books on the subject, in our period. The first, of 1889, was by Ramon y Cajal.[105] This is a modest volume from a worker famous for his histological techniques. The other book is by Rio de Lara,[106] dated 1893. This is a comprehensive work, but by its publication date so were many others in other languages.

The only other book to be noted in this section is an important and rare work on microchemical tests.[107] Originally written in Danish in 1880, the only version seen is the English translation. There was certainly a German translation published shortly after the original, and translations into French and Italian are also mentioned in the preface. This may be taken as some measure of the importance attached to the book when it was first issued. In England the book has only 118 pages, but is a very clear account of the various chemicals which may be applied to plant tissues under the microscope, and the reactions observable. It is a remarkable volume for its time, and also for its place of origin.

The books considered in this highly selective section support the impression gained from the serial literature, that microscopical technique was an English perquisite until the last third of the nineteenth century. After this time, the techniques were developed on an increasingly scientific basis in Germany, with detailed and extensive documentation as the result of their superior organization of research. On the other hand, a distinct tendency to publish regardless of merit became increasingly obvious in German literature in the last quarter of the century. By no means all the German serials are worthy in content, and the same applies to many of their books also.

References for Chapter 3

1.   C Gould—*The companion to the microscope.* (London: W Cary, editions as dated.)
2.   C R Goring & A Pritchard—*The natural history of several new, popular, and diverting living objects for the microscope.* (London: A Pritchard, 1829.)
3.   A Pritchard—*The microscopic cabinet of select animated objects.* (London: Whittaker, Treacher & Arnot, 1832.)
4.   A Pritchard—*The natural history of animalcules.* (London: Whittaker, 1834.)
5.   A Pritchard—*A list of two thousand microscopic objects.* (London: Whittaker, 1835.)
6.   A Pritchard—*Microscopic illustrations of living objects.* (London: Whittaker, 1840.)
7.   G T Fisher—*Microscopic manipulation, containing the theory and plain instructions for the use of the microscope.* (London: T & R Willats, 1846.)
8.   J Quekett—*A practical treatise on the use of the microscope.* (London: Bailliere, 1848.)
9.   R B Todd (Ed.)—*The cyclopaedia of anatomy and physiology.* (London: Longman, 6 vols, 1835–59.)
10. L Beale—*The microscope in its application to practical medicine.* (London: Churchill, 1854.)
11. J Hogg—*The microscope : its history, construction, and applications.* (London: Orr, 1854.)
12. W R Carpenter—*The microscope and its revelations.* (London: Churchill, 1856.)
13. J W Griffith—*An elementary textbook of the microscope.* (London: Van Voorst, 1864.)
14. R Beck—*A treatise on the construction, proper use, and capabilities of Smith Beck and Beck's achromatic microscopes.* (London: Van Voorst, 1865.)
15. L S Beale—*How to work with the microscope.* (London: Churchill, 1857.)
16. W L Notcutt—*A handbook of the microscope and microscopic objects with descriptive lists of upwards of 1780 objects and full directions for obtaining, preparing, and viewing them.* (London: Lumley, 1859.)
17. J W Griffith & A Henfrey—*The micrographic dictionary ; a guide to the examination and investigation of the structure and nature of microscopic objects.* (London: Van Voorst, 1856.)

18. For example, see the anonymous review in *M.M.J.* *13 :83–85 (1875)* :
    ... If a new edition were required, surely there was room for infinitely more change than has been made; if not, why impose so expensive a work upon those who already possess the second edition? ...
19. T Davies—*The preparation and mounting of microscopic objects.* (London: Hardwicke, 1864.)
20. M C Cooke—*One thousand objects for the microscope with a few hints on mounting.* (London: Warne, 1869.)
21. W T Suffolk—*On microscopical manipulation.* (London: Gillman, 1870.)
22. J H Martin—*A manual of microscopic mounting with notes on the collection and examination of objects.* (London: Churchill, 1872.)
23. J H Martin—*Microscopic objects figure and described.* (London: Churchill, 1870.)
24. See, for example, the review in *M.M.J.* *5 :182 (1871)*:
    ... It is without any aim; it can serve no purpose; and it is altogether the worst thing of the kind that we have ever seen ...
25. W Rutherford—*Outlines of practical histology.* (London: Churchill, 1875.)
26. S Marsh—*Section-cutting : a practical guide to the preparation and mounting of sections for the microscope, special prominence being given to the subject of animal sections.* (London: Churchill, 1878.)
27. G E Davis—*Practical microscopy.* (London: Bogue, 1882.)
28. A C Cole—*Studies in microscopical science.* (London: Bailliere, vol. 1, 1883; vol. 2, 1884; vol. 3, 1886; vol. 4, 1887.)
29. See, for example, the brief note in *J.R.M.S.* *5 :355 (1885)*, where it shows that there had been a break in publication, and microscopists were urged to give their support to avoid a loss.
30. E M Crookshank—*An introduction to practical bacteriology.* (London: Lewis, 1886.)
31. H Gibbes—*Practical histology and pathology.* (London: Lewis, 1887.)
32. W Fearnley—*A course of elementary practical histology.* (London: Macmillan, 1887.)
33. A B Lee—*The microtomist's vade-mecum. A handbook of the methods of microscopic anatomy.* (London: Churchill, 1885.)
34. W S Colman—*Section cutting and staining.* (London: Lewis, 1888.)
35. W Stirling—*Outlines of practical histology.* (London: Griffin, 1890.)
36. P W Squire—*Methods and formulae used in the preparation of animal and vegetable tissues for microscopical examination including the staining of bacteria.* (London: Churchill, 1892.)
37. F J Wethered—*Medical microscopy.* (London: Lewis, 1892.)
38. M I Cross & M J Cole—*Modern microscopy.* (London: Baillière, 1893.)
39. A C Cole—*The methods of microscopical research.* (London: Baillière, 1895.)
40. G Mann—*Physiological histology, methods and theory.* (Oxford: Clarendon Press, 1902.)
41. A Flatters—*Methods in microscopical research. Vegetable histology.* (Manchester: Sherratt & Hughes, 1905.)
42. H Schacht—*Das Mikroskop und seine Anwendung, insbesondere für Pflanzenanatomie und Physiologie.* (Berlin: Muller, 1851.)
41. H Schacht—*The microscope, and its application to vegetable anatomy and physiology.* Edited by F Currey. (London: Highley, 1853.)
44. M Willkomm—*Die Wunder des Mikroskops oder die Welt im kleinsten Raume.* (Leipzig: Epamer, 1856.)
45. H Welcker—*Ueber Aufbewahrung mikroskopischer Objecte nebst Mittheilungen über das Mikroskop und dessen Zubehör.* (Giessen: Ricker, 1856.)
46. A Kölliker—*Manual of human histology.* Translated and edited by G Busk & T Huxley. (London: Sydenham Society, 1853.)

47. H Frey—*Das Mikroskop und die mikroskopische Technik.* (Leipzig: Engelmann, 1863.)

48. H Frey—*The microscope and microscopical technology.* Translated and edited by G R Cutter. (New York: Wood, 1872.)

49. H Hager—*Das Mikroskop und seine Anwendung.* (Berlin: Springer, 1866.)

50. S Exner—*Leitfaden bei der mikroskopischen Untersuchung thierischer Gewebe.* (Leipzig: Engelmann, 1873.)

51. L Dippel—*Handbuch der allgemeinen Mikroskopie.* (Braunschweig: Vieweg, 1882.)

52. W Flemming—*Zellsubstanz, Kern und Zelltheilung.* (Leipzig: Vogel, 1882.)

53. C Friedlander—*Mikroskopische Technik zum Genrauch bei medicinischen und pathologischanatomischen Untersuchungen.* (Berlin: Fischer, 1883.)

54. W Behrens—*Hilfsbuch zur Ausfuhrung mikroskopischer Untersuchungen im botanischen Laboratorium.* (Braunschweig: Schwetschke, 1883.)

55. W Kükenthal—*Die mikroskopische Technik im Zoologischen Practikum.* (Jena: Fischer, 1885.)

56. F Hueppe—*Die Methoden der Bakterienforschung.* (Wiesbaden: Kreidel, 1885.)

57. W Behrens—*Tabellen zum Gebrauch bei mikroskopischen Arbeiten.* (Braunschweig: Bruhn, 1887.)

58. W Behrens, A Kossel, & P Schiefferdecker—*Das Mikroskop und die Methoden der mikrosckopischen Untersuchung.* (Braunschweig: Bruhn, 1889.)

59. A Bohm & A Oppel—*Taschenbuch der mikroskopsichen Technik.* (Munchen: Oldenbourg, 1890.)

60. A Zimmermann—*Die botanische Mikrotechnik. Ein Handbuch der mikroskopsichen Präparations-, Reaktions- und Tinktions-methoden.* (Tubingen: Laupp, 1892.)

61. A Zimmermann—*Botanical microtechnique; a handbook of methods for the preparation, staining and microscopical investigation of vegetable structures.* Translated by J E Humphrey. (New York: Holt, 1893.)

62. S Apáthy—*Die Mikrotechnik der thierischen Morphologie; eine kritische Darstellung der mikroskopischen Untersuchungsmethoden.* (Part I, Braunschweig: Bruhn, 1896; Part II, Leipzig: Hirzel, 1901.)

63. B Pollack—*Die Farbetechnik der Nervensystems.* (Berlin: Karger, 1897.)

64. A Fischer—*Fixirung, Färbung und Bau des Protoplasmas. Kritische Untersuchungen über Technik und Theorie in der neueren Zellforschung.* (Jena: Fischer, 1899.)

65. P Ehrlich, R Krause, M Mosse, H Rosin, & C Weigert—*Encyklopädie der mikroskopischen Technik mit besonder Berücksichtigung der Farbelehre.* (Berlin & Wien: Urban & Schwarzenberg, 1903.)

66. P Ehrlich *et al.*—*Enzyklopädie der mikroskopischen Technik.* (Berlin & Wien: Urban & Schwarzenberg, 1910.)

67. R Krause—*Enzyklopädie der mikroskopischen Technik.* (Berlin & Wien: Urban & Schwarzenberg, 1926.)

68. P Rothig—*Handbuch der embryologischen Technik.* (Wiesbaden: Bergmann, 1904.)

69. J W Moll—*Die Fortschritte der mikroskopischen Technik seit 1870.* (Jena: Fischer, 1908.)

70. C Chevalier—*Des microscopes et de leur usage.* (Paris: Crochard, 1839.)

71. F Dujardin—*Nouveau manuel complet de l'observateur au microscope.* (Paris: Roret, 1843.)

72. A Donné—*Cours de microscopie complémentaire des études médicales.* (Paris: Baillière, 1844.)

73. A Donné & L Foucault—*Cours de microscopie complémentaire des études médicales: Atlas éxecuté d'après nature au microscope-daguerréotype.* (Paris: Baillière, 1845.)

74. C Robin—*Du microscope et des injections dans leurs applications à l'anatomie et·à la physiologie.* (Paris: Baillière, 1849.)

75. L J Saurel—*Du microscope, au point de vue des ses applications à la connaisance et au traitement des maladies chirurgicales.* (Paris: Baillière, 1851.)

76. C Robin—*Mémoire sur les objets qui peuvent être conservés en préparations microscopiques transparentes et opaques, classés d'après les divisions naturelles des trois règnes de la nature.* (Paris: Baillière, 1856.)

77. M Michel—*Du microscope de ses applications à l'anatomie pathologique et au traitement des maladies.* (Paris: Baillière, 1857.)

78. M P Coulier—*Manuel pratique de microscope appliquée à la médecine.* (Paris: Magdelaine, 1859.)

79. A Chevalier—*L'étudiant micrographe: traité théoretique et pratique du microscope et des préparations.* (Paris: Delahaye, 1864.)

80. C Robin—*Traité du microscope.* (Paris: Baillière, 1871.)

81. L A Ranvier—*Traité technique d'histologie.* (Paris: Savy, 1875.)

82. J Pelletan—*Le microscope, son emploi et ses applications.* Paris: Masson, 1876.)

83. R Gerard—*Traité pratique de micrographie appliquée à la zoologie, à l'hygiene et aux recherches cliniques.* (Paris: Doin, 1887.)

84. R Boneval—*Nouveau guide pratique de technique microscopique appliquée à l'histologie et à l'embryogenie.* (Paris: Maloine, 1890.)

85. J H Wythes—*The microscopist; or a complete manual on the use of the microscope for physicians, students and all lovers of natural science.* (Philadelphia: Lindsay & Blakiston, 1851.)

86. J King—*The microscopist's companion; a popular manual of practical microscopy.* (Cincinnati: Rickey & Mallory, 1859.)

87. J Phin—*How to use the microscope, being practical hints on the selection and use of that instrument.* (New York: Industrial Publications, 1875.)

88. C Seiler—*Compendium of microscopical technology.* (Philadelphia: Brinton, 1881.)

89. C O Whitman—*Methods of research in microscopical anatomy and embryology.* (Boston: Cassino, 1885.)

90. S H Gage—*Notes on microscopical methods.* (Ithaca: Andrus & Church, 1886.)

91. F L James—*Elementary microscopical technology.* (St Louis: Medical Journal Co, 1887.)

92. C H Clark—*Practical methods in microscopy.* (Boston: Heath, 1894.)

93. A H Smith—*Dental microscopy.* (Philadelphia: Dental Manufacturing Co, 1895.)

94. C J Chamberlain—*Methods in plant histology.* (Chicago: University Press, 1901.)

95. op. cit., note 41 above.

96. I Hardesty—*Neurological technique.* (Chicago: University Press, 1902.)

97. op. cit., note 63 above.

98. M F Guyer—*Animal micrology.* (Chicago: University Press, 1906.)

99. H van Heurck—*Le microscope, sa construction, son maniement et son application à l'anatomie végétal et aux diatomées.* (Paris: Delahaye, 1865.)

100. H van Heurck—*The microscope: its construction and management.* Translated by W E Baxter. (London: Crosby Lockwood, 1893.)

101. P Francotte—*Manuel de technique microscopique applicable à l'histologie, l'anatomie comparée, l'embryologie et la botanique.* (Bruxelles: Lebegue, 1886.)

102. A Garbini—*Manuale per la tecnica moderna del microscopio nelle osservazione zoologiche, istologiche ed anatomiche.* (Verona: Goldschagg, 1885.)

103. D Carazzi & G Levi—*Technica microscopica: guida practica.* (Milano: Società Editrice Libraria, 1889.)

104. L Maggi—*Tecnica protistologica.* (Milano: Hoepli, 1895.)

105. S Ramon y Cajal—*Manual de histologia normal, a de tecnica micrografica.* (Madrid: Moya, 1889.)

106. D L Rio de Lara—*Manual de técnica micrografia general.* (Madrid: Moya, 1893.)

107. V A Poulsen—*Botanical micro-chemistry; an introduction to the study of vegetable histology.* Translated by W Trelease. (Boston: Cassino, 1884.)

4. A Survey of Substances used in Microtechnique 1830–1910

Introductory remarks

It was only gradually that specimen preparation became anything other than totally rule-of-thumb. Until the microscope was improved optically the imperfections of preparative techniques would not have been obvious.

By the mid-1880s the basis of modern methods was beginning to be recognized. Some experimental work was undertaken on the mode of action of the various reagents, and trial of new ones was less random than before. By the time of his fourth edition in 1896, Lee was able to say in his preface:

> It has seemed to me advisable ... to explain more fully the principles on which they [the more important processes] are founded, and to add in many cases a critical estimate of their rationality and practical value.

> The essential feature of the first edition was that it was an altogether exhaustive collection ... [but now] the book has been lightened by the jettison of much useless matter ... [although] recent methods, which may be considered to be still on their probation, have been treated with the accustomed fulness.[1]

Following Lee's example, we must omit useless material from this account; there are so many substances and mixtures which have been used in specimen preparation that a full account of them would be pointless. Therefore, we shall consider the first introduction of particular reagents, for obvious historical reasons: we must deal also with the first effective introduction, where the two are not synonymous. Equally important on occasion is the introduction of particular mixtures and procedures, and these also must be considered.

The preparation of microscopic mounts is a very involved procedure. Quite extraordinarily complicated living structures are subjected to a variety of processes which will hopefully present a final picture having as much resemblance to the original state as possible. It is a relatively modern concept that the prepared specimen might bear little resemblance to the original condition, but that when a tissue is prepared in a wide variety of ways, a full idea of its make-up can be obtained.

This understanding had not been gained by the early part of the nineteenth century. Many investigators used methods now known to be so damaging in their effects, that their results must necessarily have been limited. These early histologists were often vague about specifying their methods, and their reagents might have been different from what is understood by the same names today: nomenclature has altered, and impurities (which may have a marked effect on results) are almost certainly different in kind and level.

Modern technicians preparing histological slides almost always display a remarkable rule-of-thumb approach, on occasion bordering on the telepathic. The difference between them and their predecessors is that the modern worker has the advantage of knowledge of the mistakes of his earlier colleagues. It must be emphasized that making a good preparation is very difficult, being as much an art as a science and requiring constant attention to seemingly trivial details; some examples of such detail will be mentioned below, besides more obviously important matters.

Important survey literature

Special attention has been given to sources explaining the scientific background of reagents, for this is significant in the search for better techniques in spite of neglect by historians of science. Contemporary surveys of the state of the art have also been included, as they are useful for revealing a closer perspective of criticism than we can achieve today.

The classic contemporary survey of the history and use of stains is that by Gierke,[2] on which most subsequent historians have leant heavily in spite of its omissions and errors. A convenient, although shortened, translation is that by Seaman,[3] and these articles are still worth reading. More modern summaries include the book edited by Conn,[4] which is useful although confined to stains. The Ciba Zeitschrift,[5] and its shorter and later English version,[6] are useful summaries, and Baker's paper[7] is a good statement of some new work, although again limited to stains.

There are many other papers dealing with the introduction of stains, but they tend to be derivative and thus repetitive. On the other hand, a modern statement of the mechanism of fixing and staining, also by Baker,[8] is most valuable.

For convenience, the substances to be discussed have been separated into those appropriate to the various stages of preparative techniques, although a few chemicals are used in more than one stage.

Killing and preserving methods

If a tissue is not in healthy living state when it is fixed, it can reveal only abnormal details, whatever its subsequent treatment. This is now obvious, but it was not always so: in early days tissues were dissected out, and then left around for a while before anything else was done. Where other processing was minimal, the maceration thus obtained might have been actually useful, but as the optical quality of the microscope improved it became obvious that changes were occurring in such circumstances. Some workers then dissected their tissue specimens from living animals, and whatever one may think of this from the moral standpoint, it would avoid that autolysis which so rapidly intervenes in samples of dead tissues, as it would also avoid the other dangers of evaporation, osmotic change, and attack by bacteria and fungi. Admittedly, such living tissues were usually taken from lower animals which had often been stunned, and it remains true today that the best histological preparations are made from animals which have been shot, had their necks wrung, or met some other rapid end not relying on administration of any chemicals beforehand. For this reason, human tissues are usually poorly fixed; it is only the rare case of a sudden accident occurring in suitable circumstances that provides suitable specimens of human tissue.

For some preparations, however, the use of narcotizing agents was quite acceptable; such cases involved the mounting of aquatic organisms which were prone to contract on normal fixing, and of which whole mounts of only minimal histological detail were required. Chloral hydrate,[9] chloroform,[10] cocaine,[11] and carbon dioxide[12] are among once-popular agents, as is rapid treatment with hot liquids;[13] the last two mentioned are still used today.

Preservative liquids were first mentioned in the seventeenth century,[14] but not as microscopical agents; definite investigation of such liquids was undertaken in the eighteenth century,[15] but it was not until 1833 that Jacobson[16] suggested the use of chromic acid as a hardening agent for microscopical use. This date may be said to represent the beginning of adequate fixation, although many workers continued to use the so-called indifferent liquids to contain tissues under the microscope. The name was given on account of their apparent lack of effect on the tissues. Water itself was often used, and then blood serum and amniotic fluid were in vogue until the 1860s. Osmotic changes do not occur in these latter fluids, but autolysis and putrefaction certainly do, and work carried out in such media must have revealed pathological changes by the 1870s, when the microscope had been improved to the extent of resolving sufficient detail.

Hardening and fixing agents

The processes causing most damage to the fine structure of a tissue are those of embedding (which involves not only dehydration but also high temperatures); sectioning (causing mechanical damage); and mounting (with another change of medium). Obviously, embedding was not a process used in the early part of the nineteenth century, but there was certainly a need for hardening to allow for the rigours of freehand sectioning. Probably the earliest method of hardening soft tissues was that of boiling them; this was the method used by Malpighi[17] as early as 1666 in his study of the human brain. He thickened the substance of the brain by boiling it, and then wiped ink across the cut surface to throw into visible relief the 'glands' which seem to have been the avascular areas as outlined by the cortical blood vessels. Botanists in the nineteenth century also used the process, but in their case it was to soften the tissues. Many workers in animal histology of this time, however, merely compressed the fresh tissues; this was the method used by Bowman[18] for his work on striated muscle fibres, with the addition of dilute acetic acid. In repeating this work using a microscope of equivalent aperture, one can make out most of the structures he described. Modern preparations do not reveal a great deal more.

Also in 1840, Hannover used chromium trioxide for histological purposes.[19] He had stayed with Jacobson in Copenhagen, who had shown him the gross preservative effects of this substance. Hannover tried it out on a variety of tissues and gave his results, but it did not come into widespread use for a long time. Corti used it in his work on the inner ear,[20] and it was included in Müller's well-known mixture in 1860.[21] It was not until 1878 that Mayzel found it useful for fixing chromosomes,[22] and it was left to Flemming to demonstrate its real potential from the following year.[23]

In 1846 Blanchard used another important fixative, mercuric chloride, for flatworms.[24] The substance had been used for some years as a component of preservative fluids for liquid mounts, as in Goadby's formula, but this was its first use in a concentration high enough actually to fix tissues. It was used again in 1851 by Corti, in his work on the inner ear just mentioned, and also by Remak in 1854 in his work on liver.[25] In spite of these uses, it was not until 1878 that Lang successfully drew attention to the substance,[26] and it has continued in use to the present day. Well-known mixtures containing it include Heidenhain's Susa[27] (called after the initial letters of sublimate and saure), the Zenker formula,[28] and Gilson's fluid—still useful for general histological purposes.[29]

Acetic acid had long been known as a preservative, in food pickling for example, but it was used in microscopy only as a macerating agent until 1851, when Clarke[30] discovered its special virtue in combination with alcohol. Clarke's fluid is still in use, and is probably the oldest formula to find any place in modern histological technique. The substance was much used, alone and in combination, for nuclei in the 1880s; a further significant formula is that of Brasil, who combined it with formaldehyde for general histological work.[31]

Alcohol is another fixative long recognized as a preservative, and so used by Baker in 1743 for preserving *Hydra*.[32] It was also used, essentially for hardening alone, by Vicq d'Azyr in 1786 in his work on the brain:[33] another similar use was that by Reil in 1809.[34] These uses were not essentially as fixative, and it was not until Clarke (quoted above) mixed it with acetic acid that it was available as a fixing agent as such. Another useful mixture is that of Carnoy,[35] which also contains chloroform, and in his second formula is still used for mitotic studies.

Potassium dichromate was introduced by Muller in 1860,[36] in a formula which also contained chromium trioxide: his second formula[37] omitted this, and formed the basis of Zenker's fluid already mentioned. Erlicki's mixture[38] is interesting as containing cupric sulphate in addition to the potassium dichromate, and is still being used to rapidly harden large-volume organs. In 1910 Regaud[39] added formaldehyde to a dichromate mixture still used for cytological studies.

A very important fixative substance was introduced by Franze Schulze (who invented chlor-zinc-iodide[40]). He sent a solution of osmic acid, now called osmium tetroxide, to his pupil Max Schultze, requesting that it be tried in histological work. In 1864 he published the first account of its use,[41] on the phosphorescent organ of a beetle: this was concerned with its blackening action rather than its powers of fixation, however. Shortly afterwards, a series of tests having been carried out in the meantime, a description of its use as a fixative was published.[42] Its use was only slowly established for it was expensive and labile, and it was not until 1882 that Flemming published his first or weak formula,[43] followed two years later by the strong version.[44] Both are still in use, and are excellent for cellular structures, giving good preservation and optical differentiation. Unfortunately, penetration is very slow as with all osmic mixtures, and only small pieces of tissue can be fixed. Another mixture containing osmium designed for cytological work, and still in use, was that of Altmann,[45] but this never gained the fame of Flemming's formulae.

As has been said, osmium tetroxide is easily oxidized as well as being very expensive. Once it had been found that all that was needed as a preservative of the solutions was a few crystals of potassium permanganate in the bottle,[46] the use of the substance was accelerated. Spoilt solutions could be regenerated by treatment with hydrogen peroxide,[47] and this discovery was also a stimulus to its use. The earlier literature has many references to the preservation, or lack of it, of these solutions, as it was a very important problem in its day.

Picric acid had its initial use in microscopy as a stain[48] rather than a fixing agent, its first use in that capacity being by Ranvier in 1875.[49] Flemming developed its use as a fixative,[50] and it was further used in the well-known mixtures of Kleinenberg[51] and Bouin.[52] This last is still a widely-used mixture for general histological work.

The final substance we must consider in some detail is formaldehyde. This is still a most important fixing agent, being a constituent of many mixtures. In 1892 Trillat[53] had described the preservative effects of the substance on albumen. Blum had also used it as an antiseptic only, until he accidentally left an anthrax-infected mouse in a dilute solution overnight, and found it thoroughly hardened the following morning. He then made a systematic survey of its effect on a variety of tissues, as a simple 4% solution, sectioning after celloidin infiltration. He found that most cells were well fixed and would accept many different stains. In retrospect it is fortunate that he embedded in celloidin and not paraffin, for formalin alone is a poor fixative for the latter agent. Blum's paper[54] is the first description of formaldehyde as a fixing agent.

A few other substances have been used as fixatives, and deserve mention. Hermann[55] used a mixture containing platinum chloride, and got excellent results, but the very high price of the reagent has prevented its more general adoption. The substance was first used by Rabl,[56] who advocated it for mitotic figures, and a few specialized formulae still use it. A much cheaper substance is iodine, used as an iodine/potassium iodide solution by Kent[57] for small organisms, where it was successful. His introduced the use of nitric acid as a fixative,[58] and it is still useful as Perenyi's formula[59] for fixing yolky eggs, which remain notoriously difficult material.

Three general techniques require mention, of which the first is to use heat to fix small isolated cells such as blood corpuscles and pus. Koch introduced the cover-glass method in 1877,[60] for use with bacterial preparations, and Ehrlich used it for blood the next year, describing it in 1879.[61] The method is still valuable and widely used. A second technique, of injecting the fixative directly into the artery of the animal, was introduced by Golgi,[62] and refined by Mann[63] who claimed that within 50 seconds every cell in the body was fixed. This claim was supported by the excellent preservation of such organs as pituitary and retina, which are both delicate and normally inaccessible, and the method remains of much value. The final general technique to be noted is what we would now call freeze-drying: it has assumed much importance in electron microscopy. Introduced by Altmann in 1889,[64] small portions of tissue were kept frozen *in vacuo* over sulphuric acid at $-20\,°C$. Mann[65] refined the method to use solid carbon dioxide mixed with alcohol to freeze the tissue very rapidly, before dehydrating it over the acid.

Work on the theoretical background of fixing methods may be said to have been based on Graham's famous paper on diffusion,[66] which established the nature of sols and gels. The nature of protoplasm was investigated by Altmann,[67] who later prepared a series of papers based on the use of his fixing solution. The main figure in the research was Flemming, who constantly stressed that it is not justified to regard as normal all that can be seen in cells which have been treated with reagents. This was a very important statement for its time,[68] and he published his mixture only because he thought it caused less distortion than other fixatives then in use. He believed that protoplasm had a basically thread-like structure. Berthold described the coagulation of plant protoplasm by alcohol and other reagents, in an important pioneering book.[69] He shook white of egg with water to make a coagulum which had a fibrillar appearance. Schwarz took this work a stage further using gelatine and peptone, and showed that colloidal bodies are precipitated as minute granules; by fixing agents the size of the granules varies in direct proportion to the strength of the fixing solutions.[70] He went on to reason, in error, that coagulation represented the separation of the firmer from the more liquid parts of the protoplasmic proteins.

Some experimental work was done on actual tissues also. In 1887 a good paper was published by Trzebiński, who had tried out a number of fixing solutions to compare their effects on the ganglion cells of spinal cord. He found that in all preparations pericellular spaces appeared, that there were vacuoles in the cell substance, and that reaction to dyestuffs varied with fixing, thus causing different results from identical original material.[71]

Butschli, in 1891,[72] published his views that all protoplasm has a foam-like structure, and went on to study a wide range of living protoplasm, and also artificial foams of soap and oil, in order to explain mitotic phenomena. This was the first attempt at a purely mechanical explanation of histological appearances. In 1899 Hardy published[73] further work on the question, and refined and extended the Butschli work on foams by using various fixatives, but most importantly by introducing subsequent embedding in paraffin. He showed that there is no basic ground substance, but that all protoplasm consists of granules or spongework, with water in between. He established that the essence of fixation is the separation of fluid from solid, and that if the solids are abundant they hang together in a network.

Some work on the penetration of fixatives had obvious practical value, such as that published by Tellyesniczky[74] in 1898: he gives lists of fixatives subdivided into various groups, and offers a great deal of practical information on their relative powers of fixation and penetration. A summary of the work of others is added to one of his own extensive researches into fixation in Mann's book,[75] which is an admirable volume. Mann began to investigate the mechanism of fixation in 1888, and is in broad agreement with Hardy's findings. It remained for Berg[76] to adopt a different approach, that of estimating the changes in nuceloproteins on fixation, as opposed to those occurring in albumen. The results are very different, reagents producing a coagulation with one while not doing so with the other: this formed the basis of work in more modern times, which cannot find a place here.

The references selected to illustrate the kind of work being carried out on the action of fixatives have of necessity been few, but have included the most significant advances. It may not be without interest that all the investigators were German, with the sole exception of Hardy (of Cambridge), who entered the field right at the end of the century.

Dehydration techniques and reagents

During the period we are dealing with, alcohol (in grades of increasing strength) was the only liquid used to remove water from tissues. Altmann's freeze-drying method[77] was an alternative which did not come into its own until long after it was first published. Acetone and methylal were suggested as alternatives to alcohol for making methylene blue preparations in 1892,[78] but other now common reagents such as dioxan and cellosolve were not introduced until 1931 and 1935 respectively. Some useful small equipment was described to facilitate dehydration, such as the sieve-dishes of Steinach,[79] which were perforated and intended to contain the object as it was moved from reagent to reagent. The small porcelain cylinders of Fairchild[80] were similar, and this kind of small accessory still finds favour, especially where very small objects are to be embedded. Cobb's differentiator[81] was intended to handle small organisms also, but was rather unwieldy, while the modified Soxhlet extractor of Cheatle's so-called rapid dehydration set-up[82] was cumbersome and quite unlikely to be any more rapid than ordinary methods.

Staining techniques : general remarks

We must remember, in this context, that in the 1870s and 1880s especially, virtually every coloured substance was tried out for its microscopical effects and reported in the literature. In 1884, Gierke said:

> From imperfect and modest beginnings it [staining of tissues] has gradually grown into favour, and during the last decade with immense rapidity, so that it is perhaps well to arrest the epidemic ambition of those, especially the younger investigators, who search the copious lists of dyes for some material that, either pure or modified, they can warmly recommend for staining, and thereby become an authority.[83]

At that date the tide of dyes was not yet even halfway towards its flood! In 1902, Mann was equally emphatic:

> Later, in 1854, another observer having been led, by a fortunate accident, to the conception of the idea that present ways may not always be the best, by a second equally fortunate occurrence was enabled to discover a reliable staining method. By publishing the latter, a seed was sown, which owing to a number of further fortunate circumstances commenced to germinate. The method of staining, once having taken root in the animal histologist, grew and grew, till to be an histologist became practically synonymous with being a dyer, with this difference, that the professional dyer knew what he was about, while the histologist with few exceptions did not know, nor does he to the present day.
>
> During this prodigally liberal production of new staining methods, the advantages accruing to histology from the use of dyes were but little, for most of the discoveries attributed to the use of special stains could have been made by a skilled microscopist on perfectly unstained sections with almost, if not quite, equal ease.[84]

With Gierke, therefore, we must say:

> ... To relate the history of all these methods would make my essay too long, and those which are but repetitions of earlier work will be omitted.[85]

There was considerable technical achievement in the nineteenth-century dyestuffs industry well before the advent of the aniline colours. As Farrar[86] has shown, there was the beginnings of a synthetic dyestuffs industry, in addition to the production of a range of natural colours and mordants, and practical expertise to use this range to produce a wide spectrum of effects: it is remarkable in retrospect that microscopists were slow to exploit the process of mordanting. A good general account of the industry is given by Holmyard,[87] and Meyer[88] provides excellent perspective by placing dyestuffs into the general history of chemistry. A valuable paper by Caro,[89] one of the protagonists of the industry, is especially useful for its help in identifying the numerous synonyms of different manufacturers. A modern account by Beer[90] presents the story in modern terms and provides useful supplementary material.

Before considering the various groups of colours, some of the more significant papers on the mechanism should be mentioned, and in this context a further quotation from Mann is pertinent:

> Notwithstanding that Ehrlich in 1879 founded the science of staining as regards anilin dyes, that Bohmer, Frey, Weigert, and Paul Meyer have explained the action of haematoxylin, that Unna proved what can be done by a systematic research, that Gierke, Griesbach, Rawitz, Benda, Michaelis, and others have also shown the way how to proceed,—notwithstanding all this, the majority of histologists still prefer to dabble in stains in an unscientific manner, owing to the baneful principle that 'Experimentation is preferable to study'.[91]

It would be easy to overestimate the effects on the progress of microscopy and histology of the publication of papers descriptive of staining mechanisms: it is very likely that most workers simply ignored them! However, one especially important paper was not widely available for a long time after it was written—the doctoral thesis of Ehrlich.[92] This reveals considerable insight into the mechanisms of staining, and is of fundamental worth in the history of microtechnique. In 1879 he noted that basic dyes show striking properties in common,[93] and in the same year noted the affinity of acid dyes for certain leucocyte granules.[94] In the following year he gave a short resume of the differences between basic and acidic dyes,[95] illustrating his remarks by reference to leucocytes: thus, before he was 26 years old, he had established the basic theoretical framework of staining.

Two possible explanations for staining action are possible, the physical or the chemical. Witt[96] supported the former, arguing that staining is simply a matter of solid solutions. Gierke had already drawn attention to the fact that most staining could be explained by physical phenomena,[97] while Rawitz[98] also supported this theory. Georgievics, in two important papers,[99,100] brought osmotic pressure into the argument in a fairly convincing way, but the most able proponent of the purely physical explanation of staining is Fischer,[101] who established important factors about staining by adsorption.

The chemical explanation was favoured by Miescher, who in 1874 isolated nucleic acid from nuclear chromatin segments, and found that it united with methyl green to form insoluble salts.[102] In 1888, Knecht showed that colour develops when a free colour base in colourless state forms a coloured salt[103] with wool. Shortly afterwards, Zacharias used mixed stains to support the theory,[104] while Weber, again working with wool, was able to stain it with both acid and basic stains, thus satisfying both the amido and the carboxylic groups.[105] A model paper by Mathews,[106] on the chemistry of cytological staining, shows that all the colour acids used by histologists react in the same way. It is true that the precise mechanism of staining is still unelucidated, in spite of much work subsequent to that quoted here. Baker provides a helpful summary of what is known,[107] and shows that the factors are much more complicated than was appreciated seventy years since. Mann's summary[108] remains useful as a convincing statement of the situation as it was at the turn of the century, and to the time of the first world war.

Natural staining substances

Even after the introduction of the synthetic dyes, some of the earlier natural colours continued to be widely used, and even today some are quite irreplaceable. As we have already seen, Leeuwenhoek used the first true stain, a solution of saffron in spirit to colour muscle fibres.[109] Such a stain is not very intense and it is unlikely that its use on such a tissue would be totally helpful. In 1758 Reichel[110] used plant extracts to colour plant tissues. It seems likely that he introduced the use of logwood for this purpose,[111] in a simple unmordanted solution. This also is not a good stain when used in that way. Hill introduced cochineal[112] for the first time only a few years later, and the use of haematoxylin and of carmine is still irreplaceable today.

Earlier workers used carmine almost exclusively. By the time Hartig made popular the process of staining in 1854,[113] although other colours such as madder, indigo and logwood were well known, so exclusive was the use of carmine in microscopy that he referred to it only as 'the dye solution'. Hartig was not, of course, the first nineteenth century user of the dye: Ehrenberg fed it to various protozoa in 1838,[114] while Göppert

& Cohn studied the cell contents of *Nitella* with its aid in 1849.[115] It was, however, Corti who used the dye as a nuclear stain for the first time in his work on the inner ear.[116] This paper was a very important one, as we have seen above, for its use of fixatives; it is no less so for its use of carmine in a proper manner for the first time. It is unfortunate that it was written in French in a new German journal: his work was ignored. It was thus left to Hartig to popularize the technique of staining with a more receptive audience. Gerlach[117] provided the final impetus to its adoption, for in 1858 he left a cerebellum section (hardened in potassium dichromate) by accident overnight in a very dilute solution of ammonia carmine—the staining was perfect and his subsequent enthusiasm and teaching ensured that staining was rapidly adopted as a research technique.

In the next year, Gerlach reported that he could not stain living animal tissues.[118] Maschke, a year later, took up the story where he had found it left by Hartig, apparently in ignorance of Gerlach's efforts. He predicted[119] that proteins would soon be recognizable by their staining reactions, and he introduced indigo as a stain. The use of carmine developed apace: Gerlach had used an acetic carmine, and the first alkaline carmine (a sodium biborate mixture) seems to be due to Thiersch.[120] Beale quoted a formula for an ammonia carmine in his third edition,[121] the first time this had been published.

An important advance was made by Schwarz in 1867, when he employed the first double staining technique.[122] This used successive solutions of picric acid and carmine, and his paper has excellent colour plates illustrating the effects obtained. In the following year, Ranvier used a single-solution double stain, or picro-carmine.[123] Formulae including carmine are very numerous in the literature, and many failed to find significant application. Among those which have survived are the lithium carmine of Orth,[124] and the alcoholic borax carmine of Grenacher.[125] Important researches using carmine include Flemming's work on *Anodonta* eggs,[126] Hertwig's on echinoderm embryology,[127] and the outstanding study of Van Beneden on *Ascaris* development.[128] The most important work on the stain itself was that of Mayer, who explored the theory of carmine staining in his 1892 paper,[129] in which he published his famous carmalum formula. He showed that commercial carmine was an impure aluminium compound of carminic acid, which explained its very variable results. His historical review of the subject, and his classification of cochineal/carmine stains, are still very valuable.

Haematoxylin, derived from logwood, is an unsatisfactory stain in the absence of a mordant, and the first mention of this colour after that of Reichel was by Quekett. The paragraph in his book[130] which deals with dyeing specimens is of fundamental importance, for it was the first proper statement of the subject, and it appeared long before the usually-accepted date of the introduction of staining techniques. From page 275:

> When very thin and transparent objects are required to be mounted in balsam, they become so indistinct . . . [that] some mode of making them dark becomes necessary; this may be effected in two ways, either by charring or dyeing; . . . Some structures, especially those of an animal nature, will not bear the charring process; to these the dyeing only is applicable, and may be effected by soaking them for a time in a decoction of fustic or logwood, after which they may be taken out and dried. A weak tincture of iodine may be employed for the same purpose.

It is unlikely that logwood used in this way would have produced very good results, and it is certainly possible that Quekett found Reichel's original description and merely suggested it without full trial. A thorough search of all the surviving Quekett preparations in the Hunterian Museum of the Royal College of Surgeons of England has revealed none at all which could have been stained with this substance. There is no mention, either in the manuscript lists, or in the printed catalogue, of logwood being used.

Waldeyer was the next to try logwood, fifteen years later, and he is usually wrongly credited with its introduction.[131] He tried to stain the axis cylinders of neurones, again without a mordant: this may have been based on Quekett's advice, for his book had been translated into German thirteen years before Waldeyer's work appeared. It was not until 1865 that Böhmer was successful in using logwood, this time with an alum mordant.[132] The use of alum was inspired by its use in the dyeing industry, and this was to be the way of using haematoxylin for the next quarter of a century.

Arnold[133] gives more precise instructions for making up an alum-haematoxylin stain than had appeared before 1872, and Tait's paper[134] three years later analysed the available stains into those which colour virtually everything, and those which are selective in action: in this context he put forward his own haematoxylin stain as a good nuclear colourant. A mixture which is still used is that of Delafield: apparently this was never published by its originator, and it was attributed to no fewer than three other workers before Prudden set out the correct authorship.[135] In 1886 Ehrlich published his acid-haematoxylin formula, still in use,[136] but it was not until 1891 that Mayer discussed the ripening of haematoxylin,[137] following the discussion of its chemical nature by Nietzki.[138] This showed that when haematoxylin is oxidized to haematin, an active staining material is produced. Other important mixtures were the iron haematoxylin of Heidenhain,[139] still used for exquisite nuclear detail; the phosphotungstic acid formula of Mallory;[140] and the several multiple staining procedures using haematoxylin and aniline dyes.

Other natural staining substances have been tried in abundance. Litmus, red cabbage, indeed any and every coloured substance has probably been suggested as a microscopical stain. We need not be concerned with these trials, for none was scientifically useful. A decided alchemical approach is notable in most of the papers which have extolled the virtues of such colours.

Aniline colours and their use as stains

The introduction of the first aniline dye by Perkin in 1856 was a major innovation, not only for microscopy but for the entire chemical industry. At last there was excellent commercial reason to stimulate original work in organic chemistry. The composition of Perkin's mauve has been the subject of much speculation, largely answered by Gurr's spectrum analysis of a part of the original sample.[141] It was not until 1862 that Beneke used what must have been Perkin's mauve in microscopy;[142] it may well have been from a different supplier (Perkin's patent was not valid in France, for example), and might thus have had different impurities and staining properties. It was not until some years after 1862 that any new dye was announced with the colour description of the stain used by Beneke, as a thorough survey of the patent and other literature has revealed, and therefore the statement of the nature of Beneke's stain may be made with some certainty.

In the next year Waldeyer used a number of dyes, including mauve but also paris blue and basic fuchsin.[143] In the same year Frey used aniline blue,[144] and Roberts used picric acid.[145] Apart from the introduction of indigocarmine by Chronszczewsky in 1864,[145] the spate of aniline dyes was over for a few years. In 1874 Ranvier used cyanin,[147] Zuppinger used iodine violet,[148] and Lieberkuhn introduced synthetic alizarin.[149] The way was now open for all to scan the makers' lists and publish a paper on the use of a colour: we must be highly selective from this date on.

Methyl violet was introduced by Cornil in 1875[150] in a paper which also described for the first time the metachromatic effects of aniline dyes. Another interesting paper in the same year was that of Hermann,[151] for it was in this that the principle of differentiating a stain was brought to the attention of histologists. However, the process had already been described by Böttcher in 1868,[152] as is often stated. This was not, though, its first description, for two years before this Schweigger-Seidel & Dogiel had decolourized carmine with acid glycerine.[153]

Ehrlich introduced at least twelve stains, ranging from safranin in 1877,[154] to methyl blue, nigrosin, and acid fuchsin in 1879, neutral red in 1893, and janus green in 1898. In retrospect, this first use of safranin seems long delayed, for it had been available since 1859: it is still a very valuable stain. Ehrlich's paper which mentions safranin also surveys the then available dyes, and makes some interesting mistakes in their composition. His work with methylene blue[155] was of fundamental importance, especially for bacteriology; his discovery the next year that addition of aniline oil to the dye solution enabled it to penetrate the membrane of the tubercle organism to provide an acid-fast colour was also major.[156]

The use of aniline oil in this way led Ziehl to include phenol in his stain in place of aniline,[157] and the year afterwards the still-current Ziehl-Neelsen stain was introduced.[158] By 1884 work on the staining of bacteria proceeded apace. Loeffler introduced his alkaline (polychrome) methylene blue,[159] and in the same year Gram published his famous method.[160] It was not until rather later that its significance in classifying bacteria into Gram-positive and Gram-negative organisms was understood. One final bacterial technique should be mentioned, the flagella-staining techniques of Loeffler. Published in two papers, one in 1889[161] and the other in the following year,[162] his procedures are still in use over eighty years later.

Other significant single stains introduced included bismarck brown in 1878,[163] malachite green in 1884,[164] light green and congo red in 1886,[165] and the fat stains Sudan III in 1896,[166] and Sudan IV in 1901.[167]

As we have seen above, the first double stain was used in 1867, and other double stains were readily proposed later. The first to counterstain haematoxylin with an aniline dye was Poole, an English schoolmaster, who used aniline blue for this purpose in 1875.[168] The still widely-used haematoxylin and eosin was first proposed in 1875,[169] with the eosin used as a separate solution as is still the case. Renaut used the two mixed together in 1879, without improving the results.[170] Another one-solution stain was the aniline blue/picric acid mixture of

Tafani,[171] and yet another was the mixture of two aniline stains used together by Schiefferdecker,[172] since 1876, as he states in the paper, he used eosin with either methyl violet or aniline green.

Histologists were not slow to see that more than two colours might be applied: the first published account of a triple stain is that of Gibbes,[173] in 1880, closely followed by Richardson the next year.[174] Griesbach in 1888 demonstrated double, triple, and quadruple staining,[175] and such effects are still of use for teaching purposes. As Lee remarked,[176] little is to be discovered merely by use of extra colours on one tissue. Much can, however, be found out by subjecting a tissue to as wide a variety of different techniques as possible,[177] especially if these are combined nowadays with phase-contrast and other interference illumination.

Certain other multiple stains have stood the test of time. Among these should be mentioned the Van Gieson stain,[178] originally worked out for nervous tissue, and later applied to connective tissue. Flemming's triple stain for cytology was published in 1891,[179] and is still used. The triple stain procedure of Mallory[180] is still valuable in giving excellent results.

A specialized group of stains—the blood stains—is important for its practical utility and for its scientific background. In their developed form they are a compound of a basic with an acidic dyestuff, nowadays used in methanol solution. Ehrlich in 1879 had employed such compounds,[181] with some understanding of their chemistry, and is the originator of this class of stain. Romanovsky's contribution in 1891 was to use the still-current eosin-methylene blue compound, but more important to demonstrate a blood parasite by this means, and thus create lasting interest in the stains, still called Romanovsky stains for this reason.[182] His method was crude and it was largely accidental that the effect was secured: it required the work of Unna[183] to show that methylene blue would 'ripen' to give an active new colour, and that of Nocht (stimulated by Unna's work) to produce a definite compound dyestuff,[184] which unfortunately he left unused. Only in the following year did Jenner[185] collect this precipitate (of a different character from that of Nocht, in ignorance of his procedure), and use it in solution in methanol. Nocht's precipitate was shortly afterwards prepared and used in the same manner by Leishmann[186] and also independently by Reuter, who used a different solvent which has not stood the test of time. It required the final work by Giemsa to make the stain reliable,[187] and work by Kehrmann[188] and by Bernthsen[189] to clarify the chemistry. This story has been followed in some detail for the light it throws on the halting progress enforced, even at the end of a period in which progress in chemistry had been breathtakingly rapid, by sheer lack of knowledge of the principles of dyestuff chemistry. There was still a large element of trial-and-error in microtechnical researches, even in Germany, and it is significant that work on the topic was now being published outside Germany.

The difficulty of specifying technical details in a proper repeatable manner loomed large in the nineteenth century. Many workers were slipshod in noting their methods, quoting neither the source of their dyes nor, often, their procedure in sufficiently full detail. A well-known example of this may be taken from Nissl's prize essay,[190] wherein while describing the preparation of chromofil bodies of nerve cells he states that the brain is to be removed and put on the desk, then one is to spit on the floor. When the spit is dry the brain is to be put into alcohol. Quite apart from the unusual nature of the timing mechanism, practical experiment has shown that the nature of the spit, its quantity, the nature of the floor, and the ambient temperature and humidity can combine to give a range of times which vary by a factor of five.

There are many literature references to such lack of information, further compounded by the large number of synonyms for particular dyestuffs. In addition, minute traces of impurity can affect dramatically the histological results obtained; this was, for example, especially the case with safranin, where samples from different makers had quite different effects. After impassioned pleas such as those of Latham,[191] papers came gradually to provide the full details required. With the formation of specialist suppliers of stains and other reagents for microscopy, the situation was further improved. Probably the most famous such supplier was the firm of Grubler, first established on the advice of Weigert in 1880. Grubler started another company in 1896, but neither firm ever manufactured dyestuffs: their contribution was to maintain a laboratory wherein the staining procedures were actually tried out on samples bought in, and to sell only those which proved to be effective.

In spite of such welcome advances, some organelles remained difficult to prepare. One such is the mitochondrion, which had been described under many different names before Benda was able to produce a procedure which rendered results easy to obtain.[192] After being refined by Meves & Duesberg seven years later,[193] the technique has remained useful to the present time. In this example, both the small size of the structures (at about the limit of the optical capacity of the microscope), and their physiological nature, combine to make their demonstration difficult. The same considerations apply with even greater force to the Golgi bodies. These are usually lost in routine fixation, and even today the light microscopist is unable to guarantee that predictable results will be obtained in their preparation, possibly on account of their largely phospho-lipid nature.[194] Controversy raged for many years as to whether or not they were artifacts, and it was not until the advent of the electron microscope that their existence was amply confirmed. Nonetheless, the search for them stimulated the development of metallic impregnation techniques, which we may next consider.

Metallic colouring

In this section we must deal with the technique of impregnation with opaque substances, some of which are not of a strictly metallic nature. The first use was by Krause in 1844,[195] who treated pieces of skin with silver nitrate solution, subsequently blackening in light. He understood the nature of the precipitate to be silver chloride and silver metal, a view which is still current. Ten years later Flinzer applied silver nitrate solid direct to the eyes of living animals;[196] the work on the cornea by His two years after this also used silver nitrate,[197] but neither realized that the method had a general applicability. Hartig used silver nitrate in botany in 1854, only one month after Flinzer,[198] but it was Recklinghausen who brought the method into wider use by his paper of 1860,[199] treating a variety of tissues with dilute silver nitrate solution followed by sodium chloride solution in the light.

Following Recklinghausen's paper there was much discussion as to whether or not the results were real or artifact. Auerbach[200] was sure they were merely deposits of silver in furrows in the epithelium; and Feltz obtained similar outlines with egg-white and a photographic paper (of the albumen type in those days, of course).[201] However, Ranvier supported the method, and substituted gold for the silver once it had been formed in the tissues.[202] Some workers tried to reduce the salt in the dark, by a variety of agents; among these, formic acid was easily used, as by Bergh,[203] and is still current. His technique had the additional advantage that counterstaining with either haematoxylin or aniline dyes was possible.

Gold salts were first used by Cohnheim in 1866,[204] in a method which was both primitive and effective—treatment with acetic acid after the gold. In 1867 Arnold used an acid gold chloride mixture,[205] which is the basis of the later method of Apáthy.[206] Bastian was responsible in 1868 for the first use of formic acid,[207] refined to its modern essentials by Löwit in 1875,[208] with further improvements by Ranvier in 1880.[209] The use of gold is also a difficult technique, for during processing the picture alters, to approach perfection and then fade away again. In addition, although preparations can be obtained which look of excellent quality when freshly made, they tend to fade rapidly. Some success was obtained with absolutely fresh tissues, and especially in work with peripheral nerves, but their keeping properties were poor and few have survived to be inspected. Apáthy's method was one of the better ones.[210]

Other agents have been used for impregnation: we have seen above (page 61) that osmium tetroxide was first used for its blackening effect rather than for its fixation. Special mention must be made of Golgi's methods, which are important in the history of histology. His first account of the potassium dichromate/silver method was published in 1873,[211] followed six years later by his mercuric chloride method.[212] Both these gave excellent results with neurones, but took a very long time to perform; his rapid method of 1886[213] took about one week. Cajal developed the work further, and in 1891 published his double impregnation method, especially for sympathetic and spinal ganglia.[214] There was a great deal of mumbo-jumbo written on these techniques of impregnation, and Hill's paper does much to clarify the various formulae.[215]

Colouring living tissues

There are two distinct phases in this process: the first dates from Trembley's work with *Hydra* in 1744, and then a gap in which attention was directed towards the usual non-vital staining techniques, following which, from about 1880, there was vigorous investigation of vital staining, stimulated by the work of Ranvier and of Certes. Trembley fed *Hydra* on various coloured substances, causing it to assume the colour of its food in true intra-vitam manner as the colour was present in granules in the protoplasm.[216] In 1778, von Gleichen used carmine particles to feed *Paramecium*, and was thus the first exponent of phagocytic colouring.[217] Ehrenberg repeated the procedure, and it was described in Pritchard's book[218] six years before Ehrenburg himself published an account in his monumental work,[219] giving full and detailed credit to von Gleichen.

The first deliberate vital staining was carried out by Chrzonszczewsky in 1864,[220] who used the process in his work on the uriniferous tubule with an ammoniacal solution of carmine as the agent. Modern vital staining relies on the ability of certain dyes to stain living cells: this is an unusual property, for most cells stain only after fixation, and most stains are in any case toxic to living cells. Brandt was able, however, to use a very dilute solution of bismarck brown in 1878[221] to colour living Protozoa; and shortly afterwards Certes used quinoline blue for the same purpose.[222] In 1885 Ehrlich found that methylene blue injected into the blood vessels of living animals could stain the nerves:[223] this discovery was to result twenty years later in a rapid development of vital staining in the modern sense.

In 1875 Ranvier described the staining of living lymphocytes by the supravital technique of colouration of the cells by phagocytosis,[224] and in 1894 Galeotti first used neutral red for this purpose,[225] thus establishing a dye which is still of first importance as a vital stain. Uhma developed the modern technique of preparing thin films of the dyestuff on cover slips, to avoid plasmolysis of the cells,[226] Pappenheim applied the method to blood,[227] and Nakanishi made other improvements.[228]

The modern development of true intravital staining took place early this century, when it was realized that acid dyes were suitable. Bouffard[229] was the first to inject benzidine dyes: shortly afterwards Goldmann (a student of Ehrlich) used a colloidal dye, pyrrhol blue.[230] Subsequent workers placed much reliance on intravital techniques, especially in conjunction with histochemical work.

Embedding, infiltration, and sectioning : introductory remarks

A distinction is to be made between embedding and infiltration, although quite often in the literature this is not done. The words of Lee cannot be bettered as making the distinction at a time when the significance was becoming obvious:

> ... Imbedding methods may conveniently be divided into two classes, distinguished by the end it is intended to compass by their employment. In the one it is merely proposed so to surround an object, too small or delicate to be firmly held by the fingers or by any instrument, with some plastic substance that will support it on all sides with firmness but without injurious pressure.... This is simple embedding ... A further object is proposed in the case of the other class of methods, which may be designated methods of interstitial imbedding or infiltration methods. In these it is proposed to fill out with the imbedding mass the natural cavities of the object ... not only each individual organ ... that may be present ... but each separate cell or other anatomical element. ...[231]

The method of simple embedding had a long history, as freehand sections were often made using some material to support delicate objects against the razor. Waxy liver from the post-mortem room was a favourite material, and is quite efficacious as a trial has revealed. Other materials included liver hardened with spirit, elder pith (still used in elementary botany classes), and carrot. Pith is especially effective if divided longitudinally, the object inserted, the bundle bound with cord, and laid in spirit for about fifteen minutes. This makes the pith swell and hold the specimen tightly. German workers of the first two-thirds of the nineteenth century held that it was not necessary to use any other methods for sectioning, and looked down on those who felt the need of a section-cutting machine. This attitude helped English workers to develop the earliest microtomes, but the German attitude changed when the advantages of proper serial sections were realized.

In this context a note on the meaning of serial sections is required. It is often said that Stilling introduced the technique in 1842, from an accidentally-frozen spinal cord sliced freehand. In a limited sense this may be true, just as a loaf of bread sliced successively would be serially sectioned. However, true serial sections are not only of a uniform thinness, they are also connected in a ribbon: they could not therefore be made until the techniques of infiltration and cutting on a suitable microtome had been invented, and this did not occur until 1882, as is described on page 81 below.

The method of simple embedding using paraffin wax is usually ascribed to Klebs in 1869:[232] he cast a cylinder of the wax to fit the well of the microtome, took the specimen from spirit and plunged it into a hole in the wax made with a hot wire. This is certainly correct so far as use of paraffin wax is concerned, but it quite overlooks a very important statement in a paper by Needham,[233] who states that tallow had been used by Quekett to help in cutting injected lung. Quekett actually injected molten tallow through the bronchi into the air cells of an injected lung, which after cooling yielded a splendid mass. Needham does not give the authority for this statement, but it does explain the otherwise inexplicable perfection of some of Quekett's lung preparations in the Hunterian collection. Needham also states that all of Topping's transparent lung preparations were made in this way. There is no reason to doubt the accuracy of these remarks, which are most important in putting back the date of the introduction of waxy embedding media by a significant number of years. A variety of other supporting media were used before the introduction of true infiltration, and must be considered briefly.

The first such use is probably that suggested by Quekett,[234] who used a block of wood as an embedding medium! This was for specimens such as hedgehog spines and porcupine quills, which were literally hammered into ready-made holes in the wood, which was then sectioned in the usual machine. Quekett also introduced the use of glue as an embedding medium, for hairs and similar objects.[235] These were made into a bundle, dipped into thick glue, and dried before being sectioned in the same way. These early processes seem not to have been widely used, in spite of their free discussion in so well-known a book.

In 1856 Griffith & Henfrey suggested the use of white wax or stearine to support small objects for freehand sectioning,[236] by melting it round them by means of a hot wire. It is not clear exactly what is meant by white wax in their book. It could mean beeswax of a purified form, or it could just possibly mean paraffin wax, which had been known for over twenty years at the time they wrote. Probably, though, paraffin would have been mentioned by name had the substance been meant. They also repeat Quekett's mention of gum arabic as an embedding medium.

In 1862 Neumann used egg albumen, hardened in strong alcohol before cutting,[237] and in 1869 Klebs suggested, besides the paraffin already mentioned, a mixture of fish glue and glycerine hardened in alcohol or chromic acid before cutting. Klebs' use of paraffin was modified by Stricker in his handbook of 1869, by mixing olive oil with the wax to modify its hardness according to the temperature at which cutting was to take place.[238]

Media which did not require dehydration were also popular, even long after true infiltration had been established. In 1873 Flemming used a solution of soap in alcohol:[239] the cooled solution solidified and was simply allowed to dry in the air before cutting. For embryological material this offered some advantages, and Salensky used a similar medium, to Polzam's formula which also contained glycerine, for his work on the gemmation of *Salpa*.[240] A similar approach was used by Kadyi,[241] who started with sodium stearate and gave very complicated instructions for preparing the mass.

Other workers returned to the use of egg albumen. Calberla's formula[242] used the yolks as well as the whites, hardening in spirit of different grades and cutting after 24 hours. Bresgen used alkaline egg albumen emulsified with melted lard,[243] and Salenka used pure white of egg only, coagulating by heat followed by strong alcohol.[244] All of these albumen media suffer from the severe defect that it is impossible to remove the albumen from the sections after cutting, which makes subsequent staining difficult and sometimes also causes confusion in interpretation. On the other hand, the method allows a tissue to retain its fatty components, and may be very helpful in allowing perfect sections to be made of otherwise difficult material.

Although Stevenson is often quoted as being the first to use a gum medium with added glycerine, in 1876,[245] we have already seen that Klebs had the basis of the medium by 1869. In his paper already mentioned, Needham goes a very important step further, by suggesting use of an air-pump to exhaust all the air from a tissue in the medium: this is the first mention of what would later be called vacuum-embedding.[246] Hertwig used a mixture of gum and sugar,[247] and Kaiser recommended his well-known jelly in 1880, hardening in 90% spirit and subsequently removing the gum medium by a fine jet of warm water.[248] Such gum/syrup mixtures were to become the usual media for the freezing microtome, as a liquid was required which did not actually freeze (this would have caused much destruction of tissues by formation of large crystals of ice), but which became tough enough to withstand the knife. Hamilton's method[249] used a long soak in syrup followed by immersion in acacia mucilage before freezing in a Rutherford instrument. Cole[250] used a mixture of the two with extra sugar for delicate structures such as the spinal cord. Once tissues were in so strong a solution they were effectively preserved and could be left a year or more if need be. Sollas[251] used gelatin instead of gum: by mounting on a slide and adding a drop of glycerine before covering, a glycerine-jelly mount was automatically secured.

Some other remarkable media were suggested. Pulped paper was advocated,[252] and vegetable wax was claimed to have advantages.[253] More remarkable was the use of shellac, as a thick solution in a groove in a piece of wood: the whole was allowed to harden in air before sectioning to obtain high quality sections of insect limbs.[254] Copal varnish was also used, applied in chloroform solution and evaporated round the object by means of a nightlight.[255] When hard enough to resist a scratch from a finger-nail, the mass was cut with a fine saw and rubbed down in the same manner as a mineral section. In spite of its unlikely nature, the process allowed the most delicate lamellae of connective tissue to be demonstrated. A final example of an extraordinary embedding material would be metal: this was used in the form of a tinfoil tube (as sold containing artists' colours) which was squeezed tightly round the object and was claimed suitable even for fresh tissues.[256]

Such was the search for sectioning media before the establishment of paraffin and celloidin, which displaced all other media and still remain in use. For hard tissues such as teeth, special methods were needed, and Quekett's description of cutting thin sections with a saw, then rubbing them down with a file and then with water of Ayr stones was the basis of all later methods. He has many highly practical points to make on the processes, and his account is still as useful as most which followed it.[257]

Special difficulties arose, and still do, when tissues having both hard and soft components were to be cut. Quekett mentions the process of decalcification with hydrochloric acid (attributing it to Carpenter),[258] and also suggests the preparation of chips of such objects as fossil bones.[259] A paper written forty years later covers only the same ground, and advocates the use of molten balsam to impregnate fully the material before grinding.[260]

Infiltration with paraffin

It is likely that the idea of treating pieces of tissue with turpentine before embedding them occurred rather earlier than the usually-quoted example of Fredericq, who took organs into alcohol, then into turpentine, and thence to paraffin wax with the intention only of preserving them.[261] There is good evidence that turpentine had been used in tentative steps towards this final infiltration stage in Stricker's advice on embedding. He suggests use of a mixture of oil and wax—beeswax is probably meant—for simple embedding. The tissue was to be dehydrated in alcohol and then placed in pure oil of cloves, which he said is far preferable to the turpentine oil that was at one time so generally employed. The object remained in this until it was clear, whereupon it was to be put into the wax mixture held in a cone of paper. It is quite clear from this that the only required final step to actual infiltration would have been to keep the cleared tissue in the molten wax for rather longer than usual. This is an important statement of current practice, but no other references before this date have come to light.[262]

In 1881 true infiltration was certainly established, when Bütschli used chloroform to dissolve the by then established paraffin wax.[263] He disliked turpentine for this purpose on account of its deleterious action on the tissues, and had tried a variety of solvents. The important step here is to have the wax actually in solution, and not merely molten as had been the case with Stricker: when the solvent evaporates a block with good cutting properties is formed. This paper was published on 28th December 1881, but another paper had been published on 12th September by Giesbrecht,[264] who described the method independently, and must have the claim to priority.

Other solvents were suggested as well as turpentine and chloroform: in addition to the oil of cloves already mentioned, creosote was put forward, and Marsh suggested ether.[265] After these initial publications, papers on modifications to the process, exploring every possible facet, were numerous. Some of them show only too plainly that the processes suggested had never been carried out by the author. One ludicrous case is where Fol[266] proposed to use a parabolic mirror. In those days centrally-heated laboratories were unknown, and in a short time the ambient temperature would affect the cutting properties of the block very markedly. He suggested that the mirror might be used to direct the rays from a lamp to a focus at the block to raise its temperature, and this could have been beneficial. He also suggested that if the block be too hot, a piece of ice be put at the focus, to send cold rays towards the block! This is so devastatingly impossible that it proves that Fol never tried it before writing his book.

Sometimes very small-seeming suggestions were the key to real progress: one such case is Gaskell's suggestion of floating the sections on warm water after cutting, to allow them to flatten, and this is still the way they are treated today.[267]

Possibly the most important facet of paraffin embedding is the ability to cut a ribbon of sections. An unequivocal statement as to the exact origin is given by Threlfall,[268] who states:

> In May 1882 W. H. Caldwell, who was investigating the embryology of marine animals by the method of embedding in paraffin and cutting sections by hand, noticed that, if the paraffin was of the right consistency in relation to the temperature at which it was cut, the section adhered to the razor at the sharp edge of the blade, and did not adhere elsewhere. By cutting a second section without moving the first from the razor, the second section welded itself to the first and pushed it across the razor.
>
> On cutting a third section under the same conditions the same phenomenon occurred, and so on, indefinitely, with the result that a ribbon of sections could be formed.

From there, Caldwell and Threlfall went on to make the first automatic microtome, as discussed on page 263 below. For two undergraduates to notice so small a detail, and subsequently to devise a machine automatically to make such a ribbon, is quite remarkable! From this event sprang much of the worth of microscopical technique as it developed from that date.

The use of celloidin

The usual attribution to Duval of the introduction of the collodion process is correct for the wet process, but as Duval himself says,[269] the material had been used by Latteux from 1877 in a manner very similar to Quekett's use of gum. A bundle of hairs was dipped into a solution of collodion, allowed to dry, and the whole sliced through. This process was naturally accompanied by very strong contraction of the mass.[270] Duval's contribution was fundamental: he found means to solidify the mass without allowing it to dry and shrink, by immersing it in strong alcohol—hence the name of wet collodion. A proprietary brand of collodion was sold by Schering of Berlin, under the name celloidin; it is possible that it may have been slightly more nitrated than ordinary collodion. This proprietary material was used by Schiefferdecker in his perfection of the process in 1882;[271] he obtained a mass which was harder than Duval's, and allowed the thinnest sections to be cut. The final major improvement was that of Viallanes in 1883, who used chloroform to harden the mass to the consistency of wax, while retaining full transparency and enough elasticity to allow even better cutting.[272]

The process still had a major drawback—it took at least a week to embed a specimen which could have been put into paraffin within two hours. A more rapid method was suggested by Meyer in 1891,[273] whereby the mass was cleared after hardening, using glycerine. Gilson's rapid method was even quicker, using clearing in cedarwood oil, after infiltration at a temperature of about 50 °C, to reduce the original thin solution to one third of its volume, a very important modification.[274]

For some especially delicate work requiring thin sections of difficult material, double embedding in celloidin followed by paraffin was described, as by Kultschizky,[275] while Field & Martin advocated simultaneous imbedding in paraffin and celloidin.[276]

The celloidin method remains the one of choice for either large sections or for delicate tissues, but it is much less widely used than the paraffin method.

Clearing agents

A clearing agent has two functions: to penetrate the tissues with a medium of high refractive index, and to remove any alcohol between dehydration and mounting. Many substances have been tried as clearing agents, and plates 17 and 18 show the dramatic effect of clearing a specimen mounted dry.[277]

The first account of using turpentine to clear an object was that of Varley in 1843,[278] rather earlier than is usually supposed. After recommending the usual method of plunging the object into hot molten balsam, he goes on to say that objects can also be immersed in turpentine until they have become quite transparent, when they are to be laid on a drop of balsam and covered by the thin glass. This is a really important statement which failed to attract much attention since it was published in an obscure and short-lived journal. The title of the paper suggests that it not merely offered advice on balsam mounting, but also on clearing an object beforehand.

The use of clearing agents was established by Lockhart Clarke in 1851.[279] He treated sections (Varley had suggested the technique for whole objects) with turpentine before mounting them in balsam. In this way he was able to work out the structure of the spinal cord, and the status of microscopical technique took a tremendous step forward.

Subsequent introduction of clearing agents took place slowly. In 1863, Kuchin used creosote to clear his sections of the spinal cord of the river lamprey,[280] and stimulated Stieda to survey the range of possible clearing agents in a paper published three years later.[281] This work drew attention to the use of such agents, as in it he surveyed most of the essential oils, dividing them into two groups. Members of the first, such as turpentine, did not clear from watery sections—a matter of some import in those days—and included 17 oils. His second group could be used for imperfectly-dehydrated sections; he included 9 oils in this oil of cloves group. It is from this paper that Stricker might have derived his ideas on the use of oil of cloves before embedding. In turn, Stieda might have been stimulated in his work by a paper which had appeared earlier in the same journal, which advocated the use of oil of cloves as being better than the turpentine used by Clark.[282] This oil is still very useful, for its property of clearing from watery alcohol, its clearing to a higher refractive index than balsam, and its brittle effect on tissues which is helpful in making minute dissections. In 1869, Bastian suggested the use of benzole and phenol for clearing,[283] and these two substances have continued in use ever since. His paper is a good account of technical methods brought out, as he says, to supplement the process of Clarke.

In 1882 a useful survey of clearing agents was published,[284] with the object of finding the best possible agent. A thorough summary of the properties of essential oils was made, and choice fell upon cedarwood oil, origanum oil, or sandalwood oil. Other agents were introduced sporadically. Oil of cajeput had been recommended for celloidin work[285] in 1897, the year before Jordan investigated the behaviour of a range of oils towards this medium.[286] Ordinary aniline oil was recommended[287] on account of its ability to clear from only 70% spirit, a property which is still useful on occasion. Methyl salicylate was suggested in 1898,[288] and in 1909 terpineol was stated to be exceptionally useful as a clearing agent.[289]

There is only a limited range of materials which can be used for clearing, and choice is governed by expense on the one hand and the refractive index being sought for on the other. In spite of the small range of oils, however, the ability to clear material is of prime importance in microscopy.

Injection techniques

The vogue for injected specimens continued for much of the nineteenth century, and was given an extra importance with provision of transparent injections mounted in balsam. Some of the best injections were made by Quekett, and examples are included in colour plate IV. As has been emphasized, such slides are compelling by their sheer beauty, but are decidedly limited in the amount of histological information they can convey. In spite of this, most manuals of microscopical technique continued to devote much space to injection methods for most of the century, and the large number of specimens which has survived testifies to the vast numbers produced. Some examples of injected specimens are shown in plates 20 and 21, in both opaque and transparent forms. Attention has been drawn to such books as that by Robin,[290] which were almost entirely devoted to injection techniques; a survey of other works of the middle third of the nineteenth century shows that the usual fraction of space given to injections was about one-sixth. The account in Beale[291] is similar to all subsequent advice, with the exception of additional formulae. By the end of the century these had been grouped into various main kinds. The most usual was a basic gelatine mixture such as that of Robin,[292] which was combined at moment of use with a colouring-mass: this could be copper ferrocyanide, some other coloured inorganic compound, or an insoluble aniline colour. A good survey of such materials is given by Lee,[293] who also quotes an albumen medium. Other workers had other favourite masses, and one based on linseed oil was claimed to fill the finest vessels;[294] as it contained about 10% of carbon disulphide it would doubtless have been quite unpleasant to work with. Other injections were prepared using a shellac base,[295] asphalt,[296] and a mixture of paraffin wax with beeswax and olive oil.[297] In the 1885 edition of Lee there are over sixty distinct formulae for injection masses. In his 1921 edition there are about twenty, and this is in a very conservative publication. By that time, all formulae for opaque masses, or those suitable only for naked-eye study, had been eliminated.

The continued popularity of such preparations may have been due to two main factors. First, many of those who advocated them were men of great authority—Robin in France, Hyrtl in Germany, and Beale in England, for example. The other factor was their continued preparation by commercial mounters, for amateur purchase: these preparations were great favourites, not too difficult to prepare once the technique had become a routine, and selling for a good price. In spite of this, in the history of histology they must be considered a side issue, as they gave so little information to the scientist.

Microdissection and maceration

The outstanding early contributions to microanatomical knowledge and technique of Malpighi, Swammerdam, and Leeuwenhoek have been mentioned earlier in this book, and are ably described by Cole.[298] The work of Hill, involving the maceration of timber, has also been described above. These were outstanding achievements, of a kind not repeated in the later history of microscopy. Nonetheless, some use of both maceration and microdissection was made in the nineteenth century, but largely for commercial reasons. The techniques of genuine microdissection were developed early in the twentieth century by Barber,[299] following the lead given by Schouten in 1899.[300] However, although the work took root as a result of these later efforts, a much earlier micromanipulation apparatus has come to light, in Schmidt's paper of 1859.[301] Schmidt also devised a good microtome, to be discussed below, but his 'microscopic dissector' is of such importance that Figure 6 includes reproductions of all his diagrams.

As Schmidt says in his paper (page 30):

> In the construction of this instrument, the principal object I had in view was, to be enabled to make the slightest motion of the microscopic needles, knives, or scissors, in different directions.

The equipment is described in the caption to the plate, but he also had a much simpler, hand-held, needle holder for use in researches on nerves: it was experience with this that led him to devise the more advanced instrument described. The needle points were sharpened on a fine Arkansas stone, and the whole instrument was carefully adjusted to work with easily-effected motions. He goes on to remark:

> ... To manage the instrument successfully, delicacy of touch and a great deal of patience are required; but it is only by the latter, combined with perseverance, energy, and close observations, that scientific facts have, or ever will be, established.

By means of this instrument he was able to observe at fairly high magnification, using a Nachet No. 3 objective, that is of about one quarter-inch focal length giving a power of about × 400 overall. He could scrape off the epithelium from the lining of small ducts as they were opened. This is a somewhat remarkable invention for its time: all other microdissection was carried out, as he says somewhat disparagingly, with hand-held needles used under only low powers, with necessarily coarse results. There is little doubt that his apparatus could have been used to pick up a single bacterium, had such an idea been possible at that time.

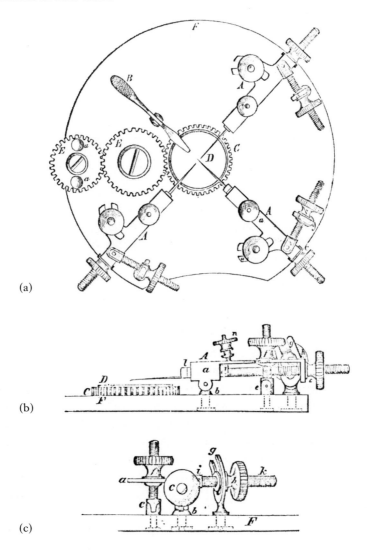

(a)

(b)

(c)

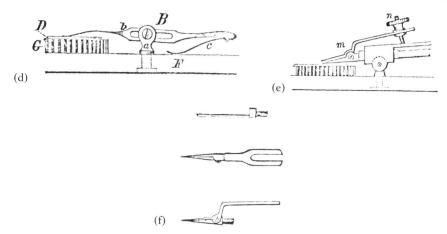

Figure 6. The Schmidt micro-dissector, 1859 (from the text figures in H D Schmidt (1859), *Amer. J. Med. Scis.   37 :13–40*).

Figure 6(a) is the plan view of the instrument, which is seen to be a plate to attach to the stage of the microscope, arranged to carry three needle-holders (A), and one specimen-holder (B). The plate carrying the specimen could be rotated by means of the knob (E) (Schmidt's text-figure 16).

Figure 6(b) is the side-elevation of a needle-holder, shown in relation to the specimen-plate. The clamp (*a*) was arranged to take interchangeable instruments. The various milled heads worked on micrometer screws to give a fine positioning of the instrument being used (Schmidt's text-figure 17).

Figure 6(c) is the end-elevation of a needle-holder (Schmidt's text-figure 18).

Figure 6(d) shows the specimen-holder. Note the fine springs which controlled the pressure exerted (Schmidt's text-figure 19).

Figure 6(e) represents the micro-scissors in the needle-holder. Their capacity for accurate work is obvious from the milled-head controls (Schmidt's text-figure 20).

Figure 6(f) illustrates the scissors, forceps, and scalpel. Each of these could be used interchangeably and accurately, and this degree of precision would not be seen elsewhere for over forty years (Schmidt's text-figures 21, 22, and 23).

The attention paid to simple dissecting microscopes by all the suppliers is a clue to the importance enjoyed by the technique. Once the process of infiltration had been implemented from the 1880s, microdissection (by which is meant a coarser form of micromanipu-lation) was less important, except in the amateur market. There was still a strong call therein for such insect dissections and disarticulations as Enock was famous for, but as far as the scientist was concerned, it required the development of true micromanipulation in the twentieth century to justify further attention.

Equally, attention to maceration as a technique was greatest earlier in the last century: today it is used only for isolation of plant elements for student courses. Before 1830 it was a major tool, often performed either by boiling or by allowing the material to rot down naturally. Max Schultze used iodized serum before compressing the tissue, and many other chemicals were used: these might have had a fairly vigorous action, as in the use of the nitric acid/potassium chlorate mixture of Kuhne,[302] and the potassium sulphocyanide of Stirling.[303] An extension of such methods was to carry them further into corrosion: here an organ injected with oily matter had its tissues dissolved away by sodium hypochlorite solution. This is the technique used by Hyrtl, and the methods were described by Altmann.[304] Again, with the coming of serial sectioning these techniques lost their earlier importance.

Mounting media : non-aqueous

It was long thought that we owe the first mountant which solidified after application to Andrew Pritchard. In 1832 he published the method, which used a drop of a gum/isinglass mixture, on a glass slide covered with a mica disc.[305] However, as Rooseboom has now shown, as early as 1795 Ypelaar had prepared mounts using venetian turpentine[306] (illustrated in plate 19): these are the earliest resin-mounted preparations ever made. This medium is still an article of commerce, although not now in use for slide preparation: it is a mixture of substances which melts easily, and solidifies again in optically-good films on cooling. This was an important innovation, which was not followed up; more than thirty years were to elapse before resinous mountants would become established.

Pritchard's 1832 method was totally overshadowed by the almost contemporaneous discovery of mounting in canada balsam. In 1835 the same author published the first description of this process, in a somewhat rare book:[307]

> There are many objects which are not sufficiently transparent to render their structure visible by transmitted light. For these, the method described in page 230 of the Microscopic Cabinet will in some cases be found useful; but the best method of mounting them, and which I have adopted with complete success, was discovered after the publication of the above work. It consists in immersing the object in Canada balsam or varnish, and pressing it between two slips of glass, so as to exclude the air-bubbles. By this treatment, many objects which otherwise possess little interest are rendered highly valuable, allowing the light to pass freely through them, exhibiting their structure, and presenting to the admiring spectator the most brilliant and superb colours . . . [pages 5–6].

This shows the date of introduction to be about 1832, and some other authors have a little to say of the event. Quekett, for example, says:[308]

> Canada Balsam.—This excellent material was first suggested by Mr J. T. Cooper, and employed about the year 1832, by Mr Bond, an ingenious preparer of microscopic objects, the first notice of it in print appears in a small book published by Mr. Pritchard, in 1835, entitled A List of Two Thousand Microscopic Objects. The older anatomists were in the habit of using varnishes of different kinds to cover their injected preparations, which, in course of time, became hard and transparent; the objects belonging to the microscopes described in page 16 are thus coated . . . [pages 275–276].

This discovery would have had considerable commercial importance, and this is borne out by Bowerbank, in a lively account of the introduction of balsam mounting:[309]

> This valuable and effective mode of mounting microscopical objects, I am informed by Mr. Topping, was originally suggested by Mr J. T. Cooper, an eminent analytical chemist, and it was first applied to the preparation of large objects for exhibition by the solar microscope by a person of the name of Newth, who was employed by the late Mr. Carpenter, the optician, of Regent Street, to exhibit them with the microscope, and who subsequently carried on a very profitable trade in objects so mounted. Mr. Bond afterwards obtained the recipe from Newth, and supplied the microscope at the Adelaide Gallery with such objects for a considerable period, but the process still remained a secret.

The story continues that Bowerbank and others were shown some of the mounts by Goadby, and upon asking how they were made, had to be told that the process was secret. However, one mount was freshly made, and some of the medium squeezed out onto the fingers of one of the party, who recognized the smell. By the time Goadby returned with more specimens the following week, the group had made its own, and much to his surprise Goadby was shown examples. These preparations were made by using two slides, thin cover glasses not being in use at that time. Examples of such early balsam mounts are shown in colour plate II.

It is likely that Pritchard derived his information from Bowerbank, for the method was made available as freely as possible in contrast to Ypelaar's use of Venetian turpentine, which was clearly kept a secret by him. This would mean that, with allowance for the intervening period, balsam mounting was introduced in 1830: this has been the reasoning behind choice of this date as a critical one in this book.

The process of applying paper covers to slides, so prominent a feature of nineteenth century preparations, seems to have originated with

Pritchard. In the same paragraph as that in which he describes use of his gum mixture for mounting,[310] he also says

> ... These sliders should have a piece of paper pasted over the talc, to protect it from injury, leaving an aperture in the centre for the object, and made of an uniform size.

In the course of the next fifty years the application of decorative, and occasionally informative, paper covers would become universal on commercial mounts, these being supplied with all manner of lithographed colour covers of considerable complexity. Some of these covers have been illustrated *in situ* on the various preparations pictured in the plates, and actual specimens were pasted into Beck's book, as already noted.[311]

Balsam mounts do not require attachment of such paper covers to ensure the adhesion of the mountant to the cover, but in spite of this the obvious decorative possibilities ensured that they would be used. An illustration of the process of attaching a cover to a slide is given by Martin:[312] see his plate 1 of page 52.

There are surprisingly few other references to balsam mounting. Varley gave concise instructions for its use in the important paper already mentioned when discussing clearing agents.[313] Here he advised use of hot balsam before going on to add the new advice on use of turpentine. It is quite unlikely that this article attracted much attention, for it was published in an obscure journal. Griffith, a well-known microscopist, writing in the same year as Varley makes no allusion to the process: He says merely that use of sliders is now quite abandoned, and leaves the size of slide to the whim of the moment:[314] he recommends the use of balsam only in molten form, and had the use of turpentine been well established he would surely have given it a mention.

Credit for use of balsam in solution, to allow it to run round the specimen without need for heat, is certainly due to Griffith, for in a footnote to another of his articles in the same volume we read:

> We are informed by our correspondent, that in the preservation of objects for the microscope in the liquid state, a solution of Canada balsam in sulphuric aether, of such a state of viscidity as is just sufficient to allow it to be laid on with a pen or stick, has answered better than any of the means published in his former paper.[315]

This implies use of the substance in deep cells, but this seems an unlikely general use on account of its great volatility and consequent shrinkage. In any case, however, the principle of solution of the balsam was established.

Less equivocally, in a paper read on 21st May 1845, Boys definitely used a solution of canada balsam in spirits of turpentine.[316] Admittedly, he used slight heat to cause the balsam to spread, and he appeared to use only dry specimens for mounting. The balsam was in solution for the principal object of avoiding entrapment of air bubbles before applying the thin cover.

Between them, these three papers place the introduction of preliminary clearing of objects, and their mounting in a solution of balsam, between 1843 and 1845. Needless to say, the method was not adopted universally or at once, and it is in this context that Clarke's paper of 1851 is important as the popularizer of the process, especially as it applied to histological sections.

Other substances were tried as solvents for the balsam. Chloroform was suggested in 1865,[317] and is still occasionally useful; benzole was used, after careful comparison with other solvents, in 1869.[318] The considerable importance of balsam mounts is that they are easy to make, and permanent records of what has been achieved: the scientist can refer back to them, and exchange them with others. Commercially, of course, this method of mounting opened up vast possibilities, which were fully exploited as time went by.

A number of other non-aqueous media have been used, either as so-called improvements on canada balsam, or to achieve high refractive indices, usually for diatom mounts. Gum dammar was introduced in 1868,[319] and this still finds some favour as it gives a refractive index marginally less than that of balsam (1·520 as against 1·524 for xylene-balsam). After some time, this medium has a tendency to become granular, which caused it to lose popularity, but a large number of mounts of the 1880s which used this medium are still in excellent order.

For some purposes, a lower refractive index is useful in enhancing visibility. One substance offering this possibility is sandarac. Gilson proposed its use in a somewhat unsatisfactory and curious manner,[320] refusing to give actual formulae on account of the great difficulty of the adequate preparation of the mixtures. He allowed them to be made up only by commercial firms who guaranteed to make them carefully. This has caused his 'euparal' mountant to be condemned scientifically, in spite of its refractive index of 1·483. It is essentially a mixture of eucalyptol, camphor, salol, and paraeldehyde with the sandarac, and has the advantage that objects can be mounted in it from 90% alcohol: it is still available commercially.

In the search for media of higher refractive indices some curious recipes have been suggested. Following Stephenson's paper,[321] in which he put forward formulae for solutions of sulphur (r.i. 1·75) and of yellow phosphorus (r.i. 2·10) in carbon disulphide—although acknowledging their dangerous nature—Abbe searched for a usable substance of high refractive index, and suggested naphthalene monobromide (r.i. 1·658).[322] This was tried out by Van Heurck for diatoms, and proved useful in allowing increased resolution. Flesch tried this medium for histological work shortly afterwards,[323] but found it disappointing: the enhanced detail visible in freshly-mounted specimens is lost after a short time. Styrax was introduced by Van Heurck as a result of work initiated by these researches, as a solid medium free from the disadvantages of a fluid mountant;[324] it was of course intended only for diatoms which do not suffer from being heated during mounting. Its r.i. when carefully prepared is 1·63, and it was widely used for diatom work for many years. One mountant which did not find widespread favour was that of Thompson,[325] which consisted of flowers of sulphur, liquid bromine, and arsenious acid heated together to the consistency of toffee! The last significant paper on high refractive index mountants was published in 1898, and considered a range of possible media, including m-cinnamene, quinidine, phenyl thiocarbamide, α-monobromonaphthalene, and piperine.[326]

Cell mounts

By 1830, as we saw on page 24 above, most preparations were put up in a fluid medium, and there were many examples of media suggested in the literature between the beginning of the nineteenth century and about 1870. The main difficulty in making such a preparation lies in sealing the cell, and thus there were many formulae for varnishes also. By use of one medium, though, this difficulty was avoided: this was a saturated solution of calcium chloride, which being highly deliquescent comes into equilibrium with the atmosphere without need of any sealing of any kind. Harting invented the medium in 1841,[327] and, as Rooseboom states,[328] in the Leiden museum there still remain examples of mounts made in 1848 using this medium.

In 1831 Varley described the making of a 'cell' (the name by which they are still known) for holding fluid mounts,[329] and in the following year Pritchard suggested the use of mercuric chloride as a mountant.[330] This much predates its employment by Blanchard[331] as what we would now call a fixing agent: Pritchard's use is set out in vague terms, but it is clear that it was that of a preservative liquid for aquatic organisms, as an antibacterial and antifungal agent, and not as a fixative.

Probably the best-known of the fluid media were those devised by Goadby, who was awarded the Society of Arts gold medal in 1841.[332] His fluids contained mercuric chloride and common salt, in different concentrations for different applications, and sometimes with addition of arsenious acid. In each formula the concentration of mercuric chloride is too small to act as anything other than a preservative fungicide.

In Germany, although both the Varley and the Griffith papers had been translated into German,[333] little use was made of English techniques before 1850. In 1843 Oschatz published some methods of making preparations,[334] using rather concentrated solutions of sucrose and acetic acid. The next year Moleschott described Harting's method of calcium chloride mounting,[335] which brought the process to a much wider potential audience than its original Dutch publication had achieved. It is interesting that, in the journal concerned, the paper immediately following is by von Mohl comparing the methods of Oschatz and Harting;[336] the latter is judged superior, and this was of some importance as von Mohl shortly afterwards published his well-known textbook,[337] which greatly influenced botanical microtechnique on the Continent. It may be of interest to record that his book, published about the same time as that by Quekett, had only 27 pages (of 277) on microtechnique, whereas the English book had proportionally five times as much. Recognition of English methods on a wider scale seems to have been started by Welcker's publication,[338] the importance of which can hardly be overstated.

In addition to preservative liquids relying on metallic compounds for their effect, some other formulae were published, of which a few examples will illustrate the approaches used. Creosote was used in aqueous solution in the well-known Thwaites' fluid,[339] and chloral hydrate was introduced in 1875 by Lavdowsky:[340] it still finds application for some purposes. Quekett's fluid was wood naphtha diluted with water.[341] Rather more important than other aqueous solutions was the introduction of glycerine in 1848:[342] three years later Deane described the first glycerine-jelly, as well as other media, including one combining glycerine and honey.[343] Some of these mountants are still occasionally used. For his well-known insect preparations Enock used 70% alcohol, whereas Clarke & Page used a formalin solution. A selected group of fluid mounts is shown in plate 22.

Specimens mounted dry continued to be produced for some special purposes: a selected group is shown in plate 23. Quekett's chapter on such mounting contained all that needed to be said.[344] He recommended this style for bone and tooth sections, and there is no doubt that the method can be successful: many commercial preparations were mounted thus, the cover being held on with gummed paper. The use of discs of cork for opaque mounts was also common, and some instrument outfits were provided with special holders to carry them in the optical axis. Edwin Quekett adopted a plan for the use of pill boxes, which had the object glued to the bottom, to be covered by the lid when not in use. Such boxes were provided with a special sprung holder for attachment to the microscope. Later in the century the pill box was glued to an ordinary slide; this allowed easy storage while overcoming a major difficulty of dry mounts—the tendency of the underside of the cover to become dirty and obscure the view of the prepared object. Another corrective was to use a wooden slide with cavity covered by an ivory disc, rotated out of the way during viewing.

Some other elaborate slides carried small objects on pins, to allow them to be rotated during observation: this was a helpful way to mount the tests of Foraminifera, for example. Perhaps the most elaborate dry mounts ever made were those which arranged as many as five hundred scales from butterfly wings into a picture of a vase of flowers or a boxing match. Scientifically quite worthless, these are nowadays expensive collector's items.

Where dry cells or wet mounts were to be finished with a ring of cement, the opportunity to decorate this was often taken by the amateur. A major difficulty in putting up wet mounts was lack of reliability of the cement, which had to contain perhaps chemically active substances of widely varying pH, with volume-altering according to the ambient temperature. Marine glue was the usual basis of the cements, apparently made of a mixture of shellac and indiarubber dissolved in mineral naphtha.[345] For shallow turned cells gold size was used; it was made by boiling linseed oil with red lead and umber for three hours, and then again with white lead. As may be imagined, the gold size supplied commercially varied in quality. Asphalt varnish was also an article of commerce, being a solution of asphalt in a mixture of boiled linseed oil and turpentine; it was used to seal covers into place. Brunswick black was a similar mixture with addition of lamp black and some india-rubber dissolved in naphtha.[346] Stieda's white cement[347] was another preparation used to finish a cover. Marsh suggested use of gelatine solution hardened by dichromate after application,[348] with a final cover of Bell's cement—a commercial formula of undisclosed composition.

One final group of mounting substances remains to be discussed— those used to fasten serial sections to the slide before removal of the embedding medium. Threlfall's original mixture was a thin solution of india-rubber in naphtha,[349] but this did not gain wide acceptance because the nature of naphtha as supplied commercially varied widely. Collodion was used by Schällibaum,[350] mixed with clove oil; the sections were floated on the previously-coated slide and warmed *in situ*. Shellac was suggested by Giesbrecht,[351] with a similar procedure. Gum arabic was also put forward,[352] as was quince mucilage,[353] but Mayer's albumen[354] was found suitably for both paraffin and celloidin sections, and is still used almost exclusively.

Histochemical tests

Some works written in the first few years of the nineteenth century showed that interest in the application of chemical tests to microscopical specimens was gaining momentum. One such example is the book by Girod-Chantrans,[355] who used a variety of tests and reagents to help in classifying plants. This was certainly an imaginative approach, and one which was far ahead of the times. Probably the first account of a definite histochemical test was that by Link in 1807,[356] who used iron sulphate solution to obtain a colour reaction with tannin. This was not a deliberate attempt to locate the site of the tannin, but only to try to discover if plant vessels opened into cells. The starch–iodine reaction was discovered by Colin & Claubry in 1814,[357] but was first applied microscopically by Raspail in 1825.[358] This paper marks the foundation of histochemistry, and is important also in the history of sectioning (see page 135 below). Raspail deliberately set out to apply what we would now call histochemical tests to his researches, and three years later he used his sugar-with-sulphuric-acid test on animal and plant cells as the basis of an ingenious classification of their properties.[359] In the following year he used three tests which are still significant: the xanthoproteic reaction, the aldehyde test for proteins, and the test which is now called the Liebermann test in ignorance of its true discoverer.[360] This is extraordinarily impressive work for that date, all neatly put together in his famous essay published in 1830.[361] Schleiden used sulphuric acid and iodine to give the blue reaction from plant cell walls in 1838,[362] and von Mohl gave a good account of the use of iodine and its reactions with plant tissues two years later.[363] The famous chlor-zinc-iodide reaction of Schulze was published in 1850,[364] and thereafter much interest was generated in histochemical tests by the botanists, but rather less by animal investigators.

Vogel had identified iron in tissues by use of ammonium sulphide,[365] and in 1867 Perls introduced his method for demonstrating iron by the prussian blue reaction.[366] A different method was used by Beale in 1861, who dissolved unwanted tissues from round the nerve fibres in which he was interested by means of gastric juice.[367] This was later developed into the technique of enzymic analysis.[368] In 1868 Klebs showed that pus gave a blue colour with tincture of guaicol,[369] and in the same year Heidenhain isolated what we now know to be ribonucleic ácid.[370] Forty-four years after it was invented, in 1888, Leitgeb used Millon's test in histochemistry,[371] and Miescher isolated nuclear chromatin by using its selective affinity for methyl green in 1878.[372] Molisch stained tissue iron-red by means of the thiocyanate re-action,[373] and phosphate was demonstrated by Lilienfeld & Monti in 1892.[374]

Some of the highlights of the history of histochemistry have been mentioned above, but the evolution of such methods is a long and detailed story in itself. For further details the introductory chapter of Pearse[375] should be consulted, together with Sandritter,[376] which in spite of confining its details to German workers does provide good references and useful short summaries. Petersen's memorial speech is another excellent source of references,[377] and a helpful synthesis of matters concerned, especially with cell theory.

It is certainly clear that histochemistry suffered a partial eclipse as a research tool with the advent of serial sectioning techniques; although some good work was done in the last quarter of the nineteenth century, the techniques whereby microscopic structures known only empirically by means of staining could be chemically interpreted by means of histochemistry did not emerge on an organized basis until the beginning of the twentieth century.

References for Chapter 4

1. A B Lee—*The microtomist's vade mecum.* (London: Churchill, 4th edn, 1896.) See pp. v, vii, and viii.
2. H Gierke (1884/5)—Farberei zu mikroskopischen Zwecken, *Z. wiss. Mikr.* *1* :62–100, 372–408, 497–557, and 2 :13–36, 164–221.
3. W H Seaman (1885/6)—Staining tissues in microscopy, *Amer. Monthly. Mic. J.* 6 :65–68, 89–94, 106–107, 131–133, 152–156, 210–216, 234–236, and 7 :13–15, 31–35, 53–54, 70–73, 97–99, 150–152.
    The importance of these articles was clearly realized at the time, for an editorial note (6 :76) says:
    > (These articles are) the most exhaustive study of the staining agents used in microscopy that can be imagined.
4. H J Conn—*The history of staining.* (New York: Biotech, 2nd edn, 1948.)
5. Anon—Die Entwicklung der histologischen Farbetechnik, *Ciba Zeitschrift Basel,* 88 (1943).

6 H J Conn (1946)—Development of biological staining, *Ciba Symposia* 7 :270–300.

7. J R Baker (1943)—The discovery of the uses of colouring agents in biological microtechnique, *J.Q.M.C.* 24 :256–275.

8. J R Baker—*Principles of biological microtechnique.* (London: Methuen, 1958.)

9. A Foettinger (1885)—Renseignements techniques. 1. De l'emploi de l'hydrate de chloral pour l'étude et pour la conservation des animaux inferieurs, *Arch. de Biol.* 6 :115–125.

10. A Korotnev (1884)—Zur Histologie der Siphonophoren, *Neapel. Stat. Mitth.* 5 :229–288.

11. C F Rousselet (1894)—On a method of preserving Rotatoria, *J.Q.M.C.* 12 :205–209.

12. H Fol (1885)—Zur Mittelmeerfauna, *Zool. Anz.* 8 :667–670.

13. Lee, op. cit. note 1 above: see pp. 11–12.

14. R Boyle (1666)—Of preserving birds taken out of the egg, and other small faetus's, *Phil. Trans. R.S.* 1 :199–201.

15. J Pringle (1750)—Some experiments on substances resisting putrefaction, *Phil. Trans. R.S.* 46 :480–488.

16. J Jacobson (1833)—Versuche über technische und medicinische Anwendung der Chromoxyde und der Chromsalze, *Liebig, Annal.* 7 :329–331.

17. M Malpighi—*De viscerum structura exercitatio anatomica.* (Bologna: Montius, 1666.) See pp. 51–54. For a convenient extract in English see: E Clarke & C D O'Malley—*The human brain and spinal cord.* (Berkeley: University of California Press, 1968.) See pp. 417–418.

18. W Bowman (1840)—On the minute structure and movements of voluntary muscle, *Phil. Trans. R.S.* 1840 :457–502.

19. A Hannover (1840)—Die Chromsäure, ein vorzügliches Mittel bei mikroskopischen Untersuchungen, *Müller, Archiv.* 1840 :549–558.

20. A Corti (1851)—Recherches sur l'organe de l'ouie des mammifères, *Z. wiss. Zool.* 3 :109–169.

21. H Müller (1860)—Anatomische Untersuchung eines Microphthalmus, *Verh. phys. med. Ges. Wurzburg* 10 :138–146.

22. W Mayzel (1879)—Ueber die Vorgänge bei der Segmentation des Eies von Würmern (Nematoden) und Schnecken, *Carus, Zool. Anz.* 2 :280–282.

23. W Flemming (1879)—Ueber das Verhalten des Kerns bei die Zelltheilung, und über die Bedetung mehrkerniger Zellen, *Virchow, Archiv.* 77 :1–29.

24. E Blanchard (1846)—Recherches sur l'organisation des Vers, *Proc. Verb. Soc. Philom. Paris* 1846 :62–65, 67–71.

25. R Remak (1854)—Ueber vielkernige Zellen der Leber, *Muller, Archiv.* 1854 :99–102.

26. A Lang (1878)—(Ueber Anatomie und Histologie mariner Dendrocoelen), *Schweizer Naturf. Gesell. Verhandl.* 61 :113–115.

27. M Heidenhain (1916)—Ueber neuere Sublimatgemische, *Z.w.M.* 33 :232–234.

28. K Zenker (1894)—Ueber Chromkali-Sublimat-Eisessig als Fixirungsmittel, *Munch. Med. Woch.* 41 :532–534.

29. Formula quoted in A van Gehudten & C Nelis (1898)—Quelques points concernant la structure des cellules des ganglions spinaux, *Ac. Med. Belg. Bull.* 12 :336–346. See p. 339.

30. J L Clarke (1851)—Researches into the structure of the spinal cord, *Phil. Trans. R.S.* 141 :601–622.

31. L Brasil (1905)—Nouvelles recherches sur la reproduction des grégarines monocystidées, *Arch. Zool. Exp.* 4 :69–99.

32. H Baker (1743)—Some observations on a polype dried, *Phil. Trans. R.S.* 42 :616–619.

33. His use of spirit is accessibly described in: J L Moreau—*Oeuvres de Vicq-d'Azyr*. (Paris: Duprat-Duverger, 1805.) See vol. 6, p. 24.

34. J C Reil (1809)—Untersuchungen über den Bau des grossen Gehirns im menschen, *Reil, Archiv. Halle   9 :136–146.*

35. J B Carnoy (1887)—Quelques observations sur la division cellulaire chez les animaux, *Cellule   3 :225–324.* See p. 227.

36. loc. cit. note 21 above.

37. H Muller (1860)—Ueber glatte Muskeln und Nervengeflechte der Chorioidea im menschlichen Auge, *Verh. phys. med. Ges. Wurzburg   10 :179–192.*

38. A N Erlicki (1877)—(untitled note), *Warschau med. Zeits   22 :15.*

39. C Regaud (1910)—Études sur la structure des tubes séminifères et sur la spermatogénèse chez les mammifères, *Arch. d'Anat., micr.   11 :291–431.* See pp. 293–294.

40. A E Fürnrohr (1850)—Verhandlungen der Section fur Botanik, Land- und Forst-wissenschaft bei der XXVII Versammlungen deutscher Naturforscher und Aertze in Griefswald, *Flora   8 :641–692.*

41. M Schultze (1864)—Ueber den Bau der Leucht-organe der Männchen von Lampyris splendidula, *Sitz. Ber. Nied. Gesell. Bonn   1864 :61–7.*

42. M Schultze & M Rudneff (1865)—Weitere Mittheilungen über die Einwirkung der Ueberosmiumsäure auf thierische Gewebe, *Arch. mikr. Anat.   1 :299–304.*

43. W Flemming—*Zellsubstanz, Kern und Zelltheilung.* (Leipzig: Vogel, 1882). See p. 381.

44. W Flemming (1884)—Mittheilungen zur Färbetechnik, *Z.w.M.   1 :349–361.*

45. R Altmann—*Die Elementarorganismen und ihre Beziehungen zu den Zellen.* (Leipzig: Veit, 2nd edn, 1894.)

46. C J Cori (1889)—Beitrag zur Conservirunstechnik von Thieren, *Z.w.M.   6 :437–442.*

47. C L Bristol (1893)—On the restoration of osmic acid solutions, *Amer. Nat.   27 :175–176.*

48. W Roberts (1863)—On peculiar appearances exhibited by blood-corpuscles under the influence of solutions of magenta and tannin, *Proc. R.S.   12 :481–491.*

49. L Ranvier—*Traité technique d'histologie.* (Paris: Savy, 1875.) See p. 136.

50. loc. cit. note 23 above. See p. 103.

51. N Kleinenberg (1879)—The development of the earth-worm, Lumbricus trapezoides, Duges, *Q.J.M.S.   19 :206–244.* See p. 208.

52. P Bouin—*Phénomènes cytologiques anormaux dans l'histogénèse et l'atrophie expérimentale du tube séminifère.* (Nancy: Thèse pour le doctorate en médecine de la Faculté de Médecine, 1897.) See p. 19.

53. J A Trillat (1892)—Sur les propriétés antiseptiques de la formaldehyde, *Compt. Rend. A.C. Paris   114 :1278–1281.*

54. F Blum (1893)—Der Formaldehyd als Härtungsmittel. Vorläufige Mittheilung. *Z.w.M.   10 :314–315.*

55. F Hermann (1889)—Beiträge zur Histologie des Hodens, *Arch. mikr. Anat.   34 :58–106.*

56. C Rabl (1885)—Ueber Zelltheilung, *Morphol. Jbuch.   10 :214–330.* See p. 216.

57. W S Kent—*A manual of the infusoria.* (London: Bogue, 1880–1882.) See p. 114 of vol. 2.

58. W His (1877)—Neue Untersuchungen über die Bildung des Hühnerembryo, *Archiv. Anat. Physiol.   1877 :112–187.* See p. 115.

59. J Perenyi (1882)—Über eine neue Erhärtungsflüssigkeit, *Zool. Anz.   5 :459–460.*

60. R Koch (1877)—Untersuchungen über Bacterien, VI. Verfahren zur Untersuchung, zum Conserviren und Photographiren der Bacterien, *Cohn, Beitr. Biol. Pflanz.   2 :399–434.*

61. P Ehrlich (1879)—Beiträge zur Kenntnis der granuliren Bindegewebszellen und der eosinophilen Leukocythen, *Arch. Anat. Physiol.   1879 :166–169.*

62. C Golgi (1886)—Sulla fina anatomia degli organi centrali del sistema nervoso, *Arch. Ital. Biol.   7 :15–47.* See p. 30.

63. G Mann (1894)—Ueber die Behandlung der Nervenzellen fur experimentellhistologie Untersuchingen, *Z.w.M.   11 :479–494.*

64. R Altmann (1889)—Ueber die Fettumsetzungen im Organismus, *Arch. Anat. Physiol.   1889 Suppl : 86–104.*

65. G Mann—*Physiological histology.* (Oxford: Clarendon Press, 1902.) See p. 139.

66. T Graham (1861)—Liquid diffusion applied to analysis, *Proc. R.S.   11 :243–247.*

67. R Altmann—*Studien über die Zelle.* (Leipzig: Felix, 1886.)

68. op. cit. note 43 above.

69. G Berthold—*Studien über Protoplasmamechanik.* (Leipzig: Felix, 1886.)

70. F Schwarz—Die morphologische und chemische Zusammensetzung des Protoplasmas (1887), *Cohn, Beitr. Biol. Pflanz.   5 :1–244.*

71. S Trzebiński (1887)—Einiges über die Einwirkung der Härtungsmethoden auf die Beschaffenheit der Ganglienzellen im Rückenmark der Kaninchen und Hunde, *Virchow, Archiv.   107 :1–17.*

72. O Butschli (1891)—Uber die Struktur der Protoplasmas, *Verh. deutsch. zool. Ges.   1891 :14–20.*

73. W B Hardy (1899)—On the structure of cell protoplasm, *J. Physiol.   24 :158–210.*

74. K Tellyesniczky (1898)—Ueber die Fixirungs- (Härtungs-) Flüssigkeiten, *Arch. mikr. Anat.   52 :202–247.*

75. op. cit. note 65 above.

76. W Berg (1903)—Beiträge zur Theorie der Fixation mit besonderer Berücksichtung des Zellkerns und seiner Eiweisskörper, *Arch. mikr. Anat.   62 :367–430.*

77. op. cit. note 64 above.

78. G H Parker (1892)—A method for making paraffine sections from preparations stained with Ehrlich's methylene blue, *Zool. Anz.   15 :375–377.*

79. E Steinach (1887)—Siebdosen, ein Vorrichtung zur Behandlung mikroskopischer Präparate, *Z.w.M.   4 :433–438.*

80. D G Fairchild (1895)—A perforated porcelain cylinder as washing apparatus, *Z.w.M.   12 :301–303.*

81. N A Cobb (1890)—Two new instruments for biologists, *N.S. Wales Linn. Soc. Proc.   5 :157–167.*

82. G L Cheatle (1893)—A rapid method of dehydrating tissues before their infiltration with paraffin, *J. Path. Bact.   3 :147–148.*

83. Seaman's translation of Gierke, op. cit. note 3 above. See p. 65.

84. Mann, op. cit. note 65 above. See p. 190.

85. op. cit. note 83 above.

86. W V Farrar (1974)—Synthetic dyes before 1860, *Endeavour   33 :149–155.*

87. E J Holmyard—Dyestuffs in the nineteenth century. Chapter 12 *in* C Singer, E J Holmyard, A R Hall & T I Williams (eds)—*A history of technology,* vol. V. (Oxford: Clarendon Press, 1958.)

88. E von Meyer—*A history of chemistry from earliest times to the present day.* Translated by G M'Gowan. (London: Macmillan, 1891.) See pp. 515–520.

89. H Caro (1892)—Ueber die Entwickelung der Theerfarben-Industrie, *Berlin dt. chem. Ges.   25 :955–1105.*

90. J J Beer—*The emergence of the German dye industry to 1925.* (Urbana: University of Illinois Press, 1959.)

91. Mann, op. cit. note 65 above. See p. 191.

92. P Ehrlich (1878)—Beitrage zur Theorie und Praxis der histologischen Farbung,

Leipzig Universitat Inaugural Dissertation 26 June 1878. Translated and reprinted in: F Himmelweit—*The collected papers of Paul Ehrlich.* Volume 1. (London: Pergamon, 1956.) See pp. 29–64 for the original, and pp. 65–98 for the translation.

93. P Ehrlich (1879)—Beiträge zur Kenntnis der granulirten Bindegewebszellen und der eosinophilen Leukocythen, *Arch. Anat. Physiol.   1879 :166–269.*

94. P Ehrlich (1879)—Ueber die specificschen Granulationen des Blutes, *Arch. Anat. Physiol.   1879 :571–579.*

95. P Ehrlich (1880)—Methodolische Beiträge zur Physiologie und Pathologie der vershiedenen Formen der Leukocyten, *Zeits. klin. Med.   1 :553–560.*

96. O N Witt (1891)—Theorie der Färbeprocesses, *Färberzeitung   1891 :404–409.*

97. op. cit. note 2 above.

98. B Rawitz (1897)—Bemerkungen über Mikrotomschneiden und über das Färben mikroskopischer Präparate, *Anat. Anz.   13 :65–80.*

99. G C T von Georgievics (1894)—Ueber das Wesen des Färbeprocesses, *Wien, Ak. Sber.   103 :589–601.*

100. G C T von Georgievics & E Lowy (1895)—Ueber das Wesen des Färbeprocesses. Vertheilung von Methylenblau zwischen Wasser und mercerisirter Cellulose, *Wien, Ak. Sber.   104 :309–314.*

101. A Fischer—*Fixirung, Färbung und Bau des Protoplasmas.* (Jena: Fischer, 1899.) See p. 331.

102. J F Miescher (1874)—Die Spermatozoen einiger Wirbelthiere, ein Beitrag zur Histochemie, *Basel, Verh. nat. Ges.   6 :138–208.*

103. E Knecht (1888)—Zur Kenntnis der chemischen Vorgänge, welche beim Färben von Wolle und Seine mit den basischen Theerfarben stattfinden, *Berlin, Chem. Ges.   21 :1556–1558.*

104. E Zacharias (1893)—Ueber Chromatophile, *Ber. d. deuts. Botan. Gesell.   10 :188–195.*

105. C O Weber (1894)—Substantive dyes and substantive dyeing. A contribution to the theory of dyeing, *J. Soc. Chem. Ind.   13 :120–127.*

106. A P Mathews (1898)—A contribution to the chemistry of cytological staining, *Amer. J. Physiol.   1 :445–454.*

107. Baker, op. cit. note 8 above.

108. Mann, op. cit. note 65 above. See pp. 369–370.

109. A v Leeuwenhoek—*Epistolae physiologicae super compluribus naturae arcanis.* (1719). In volume III of *Opera omnia.* (Lugduni Batavorum: Langerak, 1722.) The stain was used in 1714.

110. G C Reichel—*De vasis plantarum spiralibus.* (Lipsiae: Breitkopf, 1758.)

111. G Holzner (1884)—Zur Geschichte der Tinctionen, *Z.w.M.   1 :254–256.*

112. J Hill—*The construction of timber from its early growth ; explained by the microscope, and proved from experiments, in a great variety of kinds.* (London: printed for the author, 1770.)

113. T Hartig (1854)—Chlorogen, *Bot. Zeit.   12 :553.*

114. D C G Ehrenberg—*Die Infusionsthierchen als vollkommene Organismen. Ein Blick in das tiefere organische Lebel der Natur.* (Leipsic: Voss, 1838.)

115. H R Göppert & F Cohn (1849)—Uber die Rotation des Zellinhaltes von *Nitella flexilis, Bot. Zeit.   7 :681–691.*

116. Corti, op. cit. note 20 above.

117. J von Gerlach—*Mikroskopische Studien aus dem Gebiete der menschlichen Morphologie.* (Erlangen: Enke, 1858.)

118. J von Gerlach (1859)—Ueber die Einwirkung von Farbestoff auf lebende Gewebe, *Mitt. Phys. Med. Soc. Erlangen   1 :5–12.*

119. O Maschke (1859)—Pigmentlösung als Reagens bei microscopischphysiologischen Untersuchungen, *Erdm. J. Prak. Chem.   76 :37–53.*

120. M Schultz (1865)—Injectionsmassen von Thiersch und Müller, *Arch. mikr. Anat.* *1 :148–150.*

121. L S Beale—*How to work with the microscope.* (London: Harrison, 3rd edition 1865.) See pp. 195–214, and especially p. 201. Page v of the preface says:
In this edition the details of the particular method of preparation carried out by the author in his investigations with the aid of the highest magnifying powers yet made, are for the first time published.

122. E Schwarz (1867)—Uber eine Methode doppelter Färbung mikroskopischer Objecte, und ihre Anwendung zur Untersuchung der Musculatur des Milz, Lymphdrusen und anderer Organe, *Sitz. Akad. W. math. naturw. Cl.* *55 :671–691.*

123. L Ranvier (1868)—Téchnique microscopique, *Archives de physiol.* *1 :319–321.*

124. J Orth (1883)—Notizen zur Färbetechnik, *Berlin klin. Wochenschr.* *20 :421–422.*

125. H Grenacher (1879)—Einige Notizen zur Tinctionstechnik, besonders zur Kernfärbung, *Arch. mikr. Anat.* *16 :463–471.*

126. W Flemming (1874)—Ueber die ersten Entwicklungserscheinungen am Ei der Teichmuschel, *Arch. mikr. Anat.* *10 :257–292.*

127. W A O Hertwig (1876)—Beitrage zur Kenntniss der Bildung, Befruchtung und Theilung des thierischen Eies, *Morph. Jarhb.* *1 :347–434.*

128. E van Beneden (1883)—Recherches sur la maturation de l'oeuf, et la fécondation, *Arch. de Biol.* *4 :265–640.*

129. P Mayer (1892)—Ueber das Färben mit Carmin, Cochenille und Hämatein-Thomerde, *Mitt. Zool. Stat. Neapel.* *10 :480–501.*

130. J Quekett—*A practical treatise on the use of the microscope, including the different methods of preparing and examining animal, vegetable and mineral structures.* (London: Baillière, 1848.)

131. W Waldeyer (1863)—Untersuchungen über den Ursprung und den Verlauf des Axencylinders bei Wirbellosen und Wirbelthieren, sowie über dessen Endverhalten in der quergestreiften Muskelfaser, *Henle u. Pfeufer, Zeits. ration. Med.* *20 :193–256.*

132. F Böhmer (1865)—Zur pathologischen Anatomie der Meningitis cerebro-medularis epidemica, *Aerztl. Intelligenzb. Munchen* *12 :539–550.*

133. A Arnold (1872)—Haematoxylin as a staining material for animal tissues, *Lens* *1 :172–173.*

134. L Tait (1875)—On the freezing process for section cutting: and on various methods of staining and mounting sections, *J. Anat. Physiol.* *9 :250–258.*

135. J M Prudden (1885)—[untitled note, in general enquiry section] *Z.w.M.* *2 :288.* Delafield was from New York.

136. P Ehrlich (1886)—[untitled note, in general enquiry section] *Z.w.M.* *3 :150.*

137. P Mayer (1891)—Ueber das Farben mit Hämatoxylin, *Mitt. Zool. Stat. Neapel.* *10 :170–186.* Unna, who had discovered the ripening of haematoxylin independently, did not publish until the next year.

138. R Nietzki—*Chemie der organischen Färbstoffe.* (Berlin: Springer, 1889). See pp. 215–217 for the discussion of haematin.

139. M Heidenhain—*Ueber Kern und Protoplasma.* (Leipzig: Engelmann, 1892.) See p. 118 for the formula.

140. F B Mallory (1897)—On certain improvements in histological technique, *J. Exp. Med.* *2 :529–533.*

141. A sample of the original laboratory-batch (no less!), in ownership of Imperial College, London, is on display in the Science Museum. Gurr was allowed to take the spectral characteristics of a minute sample, which is freely soluble in spirit and of high tinctorial value. Its absorbance peaked at 548 nm. See pp. 123–124 of his book:

E Gurr—*Synthetic dyes in biology, medicine and chemistry*. (London: Academic Press, 1971.)

142. F W Beneke (1862)—[note without title], *Correspbl. d. Vereins f. gemeinschaflt. Arbeiten* 59 :980. He dissolved commercial lilac-anilin in acetic acid, to obtain a clear solution.

143. Waldeyer, loc. cit. note 131, above.

144. H Frey (1863)—Die Lymphwege einer Peyer'schen Plaque beim Menschen, *Virchow, Archiv.* 26 :344–357.

145. Roberts, loc. cit. note 48 above.

146. N Chrzonszczewsky (1864)—Zur Anatomie der Niere, *Virchow, Archiv.* 31 :153–199.

147. L A Ranvier (1874)—Des applications de la purpurine à l'histologie, *Archives de Physiol.* 1 :761–773.

148. H Zuppinger (1874)—Eine methode, Axencylinderfortsätze der Ganglienzellen des Rückenmarkes zu demonstriren, *Arch. mikr. Anat.* 10 :255–256.

149. N Lieberkuhn (1874)—Ueber die Entwirkung von Alizarin auf die Gewebe des lebenden Körpers, *Marburger Sitzungsber.* 1874 :33.

150. M V Cornil (1875)—Sur la dissociation du violet de méthylaniline et sa séparation en deux coleurs sous l'influence de certain tissus normaux et pathologiques, en particulier par les tissus en dégénérescence amyloide, *Comp. Rend. Acad. Sci.* 80 :1288–1291.

151. E Hermann (1875)—Uber eine neue Tinctionsmethode, *Tagebl. d. 48 Naturfvers. Graz* 1875 :105.

152. A Böttcher (1868)—Uber Entwicklung und Bau der Gehörlabyrinths nach untersuchungen an Säugetheiren, *Robin, J. Anat.* 8 :439–447.

153. F Schweigger-Seidel & J Dogiel (1866)—Ueber die Peritonealhöhle bei Fröschen und ihren Zusammenhang mit dem Lymphgefäss-systeme, *Ber. Math. Phys. Leipzig* 18 :247–255.

154. P Ehrlich (1877)—Beiträge zur Kenntniss der Anilinfärbungen und ihrer Verwendung in der mikroskopischen Technik, *Arch. mikr. Anat.* 13 :263–277.

155. P Ehrlich (1881)—Uber das Methylenblau und seine klinishbakterioskopische Verwerthung, *Zts. klin. Med.* 2 :710–713.

156. P Ehrlich (1882)—[note without title], *Deuts. Med. Wochnsch.* 8 :269–270.

157. F Ziehl (1882)—Zur Färbung des Tuberkelbacillus, *Deutsche Med. Woochnsch.* 8 :451.

158. F Neelsen (1883)—Ein casuistischer Beitrag zur Lehre von der Tuberkulose, *Centbl. med. Wiss.* 21 :497–501.

159. F Loeffler (1884)—Untersuchungen über die Bedeutung der Mikroorganismen für die Enstelung der Diptherie beim Menschen, bei der Taube und Beim Kalbe, *Mitt. d. K. Gesundheitsamte* 2 :421–499.

160. C Gram (1884)—Uber die isolirte Färbung der Schizomyceten in Schnittund Troken-präparaten, *Fortsch. d. Med.* 2 :185–189.

161. F Loeffler (1889)—Eine neue Methode zur Färben der Mikroörganismen im besonderen ihrer Wimperhaare und Gesseln, *Centbl. f. Bakt.* 6 :209–224.

162. F Loeffler (1890)—Weitere untersuchungen über die Beizung und Färbung der Giesseln bei den Bakterien, *Centbl. f. Bakt.* 7 :625–639.

163. C Weigert (1878)—Bismarckbraun als Färbemittel, *Arch. mikr. Anat.* 15 :258–260.

164. E v Beneden & C Julin (1884)—La spetmatogénèse chez l'Ascaride mégalocéphale, *Brux. Acad. Bull.* 7 :312–342.

165. H A Griesbach (1886)—Weitere Untersuchungen über Azofarbestoffe behufs Tinction menschlicher und thierischer Gewebe, *Z.w.M.* 3 :358–385.

166. L Daddi (1896)—Nouvelle méthode pour colorer la graisse dans les tissus, *Arch. Ital. Biol.* *26 : 143–146.*

167. L Michaelis (1901)—Uber Fett Färbstoffe, *Virchow, Archiv.* *164 : 263.*

168. W H Poole (1875)—A double staining with haematoxylin and aniline, *Q.J.M.S.* *23 : 375–377.*

169. A Wissowzky (1876)—Ueber das Eosin als reagenz auf Hämoglobin und die Bildung von Blutgefässen und Blutkörperchen bei Säugethierund Hühnerembryonen, *Arch. mikr. Anat.* *13 : 479–496.*

170. J L Renaut (1879)—Sur l'éosine hématoxylique et sur son emploi en histologie, *Comp. Rend. Acad. Sci.* *88 : 1039–1042.*

171. A Tafani (1878)—Nuovo metodo per colorire i preparati microscopici mediante una soluzione picro-anilinica, *Sperimentale* *41 : 52–57.*

172. E F P Schiefferdecker (1878)—Kleinere histologische Mittheilungen, *Arch. mikr. Anat.* *15 : 30–40.*

173. H Gibbes (1880)—On the double and treble staining of animal tissues for microscopical investigation, *J.R.M.S. (O)* *3 : 390–393.*

174. B W Richardson (1881)—Multiple staining of animal tissues with picro-carmine, iodine, malachite-green dyes, and of vegetable tissues with atlas-scarlet, soluble blue, iodine, and malachite-green dyes, *J.R.M.S.* *2 : 868–872.*

175. H A Griesbach (1889)—Microscopical staining, *Amer. Monthly M.J.* *10 : 30–33.* (An account of a demonstration in Würzburgh in 1888.)

176. Lee, op. cit. and extract, above.

177. For a convenient colour plate showing eight different staining methods applied to one organ, see the frontispiece of:
    W H Freeman & Brian Bracegirdle—*An advanced atlas of histology.* (London: Heinemann, 1976.)

178. J van Gieson (1889)—Laboratory notes of technical methods for the nervous system, *New York Med. J.* *50 : 57–60.*

179. W Flemming (1891)—Uber Theilung und Kernformen bei Leukocyten und über deren Attractionssphären, *Arch. mikr. Anat.* *37 : 249–298.*

180. F B Mallory (1900)—A contribution to staining methods, *J. Exp. Med.* *5 : 15–20.*

181. P Ehrlich (1879)—Methodolische Beiträge zur Physiologie und Pathologie der verschiedenen Formen der Leukocyten, *Zts. klin. Med.* *1 : 553–560.*

182. D Romanovsky (1891)—Zur Frage der Parasitologie und Therapie der Malaria, *St. Petersbg. Med. Wchnschr.* *16 : 297–302, 307–315.*

183. P G Unna (1891)—Uber die Reifung unserer Färbstoffe, *Z.w.M.* *8 : 475–487.*

184. B A E Nocht (1898)—Zur Farbung der Malariaparasiten, *Centrbl. Bakt. Bakt.* *24 : 839–844.*

185. L Jenner (1899)—A new preparation for rapidly fixing and staining blood, *Lancet 1899 part* *1 : 370.*

186. W B Leishman (1901)—A simple and rapid method of producing Romanowsky staining in malarial and other blood films, *B.M.J.* *1901 : 757–758.*

187. G Giemsa (1902)—Färbemethoden für Malariaparasiten, *Centrbl. Bakt. Bakt.* *31 : 429–430, & 32 : 307–313.*

188. F Kehrmann (1906)—Uber Methylen-azur, *Ber. deut. Chem. Gesell.* *39 : 1403–1408.*

189. A Bernthsen (1906)—Uber die chemische Natur des Methylenazurs, *Ber. deut. Chem. Gesell.* *39 : 1804–1809.*

190. F Nissl (1884)—Resultate und Erfahrungen bei den Untersuchungen der pathologischen Veränderungen der Nervenzellen in der Grosshirnrinde, (unpublished prize essay, medical faculty of Munich). Reported in: W Haymaker & F Schiller—*The founders of neurology.* (Springfield: Thomas, 2nd edn, 1970.) See p. 354.

191. V A Latham (1894)—A plea for the study of re-agents in micro work, *Proc. Amer. M.S.* *15 :209–211*.

192   C Benda (1901—Die Mitochondriafärbung und andere Methoden zur Unter-suchung der Zellsubstanzen, *Anat. Anz. Ergänzhft.* *19 :155–174*.

193. F Meves & J Duesberg (1908)—Die Spermatozytenteilungen bei der Hornisse (Vespa crabris L), *Arch. mikr. Anat.* *71 :571–587*.

194. For a useful review of the subject, see H Kirkman & A E Severinghaus (1938)—A review of the Golgi apparatus, *Anat. Record* *70 :413–431, 557–573, & 71 :79–103*.

195. C Krause (1844)—see the article 'Haut' in: R Wagner—*Handwörterbuch der Physiologie*. (Braunschweig: Vieweg, 1844.)

196. M C A Flinzer—*De argenti nitrici usu et effectu praesertim in oculorum morbis sanandis*. (Lipsiae: Diss. 1854, bei Coccus gearbeitet.) The original is not available, but is described in detail by: F Stadtmüller (1921)—Historiche Darstellung zur Dentung des Wesens der Silbermethode an nicht fixierten Objecten, *Anat. Hefte* *59 :77*.

197. W His—*Beiträge zur normalen und pathologischen Histologie der Kornea*. (Basel: Schweighauser, 1856.)

198. T Hartig (1854)—Ueber das Verfahren bei Behandlung des Zellkerns mit Färbstoffen, *Bot. Zeit.* *12 :681*.

199. F v Recklinghausen (1860)—Eine Methode mikroskopische hohle und solide Gebilde von einander zu unterschneiden, *Virchow, Archiv.* *19 :451–452*.

200. L Auerbach (1865)—Ueber die Feinere Structur der Saugadern und der Blut-Capillaren, *Breslau Jahrs. Schles. Gesell.* *43 :142–143*.

201. V Feltz (1871)—Étude éxpérimentale sur le passage des Leucocytes à travers les parois vasculaires et sur l'inflammation de la cornée, *Robin, J. Anat.* *7 :505–534*.

202. L Ranvier (1868)—Procédé pour faire apparaitre les noyaux dans les tissus imprégnés d'abord par le nitrate d'argent, *Archives de physiol.* *1 :666–670*.

203. R S Bergh (1900)—Beiträge zur vergleichenden Histologie, *Anat. Hefte* *14 :379–407*.

204. J Cohnheim (1866)—Ueber die Endigung der sensiblen Nerven in der Hornhaut, *Virchow, Archiv.* *38 :343–386*.

205. J Arnold (1867)—Ein Beitrage zu der feineren Structur der Ganglienzellen, *Virchow, Archiv.* *41 :178–220*.

206. S Apáthy (1897)—Das leitende Element des Nervensystems und seine topogra-phischen Beziehungen zu den Zellen, *Mitt. Stat. Zool. Neapel.* *12 :495–748*.

207. H C Bastian (1868)—On some new methods of preserving thin sections of brain, or spinal-cord, for microscopical examination, *J. Anat.* *2 :104–109*.

208. M Löwit (1875)—Die Nerven der glatten Musculatur, *Wien. Akad. Sitzber.* *71 :355–376*.

209. L A Ranvier (1880)—On the termination of nerves in the epidermis, *Q.J.M.S.* *20 :456–458*.

210. Apáthy, loc. cit. note 206 above.

211. C Golgi (1873)—Sulla struttura della sostanza grigia del cervello, *Gazz. med. Ital. Lombardo* *6 :41–56*.

212. C Golgi (1879)—Un nuovo processo di tecnica microscopica, *Ist. Lomb. Rendiconti* *12 :206–212*.

213. C Golgi (1886)—Sulla fina anatomia degli organi centrali del sistema nervoso, *Arch. Ital. Biol.* *7 :15–47*.

214. S Ramon y Cajal (1891)—Sur la structure de l'ecorce cérébrale de quelques mammifères, *Cellule* *7 :125–176*.

215. A Hill (1896)—The chrome-silver method. A study of the conditions under which

the reaction occurs and a criticism of its results, *Brain* *19 :1–42.*

216. A Trembley—*Mémoires, pour servir à l'histoire d'un genre de polypes d'eau douce, à bras en forme de cornes.* (Leide: Verbeek, 1744.)

217. F v Gleichen—*Abhandlungen über die Saamen- und Infusionsthierchen und über die Erzeugung ; nebst mikroskopischen Beobachtungen des Saamens der Thiere und verschneidener Infusionsen.* (Nürnberg: Launoy, 1778.)

218. A Pritchard—*The microscopic cabinet of select animated objects ; with a description of the jewel and doublet microscopes, test objects, &c.* (London: Whittaker, Treacher and Arnot, 1832.) See p. 234.

219. D C G Ehrenberg—*Die Infusionsthierchen als volkommene Organismen. Ein Blick in das tiefere organische Leben der Natur.* (Leipsic: Voss, 1838.)

220. N Chrzonszczewsky (1864)—Zur Anatomie der Niere, *Virchow, Archiv.* *31 :153–199.*

221. K Brandt (1878)—Mikrochemische Untersuchungen, *Arch. f. Physiol.* *1878 :563–565.*

222. M A Certes (1881)—Sur un procédé de coloration des Infusoires et des éléments anatomiques, pendant la vie, *Comp. Rend. Acad. Sci.* *92 :424–426.*

223. P Ehrlich—*Sauerstoffbedürfnis des Organismus.* (Berlin: Springer, 1885.)

224. L Ranvier—*Traité technique d'histologie.* (Paris: Savy, 1875.)

225. G Galeotti (1894)—Recherche sulla colorbilita della cellule viventi, *Z.w.M. Z.w.M.* *11 :172–207.*

226. A Uhma (1899)—Die Schnellfärbung des Neiser'schen Diplococcus in frischen nicht fixierten Präparaten, *Arch. Dermatol. Syph.* *50 :241–242.*

227. A Pappemheim (1899)—Vergleichende Untersuchungen über die elementare Zummanensetzung des rothen Knochenmarkes einiger Säugethiere, *Arch. Path. Anat. Physiol.* *157 :19–76.*

228. K Nakanishi (1900)—Vorläufige Mittheilung über neue Färbungsmethode zur Darstellung des feineren Baues der Bakterien, *Munch. Med. Wochnsch.* *47 :187–188.*

229. G Bouffard (1906)—Injection des coleurs de benzidine aux animaux normaux, *Ann. Inst. Pasteur* *20 :539–546.*

230. E E Goldmann (1909)—Die äussere und innere Sekretion des gesunden Organismus im Lichte der vitalen Färbung, *Beitr. Klin. Chir. Tubingen* *64 :192–204.*

231. A B Lee—*The microtomist's vade-mecum, a handbook of the methods of microscopic anatomy.* (London: Churchill, 1885.) See p. 162.

232. E Klebs (1869)—Die Einschmelzungs-Methode, ein Beitrage zur microscop-ischen Technik, *Arch, mikr. Anat.* *5 :164–166.*

233. J Needham (1873)—On cutting sections of animal tissues for microscopical examination, *M.M.J.* *9 :258–267.*

234. Quekett, op. cit. note 130 above. See p. 313.

235. ibid. See p. 313.

236. J W Griffith & A Henfrey—*The micrographic dictionary.* (London: van Voorst, 1856.) See p. 530.

237. E Neumann (1861)—Beitrag zur Lehre von den cavernösen Geschwülsten, *Virchow, Archiv.* *21 :280–284.*

238. S Stricker—*Handbuch der Lehre von den Geweben des Menschen und der Thiere.* (Leipzig: Engelmann, 2 vols, 1869–1872.)

239. W Flemming (1873)—Eine Einbettungsmethode, *Arch. mikr. Anat.* *9 :123–125.*

240. W W Salensky (1877)—Ueber die Knosprung der Salpen, *Morphol. Jahrb.* *3 :549–602.* See p. 558.

241. H Kadyi (1879)—Beitrag zur Kenntniss der Vörgange beim Eierlegen der Blatte orientalis. Vorläfige Mittheilung, *Zool. Anz.* *2 :632–636.*

242. E Calberla (1876)—Eine Einbettungsmasse, *Morphol. Jahrb. 2 :445–448.*

243. M Bresgen (1875)—Ueber die Musculatur der grösseren Arterien, insbesondere ihrer Tunica adventitia, *Virchow, Archiv. 65 :246–261.*

244. E Salenka (1878)—Das Mannchen der Bonellia, *Zool. Anz. 1 :120–121.*

245. J Stevenson (1876)—A ready method of preparing sections of diseased tissues for the microscope, *Edinburgh Med. J. 21 :605–607.*

246. Needham, op. cit. note 233, above. See p. 265.

247. C W T R Hertwig (1880)—Ueber den Bau der Ctenopheren, *Jenaische Zeits. 14 :313–457.* See pp. 314–315.

248. Lee, op. cit. note 231 above; 4th edition 1896, p. 97.

249. D J Hamilton (1878)—A new method of preparing large sections of the nervous centres for microscopic investigation, *J. Anat. Physiol. 12 :254–260.*

250. A C Cole—*The methods of microscopical research.* (London: Bailliere, 1884.) See p. xxxix.

251. W J Sollas (1884)—An improvement in the method of using the freezing microtome, *Q.J.M.S. 24 :163–164.*

252. B W Richardson (1882)—Description of a simple plan of imbedding tissues, for microtome cutting, in semi-pulped unglazed printing paper, *J.R.M.S. 2 :474–475.*

253. P Francotte (1887)—Notes de technique microscopique, *Bull. Soc. Belge Micr. 13 :140–144.*

254. J D Hyatt (1880)—A method of making sections of insects and their appendages, *Am. Monthly Mic. J. 1 :8.*

255. G v Koch (1878)—Notiz über dic Zooide von Pennatula, *Zool. Anz. 1 :103–104.*

256. M E O Liebreich (1892)—noted in *J.R.M.S. 1893 :801.*

257. Quekett, op. cit. note 130 above. See p. 297.

258. ibid. See p. 303.

259. ibid. See p. 300.

260. H J Johnston-Davis & G C J Vosmaer (1887)—On cutting sections of sponges and other similar structures with soft and hard tissues, *J.R.M.S. 1887 :200–204.*

261. L Fredericq (1879)—Sur la conservation des pièces anatomiques par la paraffine, *Bull. Soc. Anthrop. Paris 2 :18–21.*

262. S Stricker—*Manual of human and comparative histology.* Translated by H Power. (London: New Sydenham Society, Vol. 1, 1870.) See p. xxx.

263. O Bütschli (1881)—Modifikation der Paraffineinbettung für mikroskopische Schnitte, *Biol. Centbl. 1 :591–592.*

264. W Giesbrecht (1881)—Vorlaufige Mittheilung aus einer Arbeit über die freilebenden Copepoden des Kieler Hafens, *Zool. Anz. 4 :254–258.*

265. S Marsh—*Section-cutting.* (London: Churchill, 2nd ed. 1882.) See p. 68.

266. H Fol—*Lehrbuch der vergleichend mikroskopischen Anatomie.* (Leipzig: Engelmann, 1884.) See p. 124.

267. W H Gaskell (1890)—On the origin of vertebrates from a crustacean-like ancestor, *Q.J.M.S. 31 :379–444.* See p. 382.

268. R Threlfall (1930)—The origin of the automatic microtome, *Biol. Reviews 5 :357–361.*

269. M Duval (1879)—De l'emploi du collodion humide piur la pratique des coupes microscopiques, *Robin, J. Anat. 15 :185–188.*

270. P Latteux—*Manuel de technique microscopique.* (Paris: Coccoz, 1877.) See p. 236.

271. E F P Schiefferdecker (1882)—Ueber die Verwendung des Celloidins in der anatomischen Technik, *Arch. Anat. Physiol. 1882 :199–203.*

272. H Viallanes (1883)—Recherches sur l'histologie des insectes et sur les phénomènes histologiques qui accompagnement le developpement postembryonnaire de ces animaux, *Ann. Sci. Nat. 14 :Art 1.*

273. E A Meyer (1891)—Sur l'emploi du photoxyline pour les préparations micro- et macroscopiques, *Biol. Centrbl.* *10 :508–509.*

274. Lee, op. cit. note 231 above. See p. 111.

275. N K Kultschizky (1887)—Zur histologischen Technik, *Z.w.M.* *4 :46–49.*

276. H H Field & J Martin (1894)—Contributions à la technique microtomique, *Bull. Soc. Zool. France* *19 :49–54.*

277. For a useful discussion of clearing agents, see Lee, op. cit. note 231 above. Page 64 in 4th edn of 1896.

278. C Varley (1843)—Varley on the method of preparing and rendering transparent objects intended to be preserved in canada balsam, *London Physiological J.* *1 :31.*

279. Clark, loc. cit. note 30 above.

280. K Z Kuchin (1863)—On the histology of the river Lamprey (in Russian), *Kazan Univ. Mem. Phys-Math.* *1 :1–24.*

281. L Stieda (1866)—Ueber die Anwendung des Kreosots bei Anfertigung microscopischer Präparate, *Virchow, Archiv.* *2 :430–435.*

282. E Rindfleisch (1865)—Zur Histologie der Cestoden, *Virchow, Archiv.* *1 :138–142.*

283. H C Bastian (1869)—Notes on the mounting and tinting of sections of animal tissues for microscopical examination, *M.M.J.* *1 :94–103.*

284. F Ncelsen & P Schiefferdecker (1882)—Beitrag zur Verwendung der atherischen Oele in der histologischen Technik, *Arch. Anat. Physiol.* *1882 :204–206.*

285. J B Carnoy & H Lebrun (1897)—La fécondation chez l'Ascaris megalocephala, *Cellule* *13 :61–195.* See p. 71.

286. H Jordan (1898)—Technische Mittheilungen, *Z.w.M.* *15 :50–55.*

287. H Suchannek (1890)—Technische Notiz betreffend die Verwendung des Anilinöls in der Mikroskopie sowie einige Bemerkungen zur Paraffineinbettung, *Z.w.M.* *7 :156–160.*

288. F Guégen (1898)—Emploi du salicylate de méthyle en histologie, *Comp. Rend. Soc. Biol.* *5 :285–287.*

289. P Mayer (1909)—Uber ein neues Intermedium, *Z.w.M.* *26 :523–524.*

290. C Robin—*Du microscope et des injections dans leurs applications à l'anatomie at à la pathologie.* (Paris : Bailière, 1849.)

291. L S Beale—*How to work with the microscope.* (London : Churchill, 1857.)

292. C Robin—*Traité du microscope.* (Paris : Bailière, 1871.) See p. 30.

293. Lee, op. cit. note 1 above. See p. 272.

294. L Teichmann (1882)—Injection-mass, *J.R.M.S.* *2 :1–25.*

295. H F Hoyer (1877)—Beiträge zur anatomischen und histologischen Technik, *Archiv. mikr. Anat.* *13 :645–650.*

296. A Budge (1877)—Die Saftbahnen im hyalinen Knorpel, *Archiv. mikr. Anat.* *14 :65–73.*

297. H A Griesbach (1882)—Bemerkungen zur Injektionstechnik bei Wirbellosen, *Archiv. mikr. Anat.* *21 :824–827.*
    See also :—
    A Tulk & A Henfrey—*Anatomical manipulation ; or, the methods of pursuing practical investigations in comparative anatomy and physiology ; also an introduction to the use of the microscope, etc., and an appendix.* (London : Van Voorst, 1844.)
    The 413 pages of this general book are important for their comprehensive survey of current injection methods, in addition to other microscopical techniques.

298. F J Cole (1951)—History of micro-dissection, *Proc. R.S.* *138B :159–187.*

299. M A Barber (1904)—A new method of isolating micro-organisms, *J. Kansas Med. Soc.* *4 :489–494.*

300. S L Schouten (1899)—*Verslagen van het Geneesk. Congres*, reported in *J.R.M.S.* *1901 :331*. The equipment was devised for the isolation of single bacteria, using glass needles 5μ thick. Fuller details are given in a later paper: S L Schouten (1905)—Reinkulturen aus einer unter dem Mikroskop isolierten Zelle, *Z.w.M.* *22 :10–45*.

301. H D Schmidt (1859)—On the minute structure of the hepatic lobules, particularly with reference to the relationship between the capillary blood vessels, the hepatic cells, and the canals which carry off the secretion of the latter, *Am. J. Med. Scis.* *37 :13–40*.

302. W Kuhne (1863)—Ueber die Endigung der Nerven in den Muskeln, *Virchow, Archiv.* *27 :528–538*.

303. W Stirling (1883)—The sulphocyanides of ammonium and potassium as histological reagents, *J. Anat. Physiol.* *17 :207–210*.

304. R Altmann (1879)—Ueber die Verwerthbarkeit der Corrosion in der mikroskopischen Anatomie, *Arch, mikr. Anat.* *16 :471–507*.

305. A Pritchard—*The microscopic cabinet of select animated objects.* (London: Whittaker, Treacher & Arnot, 1832.)

306. M Rooseboom (1956)—The introduction of mounting media in microscopy and their importance for biological science, *Actes 8th Cong. Int. Hist. Scis.* *2 :602–607*.

307. A Pritchard—*A list of two thousand microscopic objects.* (London: Whittaker, 1835.)

308. Quekett, op. cit. note 130 above.

309. J S Bowerbank (1870)—Reminiscences of the early times of the achromatic microscope, *M.M.J.* *3 :281–285*.

310. Pritchard, op. cit. note 305 above.

311. Beck, op. cit. note 14 above.

312. J H Martin—*A manual of microscopic mounting.* (London: Churchill, 1872.)

313. Varley, op. cit. note 278 above.

314. J W Griffith (1843)—On the different modes of preserving microscopic objects, *Ann. Mag. Nat. Hist.* *12 :113–117*.

315. Untitled editorial footnote, *Ann. Mag. Nat. Hist.* *12 :482.* (1843.)

316. T Boys (1849)—On a method of mounting objects in canada balsam, *Trans. R.M.S.(O)* *2 :44–45*.

317. W H Heys (1865)—Some remarks on mounting microscopical preparations in canada balsam and chloroform, *Q.J.M.S.* *13 :19–21*.

318. Bastian, op. cit. note 283 above.

319. H C Bastian (1868)—On some new methods of preserving thin sections of brain, or spinal-cord, for microscopical examination, *J. Anat.* *2 :104–109*.

320. G Gilson (1906)—Un nouveau médium solidifiable pour le montage des préparations microscopiques, *Cellule* *23 :427–432*.

321. J W Stephenson (1880)—On the visibility of minute objects mounted in phosphorus, solution of sulphur, bisulphide of carbon, and other media, *J.R.M.S.(O)* *3 :564–567*.

322. Editorial note (1880)—Visibility of minute objects—new medium for mounting (monobromide of naphthaline), *J.R.M.S.* *1880 :1043–1044*.

323. M H J Flesch (1882)—Kleine Mittheilungen zur histologischen Technik, *Zool. Anz.* *5 :554–556*.

324. H v Heurck (1883)—De l'emploi du styrax et du liquidambar en remplacement du baume de canada, *Bull. Soc. Mic. Belge* *11 :134–136*.

325. Anon (1892)—The Rev. Father Thompson's high refractive medium, *J.Q.M.C.* *11 :123*.

326. H G Madan (1898)—On some organic substances of high refractivity, available for mounting specimens for examination under the Microscope, *J.R.M.S.* *18 :273–281*.

327. P Harting (1843)—Middel om mikroskopische voorwerpen te bewaren, *Hoeven en Vriese, Tijdschrift 10 :289–294.*
328. Rooseboom, op. cit. note 306 above.
329. C Varley (1831)—Improvement in the microscope, *Trans. Soc. Arts 48 :332–400.*
330. Pritchard, op. cit. note 305 above.
331. Blanchard, op. cit. note 24 above.
332. The formula was first published by D Cooper (1841)—On preservative solutions for mounting animal structures, *Mic. J. 1 :183–184;* but a clearer exposition is given in H Goadby (1845)—On the conservation of substances, *Brit. Ass. Rep. 14th Meeting York 1844 :69.*
333. C Varley (1834)—Ueber Samen, Keimung und Saft-Circulation der Chara vulgaris: nebst andern Bemerkungen, *Flora 14 :61–75.*
    J W Griffith (1843)—Ueber die verschiedenen Verfahren behufs der Aufbewahrung miksoskopischer Gegenstande, *Froriep, Neue Notizen 28 :33–38.*
334. A Oschatz (1843)—Ueber Darstellung und Aufbewahrung microscopischer Präparate, *Froriep, Neue Notizen 28 :20–23.*
335. J Moleschott (1844)—Harting'sche Methode mikroskopische Präparate aufzubewahren, nach seinem Aufsatze in van der Hoeven und de Vriese tydschrift voor natuurlyke geschiedenis en physiologie en schriftlichen Zusätzen, *Bot. Ztg. 2 :881–884.*
336. H v Mohl (1844)—Nachtrag zu Moleschott's Aufsatz, *Bot. Ztg. 2 :884–885.*
337. H v Mohl—*Mikrographie, oder anleitung zur Kenntniss und zum gebrauche des Miksoskops.* (Tubingen: Fues, 1846.)
338. H Welcker—*Ueber Aufbewahrung mikroskopischer Objecte nebst Mittheilungen über das Mikroskop und dessen Zubehor.* (Giessen: Ricker, 1856.)
339. Lee, op. cit. note 1 above. See p. 36.
340. M Lavdowsky (1877)—Zur feineren Anatomei und Physiologie der Speicheldrusen, insbesondere der Orbitaldruse, *Arch. mikr. Anat. 13 :281–364.* See p. 359.
341. Lee, op. cit. note 1 above. See p. 37.
342. R Warington (1849)—On a new medium for mounting organic substances as permanent objects for microscopic inspection, *Trans. R.M.S.(O) 2 :131–133.*
343. H Deane (1852)—On a new medium for mounting fresh or moist animal and vegetable structures, *Trans. R.M.S.(O) 3 :149–153.*
344. Quekett, op. cit. note 130 above. See pp. 284–296.
345. Lee, op. cit. note 1 above. See p. 31.
346. ibid., p. 31.
347. L Stieda (1866)—Ueber die Anwendung des Kreosots bei Anfertigung microscopischer Präparate, *Arch. mikr. Anat. 2 :430–435.*
348. S Marsh—*Section cutting, a practical guide to the preparation and mounting of sections for the microscope.* (London: Churchill, 2nd edn, 1882.) See p. 104.
349. Threlfall, op. cit. note 268 above.
350. H Schällibaum (1883)—Ueber eine Verfahren mikroskopische Schnitte auf dem Objectträger zu fixiren und daselbst zu färben, *Arch. mikr. Anat. 22 :689.*
351. W Giesbrecht (1881)—Zur Schneide-Technik, *Zool. Anz. 4 :483–484.*
352. H J Waddington (1881)—On the use of arabin in mounting microscopic objects, *J.Q.M.C. 6 :199–200.*
353. C Born & G Wieger (1885)—Ueber einen neuen Unterguss, *Z.w.M. 2 :346–348.*
354. A Andres, W Giesbrecht & P Mayer (1883)—Neuerungen in der Schneidetechnik, *Mitt. Stat. Zool. Neapel. 4 :429–438.*
355. Girod-Chantrans—*Recherches chimiques et microscopiques sur les conferves, bisses, tremelles, etc.* (Paris: Bernard, 1802.)

110 A HISTORY OF MICROTECHNIQUE

356. D H F Link—*Grundlehren der Anatomie und Physiologie der Pflanzen.* (Gottingen: Danckwerts, 1807.)

357. J J Colin & H G de Claubry (1814)—Sur les combinaisons de l'iode avec les substances végétales et animales, *Annal. de Chimie* 90:87–100.

358. F V Raspail (1825)—Développement de la fécule dans les organes de la fructification des céréales, et analyse microscopique de la fécule, suivie d'expériences propres à en expliquer la conversion en gomme, *Ann. Sci. Nat.* 6:224–239, 384–427.

359. F V Raspail (1828)—Nouveau réactif destiné, dans les analyses microscopiques, à distinguer les quantités minimes de sucre, d'albumine, d'huile et de résine, et analogie que l'on découvre par ce moyen entre les ovules des plantes et les organes femelles de la géneration des animaux pendant le temps de la gestation, *Ferussac, Bull. Sci. Math.* 10:267–272.

360. His researches are bound in with the next volume:

361. F V Raspail—*Essai de chimie microscopique appliquee à la physiologie, ou l'art de transporter le laboratoire sur le porte-objet, dans l'étude des corps organises.* (Paris: chez l'auteur, 1830.)

362. M J Schleiden (1838)—Botanische Notizen, *Wiegmann, Archiv.* 4:49–66.

363. H v Mohl (1840)—Einige Beobachtungen über die blaue Färbung der vegetabilischen Zellmembran durch Jod, *Flora* 32:609–624, 625–637.

364. Fürnrohr, op. cit. note 40 above.

365. J Vogel—*Pathologische Anatomie des menschlichen Körpers.* (Leipzig: Voss, 1845.)

366. M Perls (1867)—Nachweis von Eisenoxyd in gewissen Pigmenten, *Virchow, Archiv.* 39:42–48.

367. L S Beale (1861)—On the structure of tissues, with some observations on their growth, nutrition, and decay, *Archives of medicine* 2:179–206.

368. A C L Kossel & A P Mathews (1898)—Zur Kenntnis der Trypsinwirkung, *Zeits. Physiol. Chem.* 25:190–194.

369. E Klebs (1868)—Die pyrogene Substanz. Vorlaufige Mittheilung. *Centrbl. Med. Wiss.* 6:417–418.

370. R Heidenhain (1868)—Beitrage zur Lehre von der Speichelabsonderung, *Breslau, Studien Physiol. Inst.* 4:1–124.

371. H Leitgeb (1888)—Krystalloide in Zellkernen, *Graz Bot. Inst. Mitt.* 1888:113–122.

372. F Miescher (1878)—Die Spermatozoen einiger Wirbelthiere. Ein Beitrag zur Histochemie, *Verh. Naturf. Ges. Basel* 6:138–208.

373. H Molisch (1893)—Bemerkung über den Nachweis von maskirtem Eisen, *Deutsch. Bot. Ges. Ber.* 11:73–75.

374. L Lilienfeld & A Monti (1892)—Sulla localizzione michrochimica del fosforo nei tessuti, *Roma Acc. Lincei Rend.* 1:310–315, 354–358.

375. A G E Pearse—*Histochemistry theoretical and applied.* (London: Churchill, 2nd edn, 1960.)

376. W Sandritter (Ed.)—*100 years of histochemistry in Germany.* English edition by F H Kasten. (Stuttgart: Schattauer, 1964.)

377. H Petersen (1941)—Die Probleme der Zellenlehre und die ihrer Geschichte, *Anat. Anz.* 90:1–42.

5. A Survey of Instruments used in Microtechnique 1830–1910

Slides and covers

The early history of the slider has been traced from its first appearance (page 19 above), through its slow evolution into an all-glass slider (page 26) into a glass slide with mica cover (page 26). Two further changes provided the slide still in use; standardization of size, and the development of the thin cover-glass.

Various sizes of slide were used, as suitable pieces of glass came to hand. One of the first acts of the Microscopical Society of London was to suggest the use of standard sizes. The precise date must remain unknown, for the *Minute Book of Meetings of the Council* appears to have been written up in retrospect, and in shortened form. The first meeting, on 3rd September 1839, was recorded as being held

> ... to take into consideration the propriety of forming a Society for the promotion of Microscopical investigation, and for the introduction and improvement of the Microscope as a Scientific Instrument.

It was resolved that a Society should be formed, that a provisional Committee be appointed to carry the resolution into effect, and that Messrs Bowerbank, Lister, Loddiges, Quekett, Reade, Solly, and Ward should serve on that committee.

On 20th December 1839 a public meeting was held, at which the constitution drawn up by the provisional committee was discussed and agreed, officers and council appointed, and forty-five members admitted. The next record is of a meeting held on 29th January 1840, at which the minutes of the general meeting were read, and a further sixty-five members noted. The next paragraph reads:

> Mr Bowerbank also reported that he had purchased for the use of the Society a Diamond and cutting-board for the purpose of cutting glass for mounting microscopical objects. The size of the glasses fixed upon by the Provisional committee being 3 inches by 1, and 3 by 1½.

From this it would follow that the sizes were fixed in 1839, as the provisional committee did not exist after 20th December of that year. It may be that Bowerbank was the moving spirit behind the choice of sizes, which were larger than most instruments of the time could comfortably accommodate on their rather small stages. Little light is thrown on the question by others, the only other reference being by Cooper,[1] who said:

> ... Some months since the Council decided upon two sizes for the glasses on which the objects presented to the Society's Cabinet should be as far as practicable mounted; they are as follow:—Three inches by one inch—and three inches by one and a half inches. For the convenience of members a cutting-board and diamond is kept in the charge of the Curator, the property of the Society, which may be made use of by Members on application.

Cooper is here in contradiction with the minutes, as he wrote that in March 1841. However, he was admitted a member of the Society only on 17th February 1841, and it is possible that he was unaware of the precise dates of earlier events.

It is interesting that the Society was concerned only with securing uniformity for its own cabinet. The size did become standard in England quite rapidly, possibly on account of the influence of commercial mounters such as Topping, who adopted the size from the beginning. We have already seen (page 90) that Griffith, in 1843, left the size of slide to the whim of the observer, and the next serious attempt to establish a uniform size was made by Welcker, in 1856.[2] He lists a series of sizes, and others have been included in the list below to illustrate the lack of uniformity outside England:

76 × 25 mm (MSL size)
70 × 22 mm (von Mohl, 1840)
30 × 40 mm (von Mohl, 1840)
20 × 20 mm (Oschatz, 1851)
37 × 28 mm (Giessen, 1856)
55 × 26 mm (Frankfurt, 1857)
70 × 20 mm (Gerlach, 1857)
37 × 22 mm (von Mohl, 1857)
43 × 28 mm (von Mohl, 1857)
48 × 28 mm (Giessen, later format)
65 × 25 mm (Vienna format)

The list is by no means exhaustive, for many commercial mounters had their own sizes, especially on the continent. The 1856 Giessen format was the most-used size outside England until the later 1880s, after which the R.M.S. size took over, and is still universally used today.

Many other kinds of slide were made for special purposes, and a selection of those for dry mounts is shown in plate 23. The cavity slide, shown in drawings of the 1744 Cuff microscope, is still in use virtually unchanged from its original form. Various fairly complicated cells were made for keeping living things in controlled conditions while under the microscope, and a number of these are illustrated in plate 24. However, in terms of importance, the ordinary balsam-mounted slide far outweighs all others. The key to the production of such slides was the thin glass cover-slip, and its introduction must now be considered.

There can be no doubt that a thin glass cover slip was first used in 1789, as we saw on page 23 above. There is no doubt either that its use was limited to Ingen-Housz. Further, Chevalier's claim that he used thin glass covers from 1825 is unhelpful (see page 26), for they were used only for temporary preparations requiring the highest magnifications. The truly significant use of thin glass covers occurred in conjunction with balsam as a mountant for permanent preparations; the date for this event can be narrowed to fairly fine limits, but not accurately established.

When balsam mounting was introduced about 1830, thin covers were not used, as shown by such specimens as have survived and by the references to the mounting process discussed on page 90 above. The first reference to the commercial availability of the covers is in an obscure journal of 1843,[3] where their use was suggested to make side walls for cells. The commercial source is wrongly credited to Mr Dark of Birmingham, corrected in another footnote a few pages further on to Messrs Chance of Birmingham. That Chance Brothers did indeed offer thin cover-glasses commercially from 1840 was confirmed during the time of the Great Exhibition of 1851, where the firm had examples on exhibition:[4]

> They also exhibit some extremely thin glass, 200 to 300 to the inch, for purposes connected with the use of the microscope, and for experiments relating to the polarization of light, the want of which had formerly been found to be a great disadvantage in researches of this nature. This thin glass was introduced by Messrs. Chance as far back as the year 1840; and the Jury were informed by Mr. Ross, that by its use microscopes were made of very far higher power, than could otherwise have been produced.

The Mr Ross giving evidence was of course the famous optician, Andrew Ross. In his well-known paper describing the objective with adjusting-collar (to allow for different thicknesses of cover-glass),[5] it is implied that covers of mica had to be used for the thinnest slips:

> ... For finest definition if immersed in fluid, to be covered with thinnest possible film of talc ... [page 104].

But it was also usual for cover glasses to be used (page 105):

...the object is covered with glass 1/100 of an inch thick, and such glass can be readily supplied ...

Therefore the date of usual use of cover-slips of glass lies before 1837, possibly as early as 1835.

This supposition is strengthened by a further mention of cover glasses by Daniel Cooper in 1842,[6] and the whole entry is worth quoting:

Glass for mounting objects.—The best glass for mounting objects upon is what is termed the best flatted crown, which is sold by the superficial square foot. Messrs. Chater and Haywood of Thames Street always keep a stock, and will accommodate microscopists.

The very thin glass is to be obtained of Mr. Drake, Jermyn Street, St. James's, who has for some time turned his attention to the subject, and has manufactured an article the 1/100th of an inch in thickness, which may be obtained for three shillings and sixpence per ounce. The thicker kind, 1/50th of an inch, may be had much cheaper.

Mica, or Talc (as it was formerly called) is now almost entirely superseded by the introduction of thin glass; it is very readily scratched, and is seldom obtained free from cracks. Those in the habit of mounting objects, never recommend the use of this substance for covering objects, when the thin glass can be obtained.

It is clear that one had to cut one's own slides and covers in those days, and that making thin glass covers was an expensive undertaking. It is also clear, from practical experiment with modern cover-slips, that to make a balsam mount by immersing a dried object in molten balsam very easily results in a cracked cover. Although the covers were supplied by 1840, it may have been quite uncommon to use thin covers for anything other than fluid mounts until turpentine clearing and mounting in balsam solution was the mode. As we saw on page 91 above, this was not before 1843 at the earliest.

As the magnifying power of microscope objectives increased, so their working distance decreased, and need for thinner glasses for the compressoria so widely used between 1830 and 1850 increased. The role of this device in microscopical research was an important one, and many different designs were produced, a selection of them being shown in plate 25. Other special accessories were in use for a very long time, chief among them being the fish-plate. This was standard equipment with the Cuff instrument of 1744, and was still supplied with Powell & Lealand instruments in 1900. It is, of course, still very interesting to see corpuscles coursing along the capillaries of a frog's web or fish's tail, but this was a specimen of considerable importance in the 1840s and later, when many experiments were tried on wound healing. Quekett prepared a large series of permanent preparations of the web of a frog, injured with a leather punch, and allowed to heal for various numbers of days: these are still in the Hunterian collection. It is very likely that such work would have been paralleled by direct observations of the living web by means of a fish plate. This was probably the longest-lived accessory, the Marshall instrument in the collection in the Museum of the History of Science in Oxford having an early form of fish-plate as part of its original equipment.

Small apparatus used in microtechnique 1830–1910

Small equipment for manipulating specimens was a feature of the earliest outfits; Wilson,[7] for example, showed a soft brush for picking up objects in much the same form as used today. Small forceps and a tube for aquatic animals were also provided from early in the eighteenth century, and stage forceps were also included in such early outfits; they ceased to be sold only about 1950. These were attached to the instrument itself, and were thus different in kind from the ordinary small forceps used in making a mount. They remain valuable for exerting a constant pressure on a specimen while it is rotated under the objective. In later examples, pinned objects may be pushed into the other end of the forceps to allow inspection, while the earliest examples had only a spike distally onto which specimens were pushed bodily. Intermediate-aged designs had a disc instead of the spike, to allow attachment of objects with a spot of gum. One special version of the stage forceps was made in the form of three curved wires which opened and closed by a collar, to hold bulky objects such as mineral specimens. Examples of various types are shown in plate 26.

Other accessories which continued in use for much of the nineteenth century included dark wells, supported in a substage carrier, to contain opaque mounts for inspection by reflected light. Both substage and body slides could be used to focus, and such holders remain very versatile holders for larger stands; they have been unknown commercially since the beginning of this century. Quekett figures[8] a special holder for opaque objects, to the design of his brother Edwin, and these are occasionally seen with older instruments.

The turntable was an important invention in the days when attention was still largely directed to producing cell mounts for liquid media. The original design was due to Shadbolt about 1850,[9] but this was a somewhat unwieldy pattern having two wheels so that turning could be continued while the varnish was applied to the slide. Jackson gave the apparatus essentially its modern form,[10] omitting the second wheel on realising that an initial spin gave enough impetus to allow plenty of varnish to be put on the slide.

As might be expected, a whole gamut of, sometimes rather bizarre, accessories were suggested: these ranged from little cones to hold a housefly while watching it feed, all the way to specialized clips to hold down a cover while the balsam dried. All of these devices were essentially amateur in nature, and while they reveal a fascinating aspect of the character of our forefathers, are not relevant to the progress of histology.

Freehand blades for sectioning

Early microscope outfits included a small scalpel, often of a good blued steel, to make minute sections and dissections of specimens. Some of these blades are shown in plate 27, together with different examples of the Valentin knife. This was invented in 1838, and a somewhat obscure journal described it the following year.[11] This knife was almost the sole sectioning instrument for animal tissues for over twenty years, during which time it was highly praised,[12] and also modified.[13] A three-bladed version was made,[14] and the instrument was still available in 1890,[15] with the usual mis-spelling, as Valentine's knife.

In use, the blades are adjusted to give a small separation, and effort is made to cut the organ with one slice only; if a sawing action is used, bad tears are made in the section. Contemporary instruction in using the instrument stressed that it should be wetted before use, not be blunted on hard tissues such as cartilage, but kept really sharp. It was recommended for tissues such as kidney and liver. The separation of the blades was not a parallel separation until a version with two screws was produced; this is much more satisfactory in use. After some practice, adequate sections for low power work, or for use in the compressorium, can be obtained, but the instrument is not suitable for careful work, and it is naturally impossible to obtain serial sections.

Another type of two-bladed section instrument was available in mid-century, the so-called 'microtome' of Strauss-Durckheim. This was essentially a pair of spring-loaded scissors, with a screw stop to control length of cut. They were devised originally for making minute dissections of insects, but are not what would now be called microtomes.[16]

Most sections were made with an ordinary razor, used either with material simply supported between finger and thumb, or with material embedded in pith or carrot: ordinary razors were also used for the earliest true microtomes. Much practice is required to make sections of adequate quality (thinness and evenness) freehand, and only plant material is nowadays treated thus, in student classes. Parts of sections may be of adequate quality, but to obtain larger complete sections is impossible. However, in the earlier part of the nineteenth century there was no need for large perfect sections, and many eminent workers in histology prided themselves on their skill with a freehand razor, decrying such instruments as section-cutting machines. Much discussion took place as to whether the blade should be hollow-ground on one or both sides, and many directions were given for maintaining the keen edge required. At that date most men were skilled in sharpening razors, but ordinary shaving sharpness did not suffice for sectioning.

The development of the microtome from 1830 to 1870

It was recorded on page 47 above that the first use of the word 'microtome' was by Chevalier in 1839,[17] but this usage requires further discussion. His Chapter 11 deals with precautions to be taken during microscopical observation, and with the preparation of objects. In this largely derivative section he quotes several English authors, including the younger Adams. He mentions instruments for making sections as *le couteau micrometrique ou mieux, l'instrument microtomique*. He then goes on to mention the Valentin knife—*Dernièrement on m'a remis un nouveau microtome de l'invention de M. Valentin*. This is the first appearance in print of the actual word microtome, and although some authors have argued that Chevalier intended it only for knives of the Valentin type, careful reading of the rest of the chapter makes it clear that it was also used in the wider sense of any instrument for making thin sections for microscopical purposes. 'Microtomy' was not used until 1884,[18] and 'microtomist' was made universal in Lee's title the following year.

We may adopt a chronological approach to the development of the microtome between 1830 and 1870, as only about 25 instruments were described in these 40 years: after 1870 each main type of instrument will have to be followed separately. An attempt has been made to mention all instruments described before 1870, without regard to their importance; after that date, selection will be necessary on account of the often trivial modifications described to justify attachment of a name to a machine.

In the 1830s only three instruments are on record, and of these only one example has been found. This is the one by Pritchard, shown in plate 28. It was designed to be fastened to the bench, and for use with timber. The double-handed blade is a decided improvement on other, and later, knives, offering good control and adequate pressure to minimise chatter. This example is undated, but it was donated by the user's son, who stated that it was pre-1835: from the preparations seen, it is clear that Pritchard must have used such an instrument; nowhere does he describe it, possibly because it was a useful trade secret. It is probable that this was the fourth microtome to be designed, following those of Hill, Adams, and Custance described earlier: the Pritchard version seems not to be closely based on any of them, although it has more in common with the Adams than the others. This was certainly available commercially in only slightly modified form as late as 1870.[19]

Bowerbank described his section instrument, again only for wood, in 1834.[20] No illustration was provided, although it seems to resemble Pritchard's design from the description, but in addition had legs. Bowerbank said that the raising screw had 15 threads per inch, and that the wood to be cut should be immersed in spirit to extract the resinous matter, and then in water to remove the gums, before being cut with razor moistened with spirit. The razor was to be flat ground on the plate side, and used with a diagonal motion away from the body.

The third machine of the 1830s was that of Jackson. The only record of this is the statement of Michael[21] that Jackson had made a very serviceable cutting-machine for producing thin sections of wood many years before the beginning of the Microscopical Society. This is probably a reliable record of an 1830s device, bearing in mind that the foundation of the Society would be an event serving as a definite dateline to a microscopist old enough to remember it.

Five instruments were described in the 1840s, which is the decade during which the first microtomes of any lasting importance were used. Quekett's instrument is a direct development of the Adams design, and is illustrated on page 307 of his 1848 book (see Figure 7), although it had then been in use for some years. An automatic advance is provided, but it was intended only for wood. In spite of a thorough search at the Royal College of Surgeons, no trace of the device has been found, and it was not included in the list of the sale of Quekett's effects after his death in 1861.

Quekett also describes another timber sectioning machine, that of Topping.[22] This was a simple machine, but had the advantage of being screwed to the bench. An example of this instrument survives, and is shown in use in plate 29. In 1848 it could be bought from Mr Topping for 16/-d., including knife; and it is likely to be the same design as that

Figure 7. Quekett's microtome, from before 1848 (from Quekett's book, note 8 in text).
The well-known illustration from the well-known book; the instrument must have been in use by Quekett for some years prior to publication, to judge from his output of microscopical preparations. The design has more than a passing resemblance to the Adams machine of sixty years before.

used by Topping in preparing his superb sections. None of these was very thin, and practical test has shown that it is possible to use it to make sections of hardened organs such as liver and kidney: a hardening process had already been described by Hannover in 1840; and could have been used by Topping. The author has tested two Topping sections by simple chemical qualitative analysis, and significant quantities of chromium are present: this strongly suggests that Hannover's hardener had been used in the preparations. This is the first microtome which could have been used for animal tissues as a matter of course, having a circular well instead of the usual segmental one.

A more important microtome was described by Oschatz in 1843.[23] A diagram is given by Harting,[24] and is reproduced in Figure 8. The specimen rose with its clamp, a good point of the design, obviating any irregularities caused by adjustment of the clamp. A relatively large top

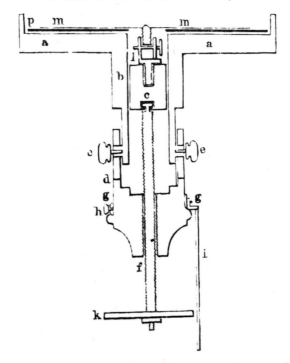

Figure 8. Oschatz microtome 1843 (from Harting's book, note 24 in text).
The large-headed screw (k) moved the specimen (l) which was held in a
small clamp. The design does not appear to be very functional.

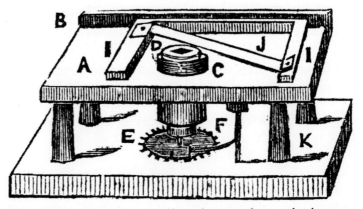

Figure 9. Fisher's illustration, 1846 (from the somewhat rare book, see note
25 in text).
So similar in appearance to the Quekett illustration that one is led to
wonder whether Fisher (who did not claim to be any kind of histologist) had
seen the famous man's section cutter, and made a sketch of it from memory.
Nonetheless, this is the first English illustration of a microtome since the
Adams reprint in 1798.

plate gave good support to the knife, and although no example of this instrument seems to have survived, it was clearly intended for animal tissues. The large head on the screw would have given finer control than most other instruments of its day, and it was to be important in stimulating others such as Welcker to design their own instruments.

The microtome illustrated by Fisher[25] is very similar to Quekett's instrument, and must surely be based on it: Fisher's picture was the first microtome to be figured in an English illustration for over fifty years, and is reproduced in Figure 9.

Of much greater significance in the history of the microtome is the Capanema instrument, as it introduced the principle of raising the specimen by an inclined plane. His paper of 1848[26] includes a good plate, reproduced in Figure 10; the essential feature of the small instrument was to move a clamp horizontally, causing it to move

Figure 10. Capanema microtome, 1848 (from the original paper, note 26 in text).

The entire instrument was only about 5 cm long. Fig. 1 is a side view showing the head of the lead-screw which moved the specimen along, thus causing it to rise vertically on account of a peg travelling in the inclined slot (f). Fig. 2 is a plan view, showing the specimen clamp with its removable handle. Fig. 3 is a lateral section showing the long lead screw: irregularities and coarseness of pitch in this would matter little provided the inclined slot was reasonably accurate, and this would be easy to achieve. Fig. 4 is an end view, showing the specimen clamp on the riser pin. The knife is seen in Fig. 5. This is a remarkably advanced design for its day, introducing a new principle of specimen advance which was not to be taken up for over a quarter of a century, by Rivet.

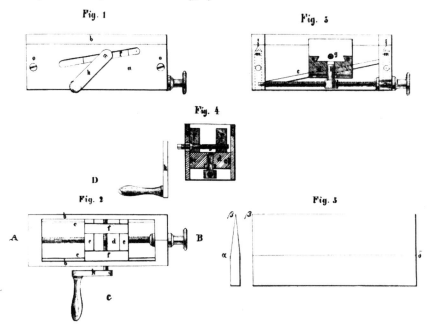

Fig. 1   Fig. 3

Fig. 4

Fig. 2   Fig. 5

vertically at the same time by means of a lateral pin engaged in an inclined slot in the side housing. A separate knife was used along the top of the housing, and this was a weak point of the design, causing lack of rigidity and control.

More instruments were described in the 1850s than any previous decade. A very simple device was suggested by Currey,[27] translator of Schacht's book on plant microscopy. The original author had suggested use of a tube, open at each end, with a divided cork containing the specimen pushed along by the finger from below. In his translation Currey suggested an improvement on this: a brass tube carrying at its closed end a screw-raised piston moving a divided cork carrying the specimen (see Figure 11). This plan was actually put forward by Ross, who made the instrument for Currey, so it should properly be called the Ross microtome. The second English edition, of two years later, omits

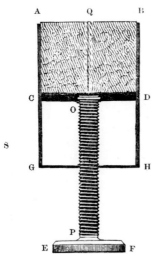

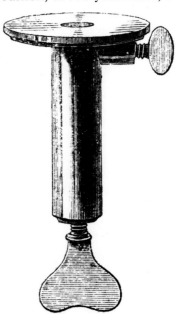

Figure 11. Currey microtome, 1853 (from the original book, note 27 in text).
This is the original hand microtome, later to be called the Ranvier microtome, although Ranvier himself never claimed to have invented it. It was actually designed by Ross, and consists of a brass tube (ABGH) with a piston (CD) raised by a screw. A piece of cork divided at (Q) contained the specimen. The lack of platform support for the knife would not have made it easy in use.

Figure 12. Hand microtome, 1855 (from the original book, note 27 in text, 2nd edition).
Lack of support for the knife was remedied by provision of a platform for the knife: this plus the very practical specimen clamp were additions to the original design, which is still available 120 years later!

both designs and includes instead what would later be called a Ranvier microtome (see Figure 12). This is likely to be a commercial development of the earlier Ross pattern, as Currey states in his preface that he had inserted the chapter containing the description of the machine, which thus did not occur in the German original. Although definitely first invented by Ross and figured in this book of 1855, the microtome was to be reinvented and popularized by Ranvier, and it is by his name that it is now known.

Welcker's well-known instrument was described in his somewhat obscure booklet of 1856,[28] and while it was based on the earlier design, of Oschatz, as Welcker acknowledges, it is much more important because this instrument, which would cut both animal and plant tissues, was to mark the beginning of the acceptance of the microtome in Germany. Welcker's original plate and a later illustration are reproduced in 13 and 14, showing the essentially simple nature of the machine. However, such was the influence of the publication, that the

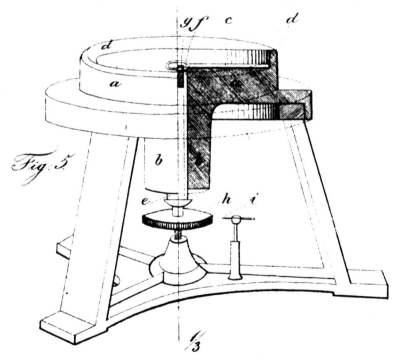

Figure 13. Welcker instrument, 1856 (from the original and scarce book, note 28 in text).
Based on the earlier Oschatz design, and with only a small diameter well, the instrument was quite solid and could be clamped to the bench—a very useful provision.

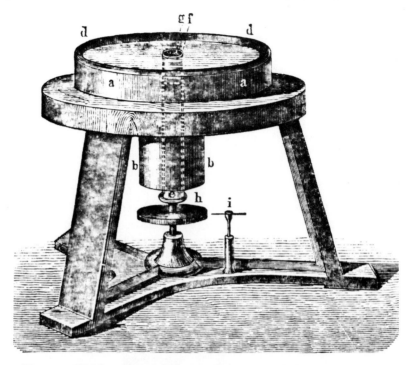

Figure 14. Welcker microtome illustrated, 1882 (from L Dippel—*Handbuch der allgemeinen Mikroskopie*. (Braunschweig, 1882)).
The illustration is taken from page 671, and shows that Welcker's instrument enjoyed some commercial sales. It was, however, omitted from the next edition, only three years later.

design became an article of commerce, and proved important in giving some academic status to an instrument which was at that date rather looked down on in Germany.

Another design was that of Gibbons,[29] described in 1856 in an Australian journal. This was not an advanced concept, being restricted to amateur production of wood sections. It is shown in Figure 15, and had an opening about half an inch in diameter, with a wheel serving the triple purpose of acting as a milled head, a ratchet, and a graduation for control of thickness.

Much more satisfactory was Smith's instrument,[30] essentially a version of the simple hand instrument shown by Currey in 1855, but with some very good modifications. With the earlier instrument, and with most successors of the same type, the specimen had to be unclamped while it was being raised (just as in the earliest Hill instrument) causing many difficulties as a result of compression and elasticity, in trying to procure consistently-thin sections. Smith realized

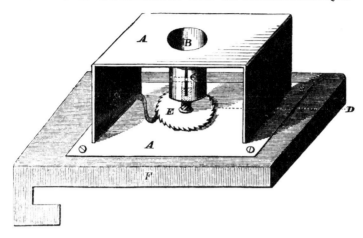

Figure 15. Gibbon's section-cutter, 1856 (from the original description, note 29 in text).
A rather crudely-designed machine, with the positive feature of hooking over the edge of the bench. The well (B) was about 13 mm in diameter.

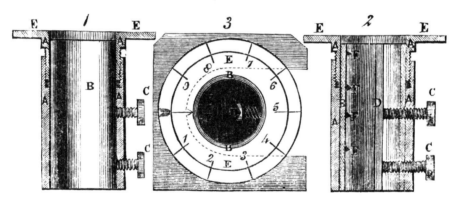

Figure 16. Smith's cutter, 1859 (from the original paper, note 30 in text).
(1) is a sectional view showing the top plate which screwed down round the fixed specimen. (2) shows the clamping bar and its protrusions which gripped the object. (3) shows the instrument slid into a socket in the bench. This is a very practical design.

this, and kept the specimen fixed all the time, arranging for the knife to descend to it rather than *vice versa*. This was done by rotating the whole top plate which carried the knife, as will be seen in Figure 16 which reproduces Smith's original woodcut. The tube was about 2 in. long, and held the stem tight all the time the knife was descending. Even more important, the entire instrument could be slid into a slot at the edge of the bench, freeing both hands to guide the blade: all the pressure was thus transmitted direct to the bench, allowing even very tough specimens to be sectioned without difficulty. This was certainly the best design which had appeared to its date.

Rather more basic was the very simple V-groove figured by Notcutt,[31] to support a stem horizontally to allow the knife to be used vertically across its end (see Figure 17). Although this is a very simple device, there is no reason to suppose that it would fail to give adequate support to thin stems, but by the time it appeared there were other much more effective machines.

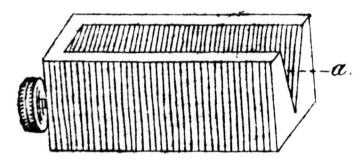

Figure 17. Notcutt's cutter, 1859 (from the original book, note 31 in text). In this very simple but quite helpful device, the specimen was simply laid in the groove, held in place by the fingers, and advanced by the screw seen on the left, ready for a vertical cut at (*a*).

One of the best of all, but one which has remained little-known, was that described by Schmidt,[32] and designed specifically for difficult tissues such as spinal cord. Figures 18 and 19 reproduce the original illustrations, and it is obvious that the instrument must have been a good practical proposition. The stage carried four clamping screws, independently movable, operating four plates of different pattern according to the nature of the tissue to be cut. Below the stage four pillars carried a screw-actuated specimen platform, with spikes to secure the material to be sectioned. This was a poor part of the design, for it would have given virtually no lateral support against compressive forces when the knife struck the object. It may be for this reason that the four plates already described were required to press on the specimen close to the plane of section. The author claimed that he had made many hundred excellent sections with the device, so it must have been quite satisfactory in use.

It was in the 1860s that the microtome became used in a variety of serious scientific investigations, and began to find respectability even among those German professors who had earlier denied the necessity for any support other than finger and thumb. The firm of Beck were

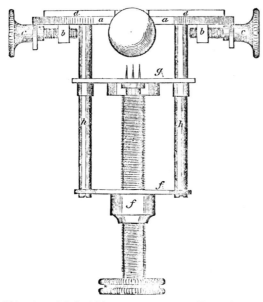

Figure 18. Side view of Schmidt's microtome, 1859 (from the original paper, note 32 in text).
The specimen stage is supported on four pillars, but the specimen itself is held by only four spikes. The rather prominent circle above the spikes is actually the head of one of the four specimen clamps.

certainly making their simple wood section machine, as shown in Richard Beck's book[33] (see Figure 20) a specimen has survived in the Science Museum, and is shown in plate 30. The device is simple, and perhaps typical of wood section-cutters offered by most English suppliers of the time, but no others are figured: they were all similarly priced, and it might be that they were made by only one or two suppliers, and factored by the rest.

Follin's instrument was described by Chevalier in 1865,[34] and is shown in Figure 21. Chevalier states that it was designed to cut timber sections (only), of definite thickness. From its construction it seems to offer no advance over previous models, but would rather have been somewhat uncertain in use on account of the height of the cutting platform over the bench. It must surely be significant that no French design had appeared before this date, and that when one did it was somewhat outmoded.

In 1866 Mouchet was awarded a French medal of honour for his wood sectioning machine, designed to secure even thickness of section.[35] He arranged a knife with strong semicircular blade to cut the timber, held in a tube below the top plate. The whole had the merit of

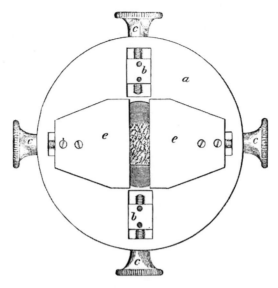

Figure 19. Plan view of Schmidt's microtome, 1859 (from the original paper, note 32 in text).

The plates (*e*) are separately adjustable to clamp the specimen, and are interchangeable according to the nature of the specimen: those shown are for liver. There is no doubt that this design was perfectly functional for soft tissues, which could not be said for any other microtome before this one: the lack of support for soft tissues at this time before the development of paraffin embedding or infiltration was cleverly overcome by these special plates.

Figure 20. Beck's section machine, 1865 (from the book by Beck, note 33 in text).

The operation of a firm blade (a carpenter's chisel specially sharpened) was opposed by counter-pressure from the hand at the base of the machine. The working outfit of the operator is worthy of note.

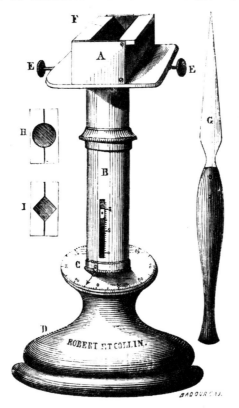

Figure 21. Follin's microtome, 1865 (from the book by Chevalier, note 34 in text).

A somewhat unconvincing design, being top heavy and lacking means to fasten it to the bench, the instrument would have to be held in one hand while the knife was wielded with the other. The plates (H) and (I) were interchangeable at (F), to hold differently-shaped specimens.

being clamped to the bench, but no illustration seems to have survived, and thus an evaluation is impossible.

A very different instrument was designed by Hensen in 1866.[36] This was for use actually on the stage of the microscope, so that fresh tissues could be examined at once. The material for which it was designed was the hearing organ of Crustacea, of a certain degree of firmness: it would not handle a wide range of tissues, but is certainly an original design (see Figure 22).

His developed his microtome in 1866, although he did not describe it until 1870.[37] He had made over 5000 sections with it by that date, and the quality of the work of His in embryology is well known; it was a matter of pride to him to have been able to make so many sections in the

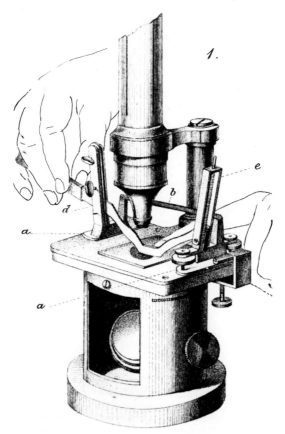

Figure 22. Hensen's microtome, 1866 (from the original paper, note 36 in text).
Designed to clamp to the microscope stage, to provide sections of fresh crustacean material for instant examination. The lever (*d*) was worked vertically under spring tension from its support (*e*). The specimen was advanced by finger pressure from behind, under visual control: the arrangement seems somewhat unsatisfactory.

time, but by 1885 it would be possible to make that number in one day! The object was carried in a clamp working in a modified substage assembly supported on a microscope foot (see Figure 23). What would have been the stage of the instrument guided the knife which moved down towards the limb, to contact a piece of metal with what could only have been unfortunate effects on the edge. The knife was supported along its length, and the material was clamped without having to be released between sections. In spite of its somewhat odd appearance, the instrument was obviously efficient, but was not sold commercially.

In the three years 1868–70, no fewer than five instruments were

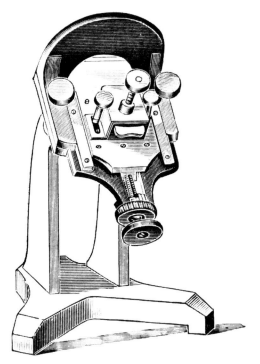

Figure 23. Microtome by His, 1866 (from the original paper, note 37 in text).
Based on the use of a modified microscope stand, the specimen was held by
the central screw directly below the stage aperture. Coarse adjustment was
obtained by the screws either side, the whole carrier being clamped.
Advance was by the large milled head, the knife being slid across the stage
above.

described. In July 1868 Luys exhibited his instrument before the
anatomical society in Paris;[38] it consisted of a heavy table on four legs,
with the central well large enough at the top to allow for insertion of
interchangeable specimen platforms (see Figure 24). This is an
interesting feature, destined to become widely adopted in the 1880s,
allowing greater flexibility in use of the microtome.

In the same year the Rivet microtome was described.[39] This design
was to become the basis of some of the most advanced microtomes ever
made, and adopted the principle of the Capanema instrument, to raise
the specimen up an inclined plane. This gave much better control of
section thickness, which was not easy to achieve with other methods
relying on screw threads, for these were not easy to obtain cut both fine
and regular in those days. An even more important advance was giving
the knife support, to allow it to be slid along under continuous pressure
from the hand. The instrument was made in hardwood, which would no

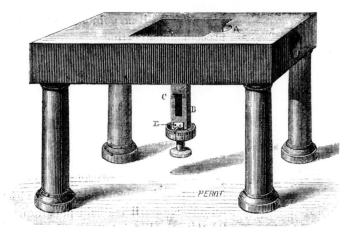

Figure 24. Luys microtome, 1868 (from C Robin—*Traité du microscope.*
(Paris: Baillière, 1871) see p. 249).
Of unusually substantial construction, which must have made for suitable
rigidity. The apparently enormous well was to receive interchangeable
specimen platforms.

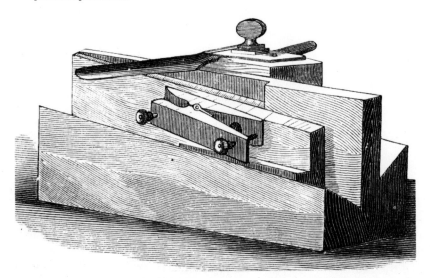

Figure 25. Rivet's microtome, 1868 (from J Gronland (1878)—Rivet's
microtome and its use, *Amer. J. Micr.* 3 *:25–29*).
The specimen advance followed the principal of Capanema, but in addition
the knife was made to traverse a definite path, to give an instrument which
would offer much more precision than any other instrument of its date.
Using both hands a series of sections could be cut quite rapidly, but a major
defect was the crudity of the object-clamp: this was only a slightly modified
clothes peg, which allowed much lateral movement when the knife hit the
specimen. A further disadvantage is the low setting of the clamp beneath
the knife: compare with the later metal design in Fig. 91.

doubt be satisfactory for a time, but obviously the knife slide would wear quite rapidly. A weak point was the object clamp, which could not have offered rigid support, as is further discussed in the caption to Figure 25.

In 1869 Hunt described his small instrument,[40] essentially a form of the hand microtome modified to carry the specimen without needing to release it every so often during cutting. This was achieved by having a second tube inside the first: this inner tube carried the tightly-clamped object, and was raised bodily by the screw at the base of the outer tube. This was a useful improvement on the usual simple microtome, originally designed by Ross, but often called the Ranvier instrument.

Hawksley's section machine was another very original design. As will be seen from Figure 26, a baseboard carried a platform running in a groove. The platform was pulled forward against spring tension, so that when it was released, it flew back again against the edge of a razor. Lateral movement of the platform allowed the thickness of the specimen to be adjusted. The process of pulling forward, adjusting the screw for thickness, and releasing, secured another section of the cephalic ganglia of beetles for which the machine was made. The design has to be mentioned as an example of that misplaced ingenuity which so often characterized inventions associated with the microscope and microtome at that date.[41]

In his paper of 1870[42] Stirling states that his microtome was devised in 1861, for Goodsir's work on the brain. The instrument is described as

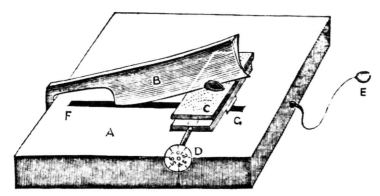

Figure 26. Hawksley's microtome, 1870 (from the original paper, note 41 in text).
The string (E) pulled the spring-tensioned stage (C) until it was caught and held. The screw (D) moved the specimen sideways on the stage to give the required thickness. The razor (B) cut a section when the stage was released to travel the length of the blade under spring tension, along the groove (G). The possibilities for lack of accuracy in this design are enormous.

an improvement of the old cutter used by botanists, and was a simple microtome provided with a clamp to fix it to the bench. The instrument was very solidly made (see Figure 27), of brass $\frac{1}{4}$ in. thick for the 4 in. diameter plate; the well was wide with a screw of 50 threads per inch. The real advance was the fitting of an accurately made piston on top of the screw, to allow the object to be raised with very little lateral motion. Specimens larger than the well were fitted into it by means of a circular gouge of exactly the right diameter, like a large cork-borer. Smaller specimens were embedded in turnip or carrot, and cut with a spirit-wetted knife. Stirling claimed that hundreds of sections per hour could be cut after practice, and that 300 per inch were easily obtained. He had tried other embedding substances, such as paraffin wax, gelatine, cheese, pith, and cork, but none was as good as the roots mentioned. This is a truly excellent paper, full of good practical advice.

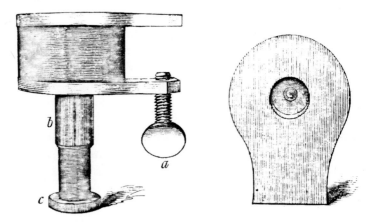

Figure 27. Stirling section cutter, 1861 (from the original paper, note 42 in text).
Basically a hand microtome with bench clamp, but with the important modification that a separate piston was fitted above the head of the screw, with attention to the clearance. When used with a suitable embedding material the machine was capable of good work. When it was described in 1870, it had been in use for nine years.

The above concludes the survey of the development of the microtome until 1870. Freezing instruments must be discussed next, for they exemplify a method which gained much popularity for twenty years, thereafter fading into a relative obscurity in the operating theatre, and post-mortem room. Following the freezers, four main groups of instrument will be discussed, before some ancillary embedding apparatus is considered.

The freezing method

The introduction of the freezing method into histology is usually attributed to Stilling in 1842, but this is not the first mention of the process. In his excellent paper of 1825,[43] Raspail showed clearly that he had used the freezing method by design. In this paper the author made four resolutions, of which the third one (page 228) is:

> D'employer les mélanges frigorifiques pour empêcher les tissus d'être défigurés et aplatis par le scalpel microscopique.

—to use freezing-mixtures to prevent the tissues from being distorted and flattened by the microscopical scalpel. This is a clear statement which shows that Raspail had already tried out the method, but his paper made little impact on the scientific world, for nowhere is there any mention of his process being applied by other scientists. The freezing method was not to become established for another forty-five years, in spite of its superiority over the somewhat crude methods in use in the first half of the nineteenth century. As has been said above, nowadays the method is used only where rapid diagnosis of tissues is required, and latterly for histochemical procedures, but as proper serial sections are not to be obtained by the process, it was abandoned for most work by 1890.

The story of Stilling's accidental (re)discovery is well known: that he left by accident a piece of spinal cord on the windowsill of his laboratory overnight, and found it frozen solid on the morning of 25th January 1842. He made a transverse section and examined it, unstained and under slight compression, at a magnification of about ×15. The radiating bundles of nerves, and the central tracts, were at once visible, and he realized that he had the key to understanding the cord structure. From there he went on to work extensively on the subject, and published a large book in 1859, which included detailed instructions for preparing the cord.[44]

In spite of all this, the freezing method was used by relatively few workers until 1871, when Rutherford published his account of a machine which made the process much simpler to use.[45] Before this time, Stilling's original freehand technique, which could produce a series of sections of more or less equal thinness if the individual was sufficiently experienced, was not adequately advantageous to be widely used. Its main advantage was that it obviated a wait of about six weeks while the organ was hardened by chromic acid or the like, and it did allow finer and larger sections to be made than could be got by use of the Valentin knife. The technique was to put the tissue in a metal crucible surrounded by a freezing mixture, and make freehand sections once the tissue was hard enough. Rutherford's contribution was to arrange for a section machine to be surrounded by a freezing bath, thus enabling the tissue to be cut in a convenient and controlled manner. The machine is illustrated in Figure 28: it is clearly a modified original Stirling. The tissues themselves were surrounded with a 0·75% salt solution, which unfortunately froze very hard and splintered as the knife hit it, but by breathing on the blade and ice while the section was being made Rutherford was able to overcome this drawback. To demonstrate the point, he killed a rabbit, removed lung, liver, intestine, muscle and kidney, and froze them at once in one mass. Sixteen minutes later they were frozen through and being sectioned. He suggested that other freezing media might be found and that the instrument could probably be developed much further.

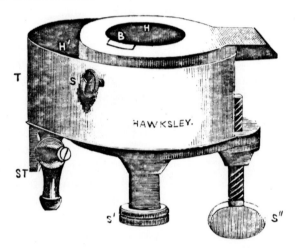

Figure 28. Rutherford's freezing microtome, 1871 (from the original paper, note 45 in text).
A version of the freezing machine seen in Fig. 27, but surrounded with an ice/salt bath. This was the machine which made universally available the cutting of frozen sections which had enjoyed only limited popularity before.

In these suppositions he was certainly correct. He himself published an account of an improved model only two years later:[46] the machine was suitable for embedded or frozen tissues, but he suggested use of gum arabic solution for the freezing medium as a decided improvement. His original woodcut is reproduced in Figure 29; and a photograph of an example of the instrument is in plate 31. The new machine had a much larger freezing bath and a larger table for the knife: the former had already been introduced by a Mr McCarthy, whose instrument had been described in the important paper by Needham already mentioned.[47] The next modification was that of Fleming, in 1876;[48] this retained the original Stirling machine, but in hand-held form with the object surrounded by scraped potato, mashed fresh brain, or other similar medium. The coolant was an alcohol/water mixture, cooled outside the microtome itself in a tin worm surrounded by a freezing mixture. This was a novel principle which obviated the need to keep a large volume of ice/salt actually round the specimen.

Figure 29. Rutherford's freezing microtome of 1873 (from the original paper, note 46 in text).
The design has been improved by a larger freezing bath, and much larger knife platform.

Also in 1875, Tait published what was essentially a modification of the Rutherford instrument, while attacking it.[49] The larger volume of coolant would be an advantage, and the other detail modifications helpful, as seen in Figure 30. In a reply to Tait published later the same year,[50] Rutherford shows that he had regarded the freezing process as being very important. He states:

When I described . . . a modification of Stirling's microtome devised by me, I was aware that I had rendered the freezing process available to the histologist in a way which would certainly prove of great service. And I was also aware that the freezing method would in the end largely supersede all other methods of imbedding the tissues for the purpose of making microscopical sections.

Tait seemed to claim to have originated the method, possibly in ignorance of earlier publications, and Rutherford was withering in his refutation of this claim. He repeated his earlier description, and condemned Fleming's modification. The original suggestion for use of a gum solution was stated to be that of his assistant, Urban Pritchard (son of the famous Andrew).

The following year an entirely new principle of freezing was adopted, that of using the latent heat of evaporation of ether. The idea originated with Hughes, whose paper included the woodcut shown in Figure 31.[51] A rubber ball was used to pump air through an ordinary glass atomiser in the freezing chamber, and thus reduce its temperature. The ether was

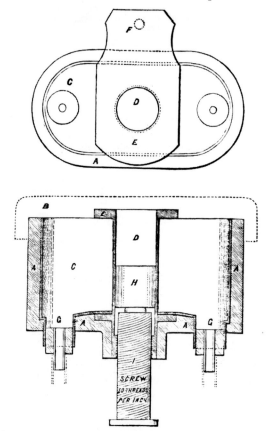

Figure 30. Tait freezing microtome, 1873 (from the original paper, note 49 in text).

A still larger freezing compartment with improved drainage was an important feature. The instrument could be actually screwed to the bench, and not merely clamped: this also was important in view of the effort required to work such a machine.

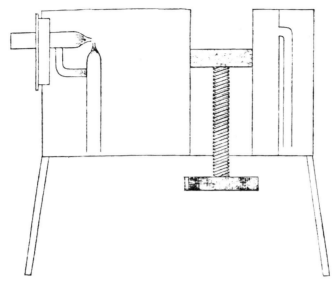

Figure 31. Hughes ether freezing microtome, 1876 (from the original paper, note 51 in text).

This was the first microtome to rely on the cooling effect of a jet of ether as it evaporated. The design was not too good, for the jet evaporated at too great a distance from the object platform, reducing the efficiency of the process and using a lot of ether.

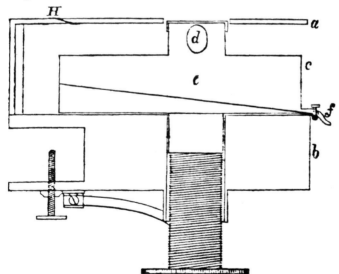

Figure 32. Lewis ether freezing microtome, 1877 (from the original paper, note 52 in text).

Apparently designed in ignorance of the Hughes machine, this was a much better device, as the atomiser was closer to the object. In both, the ether was kept in the body of the instrument.

held inside the instrument, and tissues were frozen within four minutes: its great advantage was that it could be used at any time without need to find ice. Twelve months later Lewis described his instrument,[52] developed in ignorance of the Hughes machine (see Figure 32). He subsequently claimed[53] that his device was simpler in construction and more expeditious in freezing, which was accomplished within twenty seconds as the atomiser had its nozzle more closely applied to the surface holding the tissue to be frozen. The principle gained wide acceptance, even the Rutherford machine being modified to take an ether attachment[54] (see Figure 33).

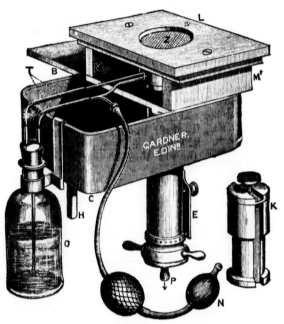

Figure 33. Rutherford microtome adapted for ether freezing, 1887 (from the original description, note 54 in text).
By 1887, commercial manufacture of microtomes was well established. All that had been done here was to add an ether platform to the original body. The ether was in a separate container. A paraffin adapter is also shown (K), and the whole outfit has a decidedly professional air compared with instruments ten years earlier.

A very simple device, hardly worthy of the name of microtome at all, was described by Urban Pritchard.[55] It consisted of a brass block on a wooden handle: the block was cooled in a freezing bath, withdrawn, and the tissue placed on it to freeze before being sectioned freehand. Much more important was the Williams instrument,[56] adapted for ice/salt freezing (see Figure 34) and having a novel knife carrier as its important

Figure 34. Williams freezing microtome, 1881 (from the original paper, note 56 in text).
This is the original design, with an ice/salt compartment and a new knife carrier, as means for adjusting thickness of section. Spare object platforms are seen alongside the instrument.

feature. This slid on the glass top plate, carrying a razor clamped on tripod feet, one of which was adjustable for height. There was thus no necessity to have a micrometer screw near the freezing mixture or under the specimen, and this was a very useful design. In the same year Groves modified the machine to accept an ether spray:[57] the ether supply was inside the instrument (see Figure 35), and Groves adapted a water-motor to work the bulb for atomising the ether. Later the same year, Groves provided a much simplified construction,[58] and the final development of this instrument occurred at the hands of Fearnley, who separated the ether container from the instrument, to allow of occasional use when need arose.[59] The original woodcut is reproduced in Figure 36, and a photograph of the instrument in plate 32. It is still easy to operate, and provides quite adequate sections with little preliminary preparation.

Taylor's freezing microtome of 1882[60] is a simple instrument provided with a supply of coolant from a remote chamber. The top of the instrument revolved to lower the plane of section, and the design was probably quite serviceable; see Figure 37. In the same year another instrument of American origin was described, by Satterthwaite &

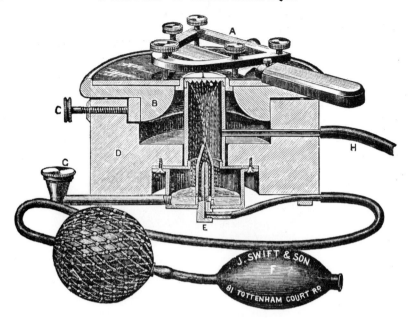

Figure 35. Groves modification of the Williams design, 1881 (from the original paper, note 57 in text).
The ice bath was omitted, and an ether spray substituted.

Figure 36. Fearnley's modification of the Groves-Williams design, 1883 (from the original description, note 59 in text).
After Groves had made the device smaller, Fearnley separated the ether supply from the body of the instrument, the first time this was done in any design.

PLATES

COLOUR PLATE I   A group of sliders illustrating typical types and the transition to glass (*Author's private collection.*)

A (241 mm long overall) and B are of hardwood for use with the solar microscope, and are the usual mica/brass circlip manufacture.

C and D are examples of opaque mounts, from 1780/90; C is for a solar microscope.

E dates from 1820/25, and is unusual in being made from two glasses cemented at their edges: the plant sections are of indifferent quality, but the slider is transitional between slide and slider.

F is a good-quality conventional slider of about 1800, containing typical insect parts. The slider is made of bone, with mica/brass circlip inserts.

G is a further example of a good-quality slider of about 1800, made in this case of wood with brass circlip/mica inserts: the timber sections are good quality and properly identified.

H is a most interesting slider, being made of glass with mica covers held separately in place with gummed paper. The date appears to be about 1830, and this example represents a transitional stage which is nearer to the slide than specimen E on this plate.

J and K are all-glass sliders with top and bottom plates of the same thickness. These are less advanced than example H, and would not allow high-power work. Both have their glasses separated by thick rag paper, which had correct thickness for separating the glasses to contain timber sections. The sliders date from about 1830.

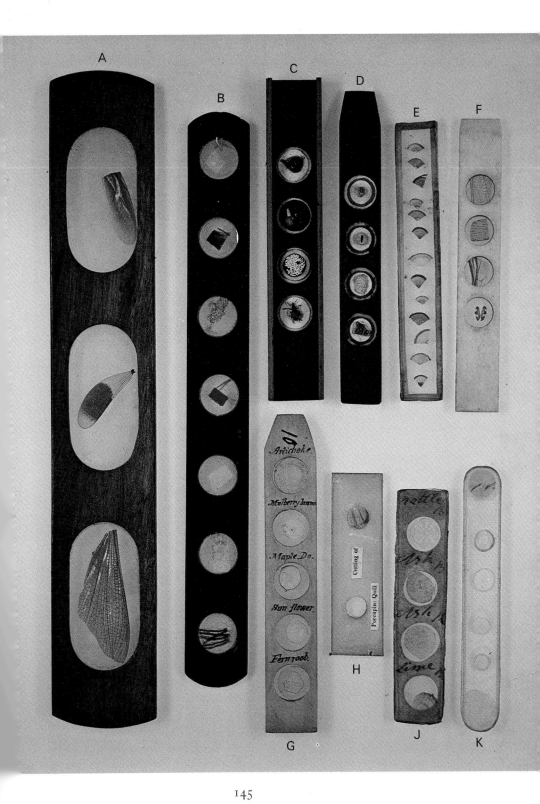

A

B

C

D

E

F

G

Artichoke.

Mulberry bark

Maple Do.

Sun flower.

Fern root.

H

Cutting of

Porcupine Quill

J

nettle

Ash

Ash

Lime

K

145

COLOUR PLATE II   Early balsam preparations
(*Author's private collection.*)

A. Male flea, mounted in molten balsam, between two equally-thick slips with ends paper bound.

B. Whole insect mounted in molten balsam between two equally-thick glasses, the top one being smaller in size. This slide may represent a stage between use of two slips and use of one slip with a thinner cover.

C. Bee sting in molten balsam between two slips of equal thickness, with paper binding.

D. An important mount of injected skin in molten balsam, between two slips, the upper one being half the thickness of the lower. Paper bands at the ends.

E. Head of *Dytiscus* in balsam between two equally-thick slips.

F. A very interesting mount of rock section on an uncovered slip. The section is cemented in place by molten balsam. Inscription:

S. Wales Oolite
with Nantite
A Ross from W Darker
April 27th 1839

G. A preparation by Topping dated 27/12/40, and thus one of his earliest. Head of fly (a preparation he was to make famous) in balsam between two paper-covered slips measuring 3×1 inches. This slide is covered with green paper, a colour he reserved for his own use: slides for sale were covered in red papers.

A

C

D

E

B

F

Topping.
27/12/10.
12.
64.

Head of Fly

G

COLOUR PLATE III   A sequence of dated preparations, 1824–1869

A. 1824   Dried injected tubercular lung by Dr James Macartney of Dublin, dated 17th August 1824: wood slip with glass cover. Tubercular patches still visible as grey nodules.

B. 1837   Dry mount of plant stem sections, paper separators, dated 4.29.1837.

C. 1842   Injected toad lung, in balsam-filled built-up cell, signed J. Quekett and dated 20th July 1842.

D. 1843   Balsam mount of infusoria dated 1843, and with thin cover. This is an important landmark, showing use of thin covers by this date.

E. 1853   T.S. chicken bone in balsam mount.

F. 1853   Dry mount of Foraminifera.

G. 1853   An important stained preparation dated 13th March 1853, of T.S. and L.S. cornea. This is the earliest stained, as opposed to injected, preparation to have come to light.

H. 1855   Dry mount of spider's web, paper-covered.

J. 1856   Salicine, a popular polarizing specimen, fully paper-covered.

K. 1857   Dated December 1857, T.S. *Acer* stem, fully paper-covered.

L. 1858   Balsam mount of *Lycopodium* spores by J Bourgogne, dated August 1858.

M. 1864   A most interesting mount, dated 30 Settem 1864, and signed Filippo Pacini. Blood corpuscles prepared to Pacini's formula, quoted on the reverse as Bicl. Mercu 1, Clor. Sodica 1, Nitrat. Potas 2, acqua ad 100.

N. 1865   Injected non-commercial mount of rabbit kidney in balsam, dated August 1865.

O. 1869   T.S. spinal cord of calf, carmine-stained, and rather roughly cut, with knife marks and wedge-shaped section. Dated October 31st 1869.

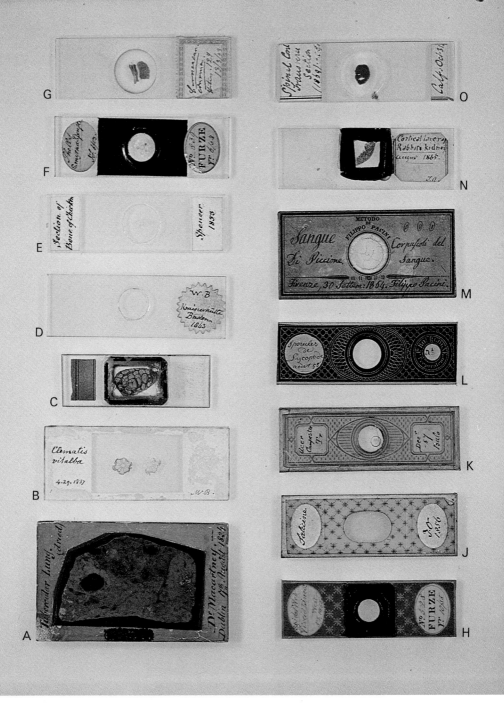

COLOUR PLATE IV   Preparations by John Quekett
(*Hunterian Museum of the Royal College of
   Surgeons of England.*)

A. An injected portion of human foetal pericardium in balsam. The slide is
   somewhat crude, being a wedge of thickness varying from 1·31 mm to
   2·25 mm.

B. A foetal foot injected whole as an opaque mount in balsam. The technique is
   described in the 1848 edition of his book, on page 260.

C. A fluid mount of a baby octopus in deep cell, with thick cover.

D. A ground section of a wombat tooth, mounted dry as advocated on page 302 of
   his book: the edges are filled with sealing wax.

E. A fine transparent injection of a foetal membrane, mounted on a slide 114 mm
   long and no less than 4·8 mm thick: the cover is about 1 mm thick.

F. A photomacrograph of specimen E—slide Dc 230. The magnification is × 35,
   and is enough to enable one to appreciate the remarkable quality of the
   preparation. The lighting was arranged to be a combination of transmitted
   and reflected so as to bring out the finest details of the vessels.

G. A photomacrograph of specimen B—slide HA 125a. The foetal foot is a magnificent specimen of the injector's art, still in excellent order, and a tribute to
   John Quekett's skill as dissector and microscopist.

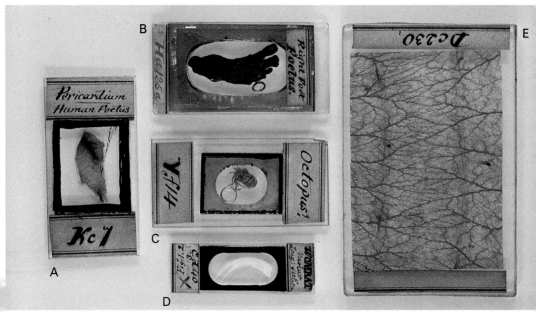

PLATE 1 An original Leeuwenhoek microscope
(*Photographed in its display case in the National Museum of the History of Science, Leiden. INV. M2a2.*)

The point designed to carry the object is seen just to the left of the plate, which has the single lens in the opening towards its top. The plate measures 22×41 mm (see page 9 of the text).

PLATE 2 Original pattern of the Hill microtome
(*Science Museum. INV. 1903–69.*)

The barrel is of the ivory construction originally described by Hill (see page 14), but this has split as is often the case with this material. This is an example of the first microtome in the world.

PLATE 3 Later pattern of the Hill microtome
(*Royal Microscopical Society. INV. 220a.*)

The barrel is of brass, signed 'Ramsden London'. The instrument is still in its original case, with spare blade. Apart from the material of the barrel, the instrument is identical with the original description.

PLATE 4 Hill microtome in use

The instrument pictured in plate 3 is shown making sections of apple-wood. Hill himself admitted that it is difficult to obtain a good edge on the blade, and that the sections had a tendency to roll. These remarks proved to be understatements: it is very difficult indeed to sharpen the blade to the edge required to avoid tearing the specimen. In use, the twig was cut from the tree, soaked in 50% spirit for three days (as was Hill's own practice), and then cut while keeping blade and specimen lubricated with more spirit. The wedge-section of the blade makes it inevitable that slices will tend to curl as they leave the knife; a microscopic examination of the edge showed that an effective angle of over 40° exists. The pointer attached to the handle lifted the small spring on the instrument, designed to reduce this curling. Much effort is needed to turn the handle to make the section.

1

2

3

4

PLATE 5   An outfit of sliders of about 1770
(*Author's private collection.*)

The pasteboard, leathercovered box (A) is typical of its kind, and held the eight sliders plus the brass slider. The sliders are of bone, with mica discs and brass circlips: all are shown with circlips visible in their cavities. The sections of timber are good enough for modern use if cleared.

The brass slider (B) was intended for support of temporary wet mounts: the glass insert is shown in position.

The typical manuscript listing of the contents of each slider is seen at (C). These are entirely usual in range of specimens and in being arranged for use with particular powers of objective, without regard for the nature of the objects.

The separately-stored ivory box is shown at (D). This is finely made, with a screw-on cap at each end (E). One side stores spare mica discs ('talcs') (F): the other keeps spare brass circlips (G). In practice it is quite easy to replace specimens between the talcs, using simple forceps of the kind universally supplied as part of the entire microscope outfit at this time.

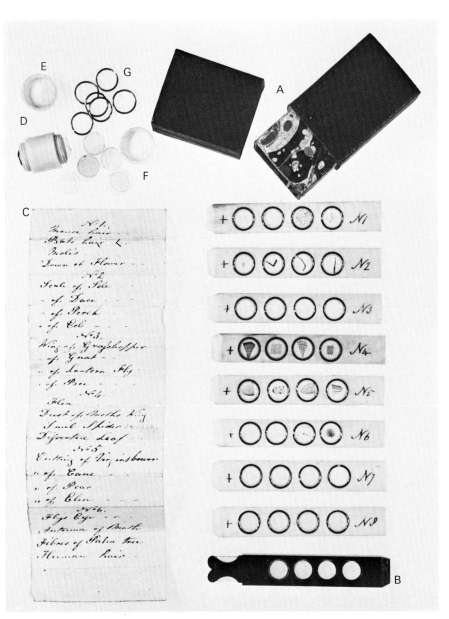

5

PLATE 6   An Adams microtome
(*Teyler's Museum, Haarlem. INV. 272.*)

Known to have been bought from Adams in September 1789, the instrument resembles the figure in Adams' book (see my Figure 5) quite closely. The angle of approach of the knife has been improved, and this makes for better cutting than the diagrammatic machine would have allowed. The specimen clamp knob is turned to the side, and this improves accessibility in use. The baseboard (overall length 221 mm) has a hole cut in it, possibly to allow drainage of spirit used to moisten the specimen during cutting.

Good design points include support of the blade at each end, thus minimizing the tendency for it to jump as it contacts the specimen, to avoid wedge-shaped sections. It is also possible to adjust the tightness of the knife in its slide, by means of the clamps visible below the plate. In use the instrument is smooth and relatively rigid. Sharpening of the blade is very easy compared with that of the Hill instrument. A further important point is that the shaped specimen well has been turned relative to the original drawing: this ensures that it is not the bark of the specimen which is contacted first. As the bark is more elastic than the centre of the twig, this avoids difficulties of compression and consequent unevenness of cut. The micrometer drive to the specimen advance is smooth and fine even today, but the instrument has certainly had very little use.

PLATE 7   A typical Ypelaar cabinet of medium size
(*Teyler's Museum, Haarlem. INV. 274.*)

Dated 1790, the cabinet (274 mm wide) has twelve drawers: four contain ivory sliders (115×18 mm), the rest ivory cells of diameters 12, 16, and 20 mm.

The timber and other plant sections were bought by Van Marum in London in July 1790. The cells were made by Ypelaar, and are typical of his products. They contain 36 marine specimens and 164 very miscellaneous objects, all of excellent quality, with well-chosen material carefully positioned and displayed (see also plate 19).

6

7

PLATE 8   Examples of Hewson's hermetically-sealed preparations
(*From the Hunterian Museum of the Royal College of Surgeons
of England.*)

Specimens from left to right:

2 'lower lip, injected shewing papillae &c'
7 'upper part of ileum, internal surface, villi shown'
29 'upper eyelid, human subject'
50 'liver, with injected vessels'
19 is shown on the reverse side, for the label, which reads

THE INTERNAL SURFACE OF THE SMALL GUT OF A DOG
PREPARED BY W HEWSON

All the tubes are flattened. Number 50 measures 77 mm long overall.

Each piece of tissue is secured by clips to a piece of metal, possibly lead, with the label scratched rather roughly on it.

PLATE 9   Photomacrograph of specimen 7, magnified × 50

The villi are clearly to be seen, with capillaries visible. The tissue is rather wavy in preservation, with much of the liquid having evaporated. At the bottom of the picture the villi are still covered, and details are surprisingly clear.

PLATE 10   Photomacrograph of specimen 2, magnified × 50

The magnification might have been a typical one in the late 1700s, although present-day definition is likely to be superior. The quality of the preparation, however, is very good: excellent detail of the capillaries has been achieved, and the whole is well preserved. Very few other liquid mounts can have survived in good order for over 200 years.

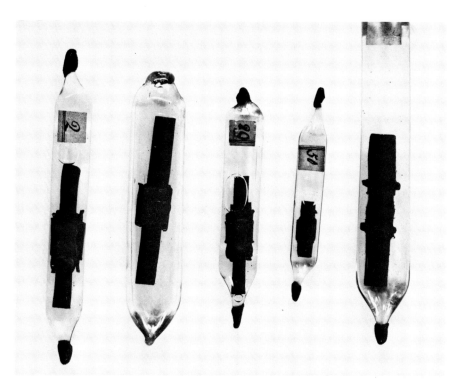

8

9

10
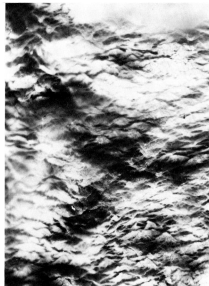

159

PLATE 11   Examples of Hewson's dried preparations
(*From the Hunterian Museum of the Royal College of Surgeons
   of England.*)

Specimens from left to right:
   M.11   muscle with attached tendon
   E.1   choroid of turkey
   G.3   foetal chorion
   K.4   kidney, horizontal section for vessels & tubes
   M.10   horizontal section of heart
   L.1   fish gill, injected
   I.9   large intestine, injected.
Specimen G.3 measures 32 mm square. The thickness of the glass varies from
1·42 mm (K.4) to 2·14 mm (L.1), with M.11 being an exceptional 4·96 mm. The
squares are generally more roughly cut than the oblongs.
   Specimen L.1 shows the area of varnish applied to the specimen. Tests showed
that the varnish is spirit-soluble, and possibly of the nature of shellac, although the
irreplaceable nature of the specimens precluded any kind of extensive test.

PLATE 12   Photomacrograph of specimen I.9, magnified × 50

A combination of reflected and transmitted lighting was used, the better to bring
out the detail. The nature of the intestine is obvious, in spite of an accumulation of
dirt, removal of which was not attempted.

PLATE 13   Photomacrograph of specimen G.3, magnified × 50

The same lighting was used, and the tissues, by their very nature, have stood up to
the somewhat drastic preparative treatment very well. The detail of the capillaries
is impressive, and the slide still gives plenty of information as to the nature of
the chorion.

11

12   13

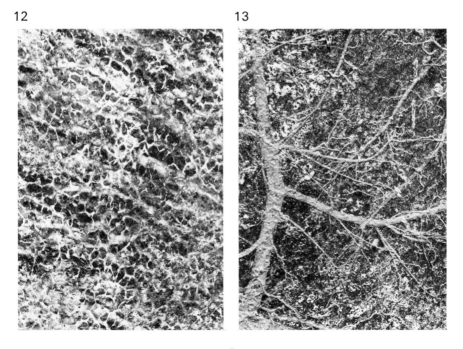

PLATE 14   Photomicrograph made with the 1826 Lister microscope (*Wellcome Museum of the History of Medicine. INV. 234/1949.*)

A good-quality modern preparation of striated muscle made with the old microscope, showing good rendering of detail of nuclei and striations at a magnification of × 400.

PLATE 15   Photomicrograph of 1849 preparation with modern microscope (*Author's private collection.*)

To contrast with plate 14, a first-quality modern microscope was used to photograph an early preparation of striated muscle: this was made by teasing out the fibres and compression, before plunging into hot Canada balsam and covering. The effect on detail is only too apparent: it was less the shortcomings of the microscope and much more those of preparative techniques which obscured histological detail at that time. The magnification is × 200, as no higher power would reveal better detail. The striations which appear to be visible are more the result of the heat treatment than any mirror of physiology!

PLATE 16   Photomicrograph of fresh compressed liver with a microscope of 1780

A rat was killed and the liver promptly removed, finely sliced with sharp scissors, and compressed in a compressorium with a drop of water. Pressure was increased until the tissues were thin enough to see through. A common three-legged microscope of about 1780 was used to view and photograph the specimen at a magnification of × 450. The illumination had to be directed somewhat obliquely to render any detail visible at all, when the globules usually reported until the late 1820s were plainly seen.

The compressorium was then transferred to a modern microscope for inspection: the liver cells were to be seen quite adequately albeit with little contrast. No globules were seen, of course.

14

15

16

PLATE 17   An uncleared timber section

A timber section was removed from its original eighteenth-century slider, and mounted dry on a modern slide, under a modern cover-glass. The photomicrograph shows the visual appearance at this stage.

PLATE 18   The timber section cleared

The section described in plate 17 was left on the microscope and no adjustment was made to any controls. Xylene was allowed to creep under the cover-glass, thus clearing the section. A second photomicrograph was made of its new appearance, on the same length of film. Both negatives were processed together, and both were printed onto different parts of the same piece of photographic bromide paper.

The only difference between the two pictures, therefore, is the effect of the clearing agent: this has dramatically increased the detail visible, demonstrating the major bar to microscopical research of using uncleared specimens.

PLATE 19   Ypelaar resin mount of about 1795
(From the collection of the Museum of the History of Science in Leiden.)

Photomacrograph of a mount put up in resin by Ypelaar as early as 1795, and thus one of the oldest resin-mounted preparations in the world. The nature of the resin has not been established beyond doubt, but its refractive index and other characteristics, plus knowledge of the availability of substances at the time, strongly suggest that it is venetian turpentine. The appearance shows that the objects were plunged into molten resin, and the clearing action is obvious in spite of the age of the preparations.

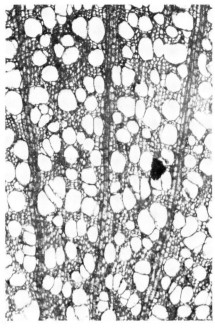

17

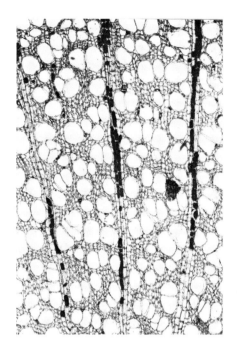

18

19A

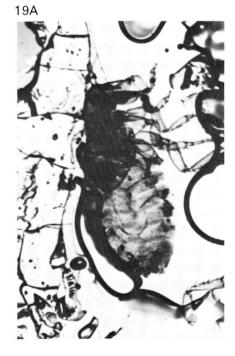

19B

PLATE 20   Injected preparations by Hyrtl
(*Hunterian Museum of the Royal College of Surgeons of England.*)

Joseph Hyrtl was one of those who kept alive the interest in injected specimens during much of the nineteenth century. By virtue of his position as professor of anatomy in Vienna from 1845 to his retirement in 1874 he had something of a monopoly on the Continent for supply of such preparations, using simple injections as well as corrosion techniques.

A. *The entire tray of forty specimens*
Each preparation is mounted in an ebony tablet, measuring 31 × 45 mm, with a cutout supporting two covers. There are four series of mounts: green covers for skin and respiratory, yellow for kidney and eyes, pink for digestive, and blue for liver and various. A label at the bottom of the tray states that the magnification used is not to exceed 30 ×.

B. *Eight selected mounts*
The front side of four mounts are shown, and the rear labels of four others. That of the small intestine of a chamois is interesting, as it is stated 'to refute the dogma of modern physiologists that the cavity of the Peyerian glands communicates directly with the chyliferous vessels'.

C. *Photomicrograph of Hyrtl's preparation of injected iris and choroid of salamander*
One has to agree that the specimen is a magnificent example of opaque injection technique. Magnification × 20.

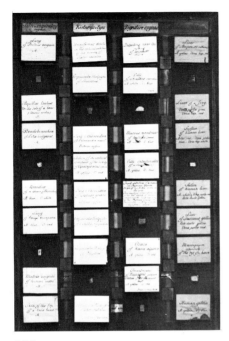

20A

20B

20C

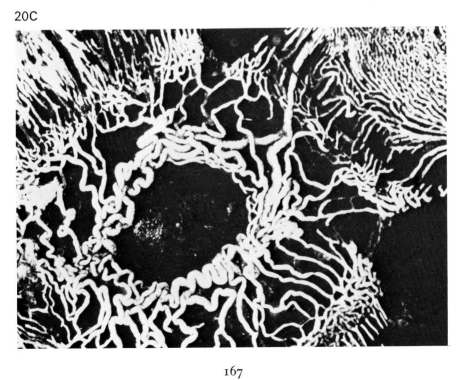

PLATE 21   Examples of transparent and opaque commercial injected mounts

A. Opaque mount of human placenta, by J Bourgogne. Typical size and finish of about 1845.

B. Transparent T.S. rabbit kidney, by Cole. A good quality large section, from about 1882.

C. Unusual preparation of injected small intestine, mounted dry.

D. Topping human kidney L.S. A good quality even section of large size.

E. German-made transparent injection of cat cerebellum, of good quality and large size, supplied by Beck in the 1860s.

F. Fluid mount of injected rabbit intestine in built-up cell.

G. Hett preparation of typical object, large intestine of American squirrel. The square cell with induced bubble is standard Hett make, of about 1850.

H. Hett circular mount, skin from scalp of 3-day-old child, well injected, from about 1855.

J. Suter mount of T.S. cat lung of about 1885.

K. Topping opaque mount of injected human skin of about 1850.

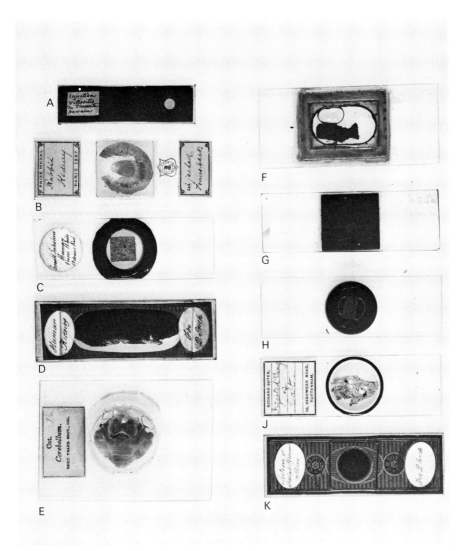

A

B

C

D

E

F

G

H

J

K

21

PLATE 22 A selection of fluid mounts
(*From the Author's collection.*)

A. Tube of the type used by Goring (see p. 24).

B. Enock dissection of abdomen of queen honey bee.

C. Clarke & Page mount of dissected genitalia from drone bee.

D. Enock entire mount of parasite from swallow.

E. Entire frog foot, injected, in cavity slide of large size.

F. Quekett fluid mount of injected foetal stomach—an exceptionally fine specimen.

G. Clark, fluid mount of *Daphnia* arranged for dark-ground illumination.

H. Baker mount, injected toad ovary for eggs.

J. Clarke & Page, *Lophopus* expanded. A good example of results from using a narcotic to prevent an aquatic organism contracting during fixation (see p. 59).

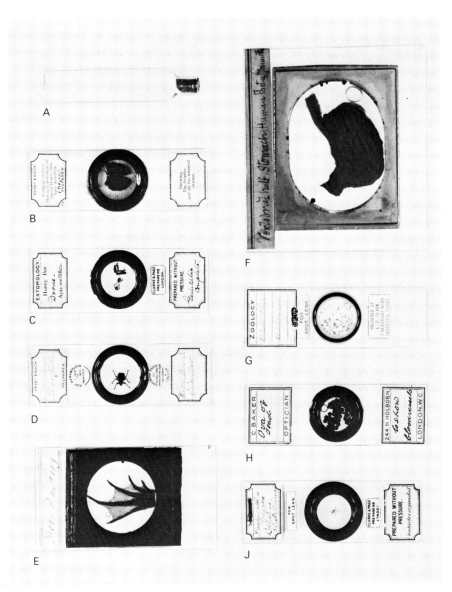

A

B

C

D

E

F

G

H

J

22

PLATE 23   A selection of dry mounts
(*From the Author's collection.*)

A. Pill-box mount containing a small dried flower, made by the author after the fashion of nineteenth-century preparations.

B. Dry mount of leaf to show hairs, under a cover-glass.

C. A mount of 48 Foraminifera from the London Clay 35 ft. below Charing Cross.

D. Typical arranged groups of Polycistina, for decorative purposes.

E. Dry pill-box mount of *Funaria* capsule: the lid is a tapered fit for secure fitting.

F. Dry pill-box mount of coral. A well conceived mount which allows routine inspection through the glass-topped pill-box lid, which can be removed totally if required. The box is bottomless, to allow for use of the lieberkuhn in illumination.

G. Rotatable ivory cover over a leaf surface contained in a cavity in a mahogany slip.

H. L.S. tooth mounted dry, sealing-wax edges, paper pasted over the cover. A standard preparation by a well-known firm, dating from about 1860.

J. Wood slip pierced to carry two foraminiferal tests mounted on rotatable pins.

K. Alpaca cloth specimen mounted in typical amateur fashion.

L. Dry cell mount of fruit, with cemented cover-glass, very dirty on its under-surface.

M. Rotating disc on mahogany slip, of mount first described in 1867.

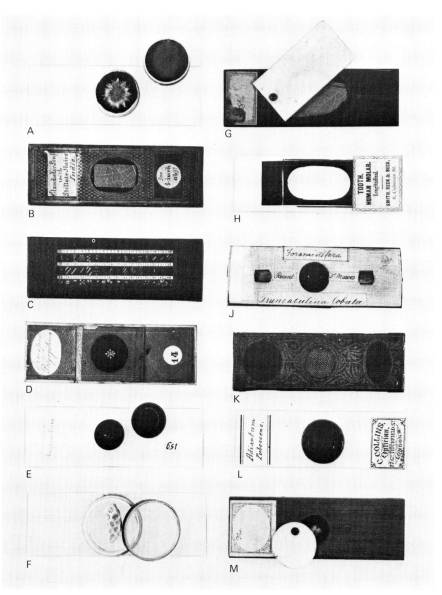

PLATE 24 A group of live cells and related apparatus
(*From the Author's collection.*)

A. Holman's siphon slide, designed to allow perfusion of the cavity with a variety of liquids. One of the brass tubes has been lost. Date about 1890.

B. Animalcule slider, dating from about 1770 and typical of the sliders provided to hold a drop of water to view small living specimens.

C. A multi-cavity slide of the form favoured for work with Protozoa in the 1870s. The ground surface allowed a cover to be kept closely in place.

D. Botterill's trough was designed for use with a vertical stage, and has the great merit of being easily taken apart for cleaning. The two vulcanite plates are held together by brass screws, to contain large cover slips separated by a rubber ring. The depth of the cell is alterable by using thicker rings, and the design remains one of the most practical ever made.

E. Live trough of large size for vertical use, cemented construction.

F. Holman's life slide designed to allow infusoria to be kept under observation to the edge of the cover at high powers.

G. Ivory wedge from live cell, designed to keep a piece of glass in position in a trough such as that shown at E, to limit the movement of a specimen.

H. Thick cell for use horizontally with larger specimens.

J. Brass life cell for use in vertical position: cemented construction.

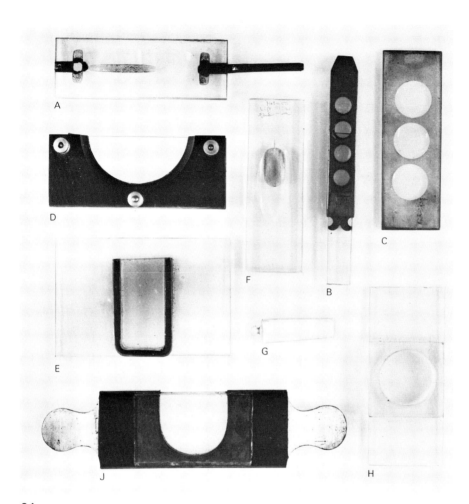

A

D

E

F

G

B

C

H

J

24

A

B

C

D

E

F

W. Watson & Sons Ltd.
London.

H

J

25

177

PLATE 26   Small equipment for specimen handling
(*From the Author's collection.*)

A. Dark wells and substage holder, from a Powell & Lealand outfit supplied in 1885.

B. Stage forceps/needle-point specimen provision from the compass-type Wilson microscope of about 1740.

C. Stage forceps from a simple microscope of about 1790; the original point is now provided with a bone tablet, white one side and black the other.

D. Stage forceps from a Powell & Lealand outfit of 1875: the large size is indicative of the size of the stand (No. 1) with which it was to be used. Note the end now carries a cork-filled box.

E. Forceps from a Cuff-type instrument of about 1770.

F. Forceps from a Culpeper-type instrument of about 1780.

G. Forceps from a Powell & Lealand outfit of 1885.

H. Specimen holder of simple forked type from an outfit of about 1790.

J. Fish tube, with its wire extractor, from about 1790.

K. Watchglass from an outfit of about 1780.

L. Live box from an outfit of about 1770.

M. Insect stage, Eltringham pattern, supplied in 1957 from a 1930s design. The cork-loaded stage rotates and inclines, on a base of standard 3×1 size: fine positioning is assured by the mechanical stage controls of the stand, giving control of viewing in all planes.

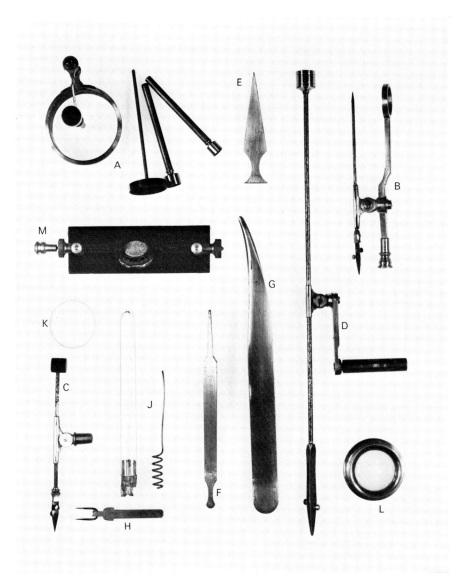

26

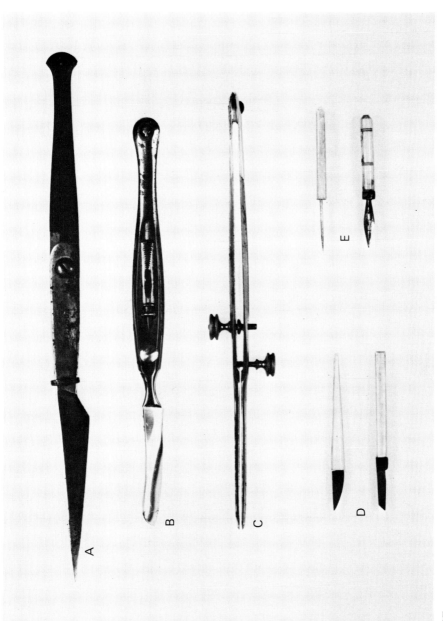

27

181

PLATE 28   Pritchard microtome, circa 1835
(*Science Museum. INV. 1876–1133.*)

The only surviving microtome of the 1830s which has so far come to light, said to have been used by Andrew Pritchard in making wood sections.

PLATE 29   Topping microtome in use
(*Wellcome Museum of the History of Medicine. INV. R.5007.*)

Screwed to the edge of the table in use, the T-shaped mahogany instrument carried a brass well with screw to raise the specimen, and a clamp for use in cutting. Surprisingly good sections can be made with this 1840s machine, supplied by Topping and also used by him in making slides.

PLATE 30   Beck microtome, 1860s
(*Science Museum. INV. 1876–1131.*)

Of simple but robust construction, the design was made by Beck for some years before 1865.

PLATE 31   Rutherford improved freezing microtome, 1876
(*Science Museum. INV. 1876–1137.*)

Complete with original felt jacket, the instrument is in mint condition, and corresponds exactly with the original description of 1873.

29

31

28

30

PLATE 32   Fearnley freezing microtome knife in use
(*Wellcome Museum of the History of Medicine.*)

The knife carrier embodied a new principle—that of keeping the specimen in constant position, and moving the knife down to it. The carrier slides across the glass top plate with entire control of angle of slice. The knob at the bottom of the picture alters the tilt of the carrier relative to two fixed feet, thus altering the height of the knife edge, and the thickness of the section. In use, the whole instrument still works very creditably.

PLATE 33   Hand microtome of the 1880s

Good design, with ample sized top plate giving good guidance to the razor in use; the screw head for raising the specimen is larger in diameter than many which were provided, and gives good control of section thickness.

PLATES 34 and 35   Cathcart microtomes

The Cathcart-Darlaston metal microtome of the 1920s represents the final stage in the development of this popular freezing instrument, based on the design introduced in 1906 to cut to a repeatable 1/1000″. The provision of a definite track for the razor carrier is a helpful feature in use, and perfectly adequate sections have been cut with this example.

The wooden model dates from about 1890. The glass slides are missing from the tops of the sides, and the ether atomizing equipment is also absent, but the general details are those of the improved model first offered in 1888.

32

33

34   35

34

35

PLATE 36   Zeiss bench microtome, 1877
(*Optisches Museum, Carl Zeiss Stiftung, Jena. INV. 341.*)

An early example of the design, preserved in excellent order. Very solid in use, and very well made.

PLATE 37   Korting microtome, 1880
(*Optisches Museum, Carl Zeiss Stiftung, Jena. INV. 338.*)

One of the first such instruments to be made, having a vertical plate 144 mm long. The example shown is preserved in excellent order, and still works very smoothly and evenly.

PLATE 38   Korting instrument, 1883
(*Optisches Museum, Carl Zeiss Stiftung, Jena. INV. 401.*)

A larger version of the instrument shown in plate 37, but not shown in any Zeiss catalogue. The plate is 251 mm long, and the object-clamp has been modified. This particular machine is interesting, as it possibly represents a development of the design by Zeiss, before they decided that microtome manufacture should be left to specialist firms.

37

36

38

PLATE 39   Cambridge rocking microtome, original pattern
(*Museum of the History of Science, Oxford.*)

This example, one of four handled, is still in its original condition, and perfectly usable. The red finish is diagnostic of this first design, first offered in 1885.

PLATE 40   Minot microtomes
(*Universiteitsmuseum, Utrecht. Photograph used by permission.*)

From left to right: the Minot-Blake design of 1899, the original Minot first sold in 1886, and the improved Minot of 1892.

PLATE 41   Cambridge automatic microtome
(*Museum of the History of Science, Oxford.*)

This machine is essentially complete and in working order: very adequate sections have been cut with it, and the speed of operation is still remarkable.

PLATE 42   'Caldwell' automatic microtome, 1883
(*From Threlfall's paper, see note 174 of chapter 5.*)

The copy instrument made on Threlfall's orders to the original design.

PLATE 43   Examples of commercial preparations (1)

A and B are mounts by West. A is an example of one of the earliest mounts to carry the maker's name, stamped on quite crudely. B is not so marked, but came from a set contained in a leather box stamped with West's name.

C to F: Pritchard mounts
C is an early example of a mica/glass test object, with handwritten maker's name. D and E are examples of typical wax-edged mounts, the smaller having the maker's name printed on the reverse. F carries three ground sections of fossil timber, and is diamond engraved with the name A Pritchard 162 Fleet Street.

G and H: Topping mounts in green paper covers
Topping kept green-covered mounts for his own use. His monogram is seen on slide G, but is covered by his address label on slide H.

I to K: Topping mounts
The name C M Topping always appeared in small circles on his mounts: the central part of the covering paper was cut away as needed.

L and M: mounts by Amos Topping
It is not clear to what extent Amos, son of C M Topping, mounted his own material. He had different paper covers from those of his father, but often did not apply paper covers at all.

N to P: factored mounts by the Toppings
O and P are C M Topping mounts factored by two different firms. N was sold by Watsons, who eventually took over C M's stock on acquiring Wheeler's business, which had been the first buyer.

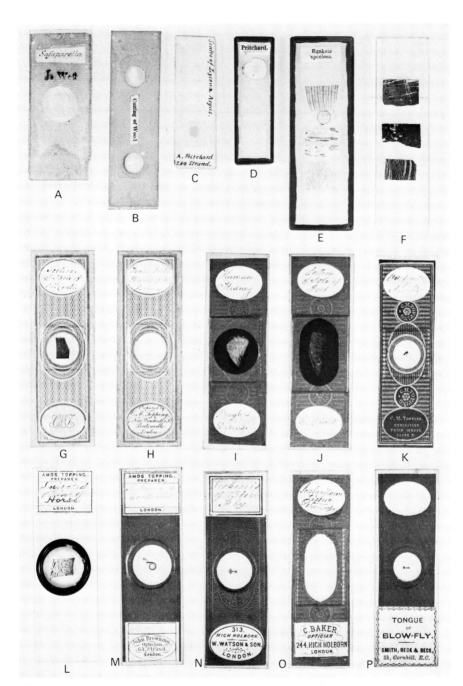

A · B · C · D · E · F
G · H · I · J · K
L · M · N · O · P

43

PLATE 44   Examples of commercial preparations (2)

A to E: mounts by J B Dancer
A is a typical mount with one label, dated 12/70, and with the usual monogram. B is probably an older preparation, perhaps from the 1860s, and is quite un-ornamented. C has a paper cover and the later label. D is a dry mount with the usual form of initials for such preparations, while E is an example of the micro-photographs for which Dancer became so well known.

F to K: slides by J T Norman
F shows the ordinary Norman green paper cover, with the monogram obscured by the lower label. G is an example of the more ornate cover, in red and gold as well as green. H is a later slide with original upper label. L shows the plain label with incorporated monogram at the corners. K is an unusual early paper-covered mount, with the name Norman printed once only next to the opening.

L to P: preparations by Richard Suter
L is a good section of rabbit duodenum carrying Suter's usual label. M is a good horizontal section of a foetal hand: the address is 5 Highweek Road instead of the usual 10. N is another good section, with later label. O is a slide of a popular subject in the 1880s—coal miner's lung section showing many coal particles. P shows a good section of a large specimen, and has a white label whereas all the rest are the usual pink.

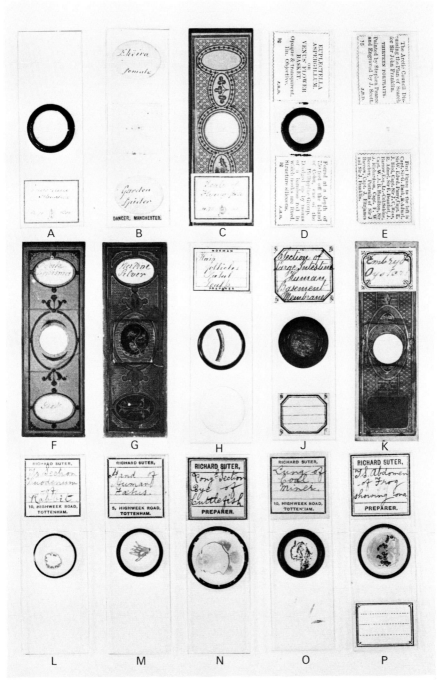

PLATE 45   Examples of commercial preparations (3)

A *to* D: *mounts by Hunter*
A is one of Hunter's earlier mounts, late 1860s, showing spicules. B is a good section of rat skin dated 1876. C, of foetal hand, shows good detail. D is a slide mounted when Hunter had gone into partnership with Sands.

E *to* K: *preparations by the Bourgogne family*
E is an early mount, pre-1852, of T.S. human cranium with manuscript label and signature. F shows the typical Bourgogne paper cover used from about 1852–55: this example is dated 1854. After 1855 the paper covers were abandoned, simple labels only being used. J Bourgogne was the senior member of the family, and G shows one of his mounts (of L.S. human finger-tip) dated 1862. Charles Bourgogne had his own labels, one of them shown at H. H(enri) also had his own, one dated March 1859 being shown at J attached to a very adequate L.S. human canine tooth. The Eugene label is seen at K, on a diatom mount.

L *to* P: *mounts by Beck*
A typical slide of the earliest period is seen at L, with the then usual black paper cover on a balsam mount of T.S. human hair. The slide M was probably bought in, as it is paper covered, unusual for slides from this maker. N to P illustrate three series of preparations from the 1880s/90s: all are of good quality.

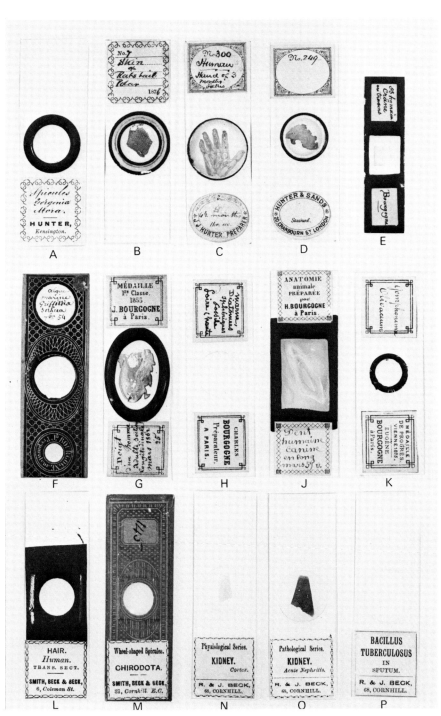

PLATE 46 Examples of commercial mounts (4)

A *to* E: *Baker preparations*
A and C carry the extra labels of the lending department, started in 1898, while B is an early preparation. D is from one of the sets of plant histology slides, and shows excellent mitotic figures. E is a good example of a mount for the amateur—a beautifully prepared caterpillar exoskeleton.

F *to* K: *mounts by Arthur Cole*
F is a diatom preparation of the late 1870s, of the type for which Cole was well known before he started to offer histological slides. G is an interesting triple mount from the 1880s, having three regions of cat spinal cord very adequately prepared. H and K are from the famous series of preparations with accompanying descriptions in his Studies in microscopical science (see page 36). H is from volume 2, and is of his usual high quality of injection: all the capillaries are properly filled, and none has burst. K is from volume 3, and is not of such good quality: the section is folded and torn, although it does show the characteristics of the disease. J is another pathological slide, of meningitis, and of better quality.

L *to* N: *preparations by Martin Cole*
L is from his student series, but of excellent quality: the developing tooth and bone in the foetal jaw section are entirely adequate even today. M is a well-stained section of bacteria *in situ* in tuberculous lung. N shows a good slide of T.S. nerve.

O *and* P: *research preparations by Arthur Cole*
O (labelled 'K' on a typical A C Cole label) shows an ulcerated joint: it is of high quality and stands oil-immersion examination even today. P is dated 1889, and has the typical Cole shield: it is labelled 'human carotid artery tied with fine silk using the first degree of force'. Clearly, Cole acted as consultant for medical research.

PLATE 47 Examples of commercial preparations (5)

A is a Walmsley mount of the 1880s, while B is another United States preparation, by the Essex Institute, made about 1875.

C is a section made by Ward of Manchester in the early 1880s: such pathological lungs attracted much interest at the time.

D and E are rock sections: D is by Murby, of late date and fine quality, while E is by How, and dates from the 1890s.

F is Thum diatom mount in naphthalene monobromide, still in excellent order. G is a typically-labelled preparation from the Naples marine zoology laboratory.

H shows a typical label of Cogit: Streptococcus from culture, about 1895.

J is one of the famous Van Heurck diatom preparations, with silvered mount for reflected light.

K is another diatom mount, by Tempère-Peragallo, dated about 1895.

L is a Möller preparation of about 1885, while M shows a Rodig mount of 1875.

N is a Klönne & Müller cheese mite preparation, dated 1886. O is a Boecker rock section, of about the same date.

P is a mount by Hensoldt, showing insect borings in fossil wood.

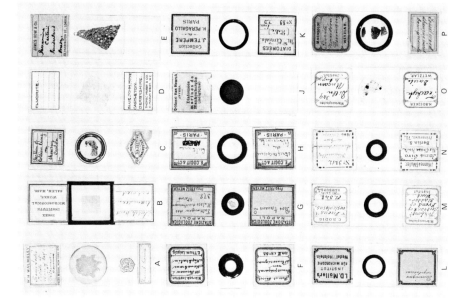

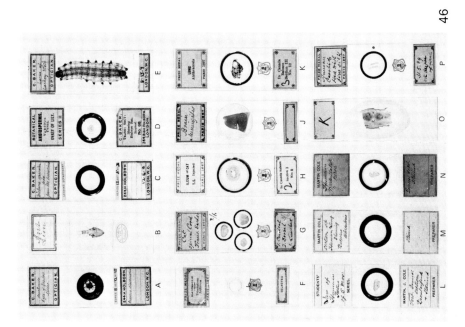

PLATE 48   Examples of commercial mounts (6)

A *to* E: *Wheeler preparations*
A is covered with Wheeler's green paper, while B and C are in the more usual plain red covers. D shows a slide mounted in plainer finish, and E is a Webb engraving.

F *to* K: *mounts from Watson*
F is a Wheeler slide with a Watson address label (compare with B). G is a typical mount for the amateur, of a disarticulated wasp beautifully prepared. H is an unusual Watson address label on a double fluid mount, while J is another Watson speciality—an arranged group of 80 diatoms in the form of a very precise star. K shows the Watson Edinburgh branch label on a good dry mount of a gnat's head.

L *to* P: *slides by Sinel and Hornell*
The Sinel mount at L is excellent quality, the larvae having all their tube feet extended, and both oral and aboral sides being shown. M is a Sinel & Hornell mount with good histological detail. N shows a Hornell slide of T.S. eye of young octopus, with good histological detail in the retina. O is a slide from Hornell's set of marine zoology: it is still good enough to be used in class. P is a preparation from Hornell's general tange, of T.S. human testis.

PLATE 49   Examples of commercial mounts (7)

A *to* G: *Flatters mounts*
A dates from about 1902, by Flatters alone. B and C show the special label used by Flatters to identify his slides: the word 'accurate' was the special one sent out with the sets of subscription mounts about 1910. Slides D to G show a succession of changes of address and company name taking place between 1918 and 1928: G is a specially interesting type slide of 20 diatoms, with accompanying list.

H: a Flögel preparation of 1883, showing carefully-selected diatoms.

J: a Thomas preparation from Chicago, showing diatoms.

K: a Voigtlander slide of human hair, of about 1885.

L *and* M: *Clarke fluid mounts*
L is by Clarke alone, M by the famous partnership with Page.

N *to* P: *histological slides by Sigmund*
P dates from about 1909, while N is from his well-known series issued with notes about 1912. Zeiss took over the sales later, their label being shown at P.

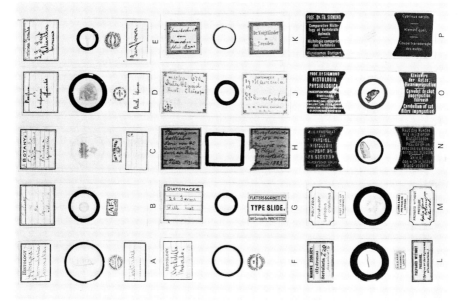

49

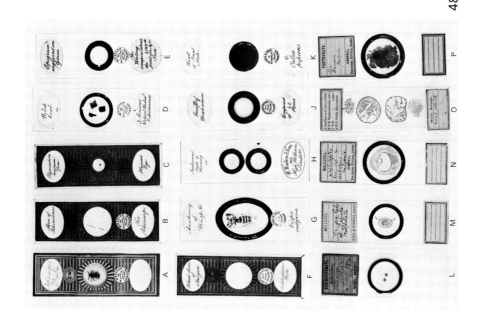

48

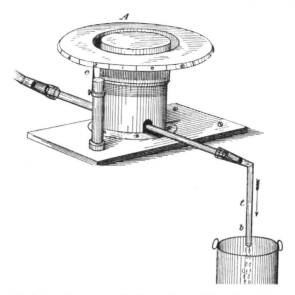

Figure 37. Taylor's microtome, 1882 (from the original description, note 60 in text).
A rather simple microtome has the interesting innovation of a supply of coolant from a remote reservoir.

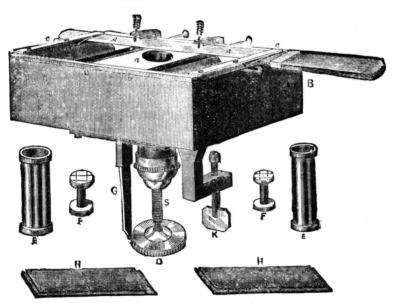

Figure 38. Microtome by Satterthwaite & Hunt, 1882 (from the original description, note 61 in text).
A modification of the Curtis machine (see Figure 59), by addition of an ice/salt bath. The substantial knife guides are a good feature of the machine, and interchangeable supports are provided for extra flexibility.

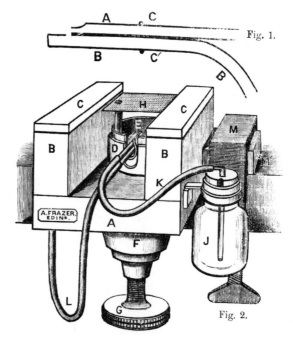

Figure 39. Original Cathcart freezing microtome, 1883 (from the original paper, note 62 in text).
The success of this instrument lay in its careful approach to the design of the ether atomiser, shown enlarged. Tube A brought ether from bottle J, and the small nozzle in tube B blew the liquid directly on to the underside of the specimen platform H. The instrument was inexpensive, and was to become a long-lasting favourite. The knife moved along glass plates C to minimize friction.

Hunt.[61] Based on the Curtis machine (see page 218 below), the large knife in sensible carrier would have been very firm in practice (see Figure 38), but it may have been overshadowed by the introduction of another instrument the following year, one destined to have wide popularity and long life—the Cathcart microtome.[62]

This was designed to use the minimum quantity of ether to maximum effect, and to be of low cost: in fact, it sold for only 15/–d., and would freeze a block of tissue $\frac{1}{4}$ in. cube inside two minutes for one farthing's worth of ether! The original illustration is reproduced as Figure 39. By 1888 the design had been improved to have a bigger head for finer adjustment, two clamps instead of one, an improved freezing-plate, and a paraffin attachment. After a few improvements of details in the intervening years, the final form was the Cathcart–Darlaston instrument,[64] introduced by Watson in 1906. From their catalogue of 1909,[63] we see that the early form was still on sale for 21/–d., an improved form

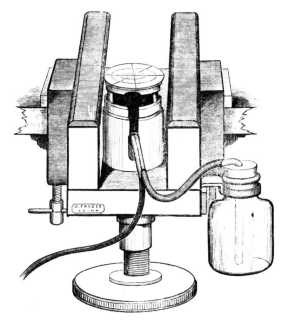

Figure 40. Improved Cathcart microtome, 1888 (from the original description, note 63 in text).
This had an even better atomising chamber and larger micrometer wheel for better control of the advance. It clamped to the bench at both sides, to withstand the considerable pressures exerted in using the machine. A paraffin stage was included in the cost of the instrument, which was listed at only 21/–d. in the 1893 Watson catalogue.

with knife-guides was listed at 27/6d., and the latest modified version at 45/–d. This was the first model to offer an automatic advance.

Many other models of freezing instruments were introduced over the years. In 1884 Gray described his ether model,[66] which was very simple in design and merely provided some guidance for the knife. Boecker's instrument was rather more involved, and not only gave a diagonal motion to the knife, but raised the specimen after each cut.[67] In 1885 Jacobs described his simple instrument, which relied on a large copper rod being suspended in an ice/salt compartment to conduct heat for a long time without further attention.[68] This was potentially a useful device, and he suggested the use of a gum/gelatine mixture with glycerine as the freezing medium, another useful advance.

The instrument of Hayes,[69] of 1887, was much more advanced, based on the use of a large V-section to guide the knife across a micrometer-raised freezing chamber (see Figure 41). The whole design shows careful application of scientific principles to the problem, and the

Figure 41. Hayes freezing microtome, 1887 (from the original description, note 69 in text).
A good design, using a massive cast-iron groove to guide the knife although this is somewhat remote from the specimen. The position of the freezing chamber could be adjusted longitudinally, and the design is quite original.

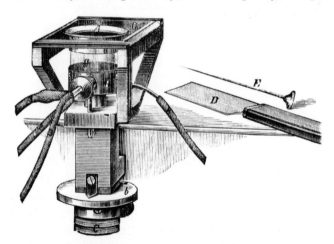

Figure 42. Jung freezing microtome, 1887 (from the original description, note 70 in text).
Stated to be based on the design of Lewis, it is somewhat outmoded for this date, in spite of being the product of a specialist maker.

equipment should have been very satisfactory in use. The Jung instrument of the same year[70] was based on the earlier machine of Lewis, and is a comparatively simple design with a separate knife which would be much less satisfactory in use than the Hayes machine (see Figure 42).

Of massive construction, the Delepine instrument was described only in 1900,[71] being made by Beck at that time, although the design had by then been in use for 18 years. There is evidence that Cook of Manchester had made earlier versions, one of which is still preserved in the department of physiology at Edinburgh.[72] The machine was thoroughly practical and of great rigidity, with clamps for both ends of the knife (see Figure 43); it was further improved the next year by being given a longer knife-carrier and rails,[73] making it one of the most satisfactory patterns ever made. Unfortunately, by 1901 the German manufacturers had a virtual monopoly of large microtomes, and it is unlikely that Beck's production runs were very large.

Figure 43. Delepine instrument, 1900 (from the original description, note 71 in text).
Of massive construction, and developed over 18 years, this instrument was one of the most advanced ever offered by a British manufacturer. One example handled proved most satisfactory in use.

Two instruments of interest for their approach to the problems of design should also be mentioned, both described in 1888. Dale's freezing microtome[74] was an attempt to provide an advance mechanism which would leave both hands free. This was achieved by attaching a cord and pulley to a ratchet which worked a piston in the usual Ranvier style, the rate of advance being determined by a locknut. Otherwise, the instrument was an ordinary ice/salt type. A similar, but larger, bath was used by Bruce in his design for a freezing instrument which would section an entire brain.[75] The principle was to have a large platform over the bath, connected with it by a number of cruciform-section pillars to ensure rapid conduction of heat from specimen to bath, to allow of relatively rapid freezing in spite of the large size of the object. The blade moved over the specimen on a plough guided by grooves, the angle of tilt being altered to lower the height of the edge. Overall, the equipment measured 22 in. by 12 in., the edge of the blade being 9 in. wide. There is no doubt that this would have been a thoroughly serviceable instrument, and it was marketed by Frazer of Edinburgh.

In 1901 a radically new departure in freezing microtomy was introduced, when a cylinder of liquefied carbon dioxide was used as a source of refrigerant,[76] a method still adopted today on account of its efficiency. The device (see Figure 44) was very simple, fitting directly on the nozzle of the cylinder, and being really only a valve with specimen support above and knife rails alongside. The specimen plate contained a spiral tube through which the gas escaped, giving very efficient cooling, but the lack of proper guidance for the knife made the machine more suitable for rapid diagnosis than original research.

Another new coolant, ethyl chloride, was proposed in 1903:[77] this was sold in glass tubes provided with thumb-operated valve, and was intended as a local anaesthetic. Katz suggested its use on tissues already hardened by formalin, but it would have been an expensive method for routine. The idea is still used today, with an aerosol can of liquefied refrigerant intended for occasional use.

Two other designs of freezing instrument should be mentioned, as examples of very different approaches followed early this century. In 1904 Osterhout published[78] a description of a very elementary freezing instrument (see Figure 45) based on a design of 1896. The special feature was the remote freezing bath, and a heavy iron blade with much attention to its sharpening being given in the paper. Ether or carbon dioxide cooling could be adopted in addition, but the design was outmoded when it was described.

A very different approach was that adopted by Krause in 1908.[79] He

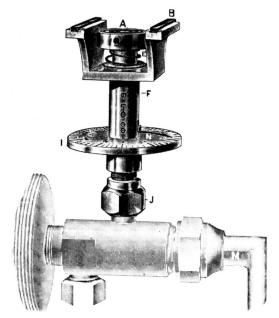

Figure 44. Carbon dioxide microtome by Bardeen, 1901 (from the original paper, note 76 in text).

Liquid carbon dioxide was vented through spiral ways in the rather crude specimen platform of the device, giving rapid cooling but little control of the blade, making it more suited to the needs of the operating theatre than the laboratory.

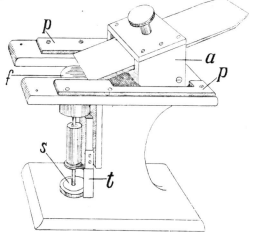

Figure 45. Osterhout instrument, 1904 (from the original paper, note 78 in text).

Coolant was brought from a remote bath, the specimen being cut by a specially sharpened blade. Designed for cytological work, the design is curiously outmoded compared with the sophisticated instruments available in 1904.

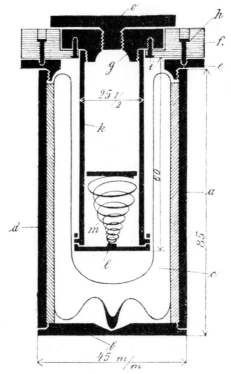

Figure 46. Krause solid carbon dioxide specimen platform, 1908 (from the original paper, note 79 in text).
a is the metal container, interchangeable with an ordinary paraffin block holder, inside which is a vacuum flask *c*. The spring-loaded internal platform pushes the solid carbon dioxide against the under side of the specimen platform *o*. This design has a great deal of merit.

used a vacuum flask in the base of a cooling platform, to hold solid carbon dioxide in contact with the plate carrying the specimen. The solid was got by allowing the gas to escape from a cylinder, and proved very satisfactory in use, the author claiming that he could obtain one hour's cooling for one penny (see Figure 46).

Other instruments were available for cutting frozen sections as an alternative to embedded ones. One such example was the so-called polymicrotome of Hailes, to be discussed in the appropriate later section. Practical trials of the freezing method have confirmed that many earlier sections cut this way must have been unsatisfactory, at least for high-power work. The average time taken to freeze in a Rutherford machine is far too long to avoid the formation of large disruptive ice crystals in the cells. With an ether machine and small pieces of tissue, the situation is rather better, as practical experiment has demonstrated. This was probably not apparent at the time of the popularity of the freezing process, although the need to cool paraffin blocks as rapidly as possible, to avoid large crystals of wax which disrupted the cutting, was reported in 1888.[80] The more rapid cooling produced by use of carbon dioxide would have been desirable in this respect, while the modern aerosol spray gives very rapid cooling and little disruption of the cell contents.

For sheer versatility, however, the usual infiltration techniques cannot be surpassed: they alone allow accurate serial sectioning. The freezing method, in spite of its great convenience and speed, did not live up to the ideals put forward by its popularizer, Rutherford.

The hand microtome

We saw on page 122 above that this instrument was first designed by Ross; popularized by Ranvier, it has survived to the present day for students of botany, and little change in its design was possible or has occurred. Its main advantage is its cheapness, for it is not notably easy to use, there being no positive control of the knife which tends to jump when it meets the specimen. In its development after 1870 there were those who preferred it not only because it was cheap, but because they thought that nothing more complicated was ever required. For example, Bird, who was secretary of the Medical Microscopical Society at the time, read a paper[81] in which he put forward the merits of embedding in pith as a corrective against those who felt that only wax could give adequate results. Pith was the only material used in the laboratory of Ranvier at the College de France, and sections of spinal cord obtained with its aid had been especially good. The Ranvier microtome, with a smaller well than the English designs, was the only kind of section instrument in use there. Bird certainly gave a convincing account of pith and its merits in that paper, remarking that it does require a microtome for good results and that many microscopists seemed to have an innate hatred of that instrument. He went on to say that there was much more in the skill of the operator than in choice of embedding medium, and in 1874 that was no doubt perfectly true. In any case, the paper is valuable in warning us against assuming that the microtome was widely or rapidly accepted, in spite of the later proliferation of designs.

Some attention was given to the hand microtome, however. Schiefferdecker[82] published a description in 1876 of a revolving-top instrument strongly reminiscent of the Smith design of 17 years earlier. An improved specimen clamp, acting like a three-jawed drill chuck of

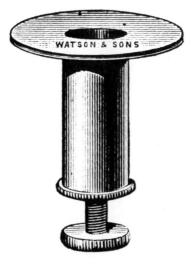

Figure 47. A hand microtome of 1909 (from the catalogue, note 65 in text). Offered at only 5/–d. for botanical work, and lacking a clamping screw for the specimen, this instrument was virtually identical with the design of half a century before.

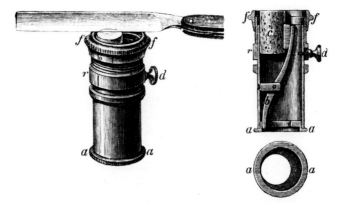

Figure 48. Reichert's hand microtome of 1885 (from C Reichert—*Illustrated catalogue and price-list of microscope stands, microtomes, etc.* (Vienna: Reichert, No. XI, 1885)).
The specimen was clamped in place, and the top plate lowered to meet it, which is a good design. The eccentric ring (*a*) operated the internal clamp, allowing rapid interchange of specimen which was embedded in cork or pith. A weak point is that there was little support for the knife.

the present day, was described by Fiori in 1899.[83] Figures 47 to 50, and plate 33 illustrate a range of hand microtomes, including one by Leitz which survived unchanged from 1907 to 1936. All the main suppliers offered them throughout the last part of the nineteenth century, the only variation being in the size and finish.

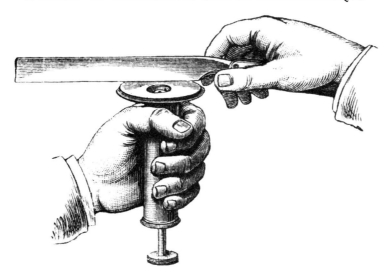

Figure 49. A hand microtome in use, 1892 (from C Reichert—*Mikroskope 1892*. (Vienna: Reichert list no. 18, 1892)).
The illustration seems a very awkward representation, as little accurate pressure could be exerted with the razor held as shown.

Figure 50. Leitz hand microtome, 1907 (from E Leitz—*Microtomes*. (Wetzlar: Leitz, 1907)).
Compared with the Watson version at 5/–d., this one at 15/–d. is certainly of superior construction. The top plate is of glass, and the micrometer head is large, with divisions of 0·01 mm. The clamp moved with the specimen, and the whole design is first-rate. Such good quality instruments for hand sectioning are no longer available, although this particular version survived to be offered in the 1936 Leitz list (No. 53 F/en), at 44/–d.

Microtomes 1870–1910 : (a) movable knife with vertical feed

We have seen that although a number of workers preferred the hand microtome, others realized the advantages of clamping this to the bench to free both hands to control the action of the knife, and this is the basis of an extensive and important group of microtomes. In this subsection it will be convenient to divide instruments with moveable knife and vertical feed into four groups: with separate blade; with attached knife but manual feed; with attached knife working in a groove; and with attached knife automatically advancing the object.

The first subgroup, with movable blade not attached to the specimen-raising system, was effectively launched by the Pritchard instrument described on page 117 above, but brought to popularity by the Stirling section cutter. The pattern remained available for many years, examples being the Cole pattern and the Army Medical Museum pattern shown in Figures 51 and 52, from an 1890 list.

Others tried other forms of the instrument; for example, Hailes suggested a modified version of the Topping machine (his own words) which took the object clamp up with the micrometer screw, thus obviating any jerking in the advance, a useful device but not original, as we have seen.[84] The top plate carried two hardened steel strips to support the blade, and an attachment allowed use of a saw for tooth and bone sectioning. The machine was available commercially from Baker, who offered it for 45/–d. in his 1887 catalogue: it was still being offered at the same price in the 1899 list, but was thereafter omitted.

Figure 51. Stirling-type bench microtome, 1890 (from the Queen catalogue, note 15 in text).
Typical of elementary section instruments, this device sold at $8.00.

Figure 52. Large Stirling-type bench microtome, 1890 (from the Queen catalogue, note 15 in text).
This is the Army Medical Museum pattern, as used by Woodward, and incorporating arrangements for taking up wear.

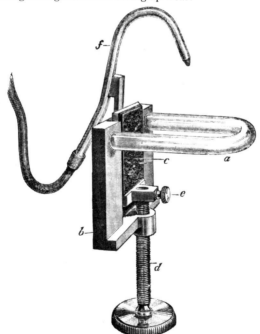

Figure 53. Roy's microtome, 1879 (from the original paper, note 85 in text). The glass horseshoe which guided the knife is labelled *a*; the tube *f* was for dripping liquid on to the knife. The design is very elementary, and would have been suitable only for very simple sectioning.

In 1879 Roy[85] suggested another very simple instrument, shown in Figure 53. The blade was slid along a glass horseshoe through the suitably embedded specimen: it is difficult to see how the instrument could have been made sufficiently robust, or how the glass could have been flat enough to allow even passage of the blade. Another instrument which seems to have offered little advantage over other models of its day was the so-called polymicrotome[86] by Hailes (not the same Hailes as in note 84). It was adaptable for both embedding and freezing, and one good point was its provision of a rapidly-set regulator for limiting the throw of the raising lever, thus allowing easily repeatable lifts. It was claimed that 200 sections per minute could be cut with this machine (see Figure 54).

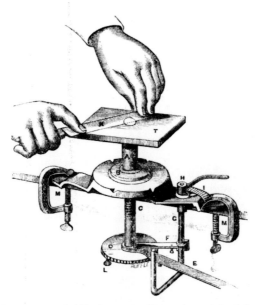

Figure 54. Polymicrotome by Hailes, 1880 (from the original description, note 86 in text).
The essential feature of this instrument is the lever (E) which raised the specimen. It could be tapped quickly between sections to raise the specimen by a fixed amount, determined by the setting of the stops (G).

Some models stood on the bench, rather than being clamped to it. The Pearson–Teesdale instrument was one such, which stood on three legs and adopted[87] the Schiefferdecker principle of keeping the object tightly clamped while lowering the blade towards it (see Figure 55). In addition, the object clamp was of the three-jawed type, and in retrospect the design seems to have been good, but there is nothing to show that it

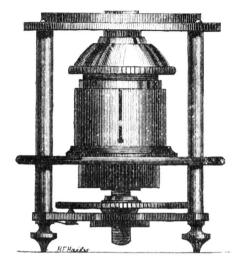

Figure 55. Pearson-Teesdale instrument, 1880 (from the original description, note 87 in text).
Notable for the three-jawed specimen clamp, and for lowering the knife to allow the sections to be made without releasing the specimen, this is a very serviceable design of a basic microtome.

was adopted commercially. Another instrument intended to be sat on the bench was marketed by Zeiss,[88] a specimen of which has been inspected (see plate 36). It was well made to a suitable design, and would have been serviceable in use.

Some tried to make what was essentially a large Stirling cutter big enough to section an entire human brain, and probably the first to do this with any degree of success was Gudden,[89] in 1875. He mentioned in his paper that he had looked at various other designs, including one by Betz,[90] and outlined the considerations for cutting an entire brain. The microtome thus designed stood on three legs and was made by Katsch of Munich. It was essentially an enlarged instrument arranged to cut under water (necessary to prevent tearing of the section), and with a long knife, necessary to cut such a large organ as the brain. By 1882 the instrument had been given four legs, and was offered commercially[91] (see Figure 56). There was at that time a great deal of interest in sectioning the entire brain. We have already seen something of this in connection with the freezing microtome, and shall see more when other forms of microtome are discussed. In 1882 Deecke[92] produced a large microtome for the same purpose, using a brass cylinder 9 in. diameter and 14 in. high, inserted in a copper basin to allow cutting under spirit; the specimen was attached by pieces of cork before being embedded in a

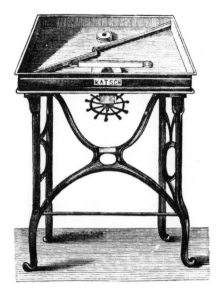

Figure 56. Katsch's large microtome, 1882 (from the original description, note 91 in text).
This is essentially a large simple microtome placed beneath a shallow tray of water in which the long knife worked, and is the commercial version of the instrument designed by von Gudden, used to make probably the first series of sections of the entire human brain.

wax and oil mixture, in much the same manner as that suggested by Betz and others long before. The wax cast was removed as cutting proceeded with a 16 in. blade using sawing cuts—the author stressing that this required some practice! The loss in a series of sections should not exceed two or three per cent; they were floated from the basin in which they had been cut, onto writing paper, and mounted on strips of glass up to 8 in. by 10 in., as required, in a manner strongly reminiscent of Gudden.

Other instruments of more conventional size and less heroic operation were also made, such as that offered by Bausch & Lomb in 1891.[93] This design was quite adequate, and was made as a result of the firm finding that some of their earlier and more complicated instruments were selling rather slowly. Somewhat similar was the instrument[94] made by Flatters, which offered the extra feature of a swivelling top plate to allow it to be made with a slight reverse taper to prevent the wax turning during cutting, always a great annoyance with these models (see Figure 57). Remarkably, perhaps, the same instrument with very minor modifications was still on sale from the successors to the company in the 1960s (see Figure 58).

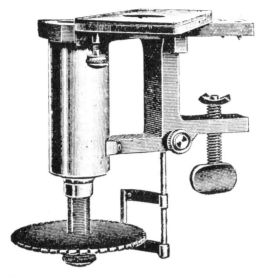

Figure 57. Flatters' section instrument, 1905 (from the original description, note 94 in text).

The large-diameter milled head and the click-stop, plus the swivelling top plate with tapered aperture, were good features of this design by a practising microtomist.

Figure 58. Flatters & Garnett microtome, 1955 (from Flatters & Garnett catalogue, 1955).

Now made from one casting, but exactly the same in principle as the instrument sold over 30 years before. In use the machine is perfectly effective when used with pith- or carrot-embedded material.

The second subgroup, with knife attached, offered greater security of action. The Curtis design of 1871[95] was good in its way, as the knife-frame held the blade just clear of the glass top plate while increasing the rigidity of the edge (see Figure 59). The author stated that he needed an

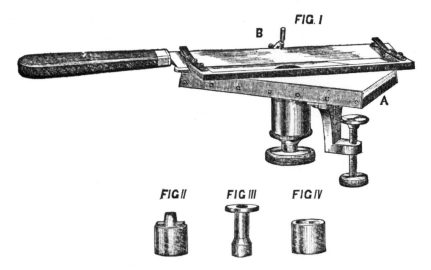

Figure 59. The Curtis microtome, 1871 (from the original paper, note 95 in text).

Essentially a simple microtome provided with a large knife platform to carry the substantial blade just clear of the glass surface, in a pivoting action. The three types of interchangeable specimen platform are also shown. This was a good earlier design, capable of making good sections of the eye.

unbroken section of considerable extent, that he needed many sections from scarce material, and that he also required sections of a whole organ: his design was intended to provide all these. He prepared whole human eyes by immersion in Müller's fluid for at least three weeks, then opened them and soaked in water for two days before hardening in alcohols before soaking in clove oil. Most interestingly for the time, and possibly following Stricker's advice (see page 80 above), he then put the tissues in paraffin wax thinned with about 8% of turpentine or benzene, kept just above the melting point for a few minutes until the excess oil of cloves was rinsed off; after this the specimen was put into a cavity melted in a block of wax. This was very close to true infiltration, and it is surprising that Curtis did not arrive at that technique. This paper is of much interest in the history of embedding, but seems never to have been discussed in this context.

Hoggan's microtome of 1874[96] was an unusual and original design, based on provision of guides for a saw. In talking of the design of his machine, Hoggan said a few words which are significant:

... now-a-days, however, anyone desirous of obtaining a name in histology has only to add a screw [to an existing machine] here, or a plate there, or take away the same, and henceforth the machine is called by his name, and the histological laity who examine the parts of the completed machine, gape in wonder at the ingenuity of the individual whose name it bears.

In spite of this acute observation of human weakness, Hoggan's own machine was only for hard materials such as teeth: he claimed that he could make twelve sections of bone ready for mounting within 40 minutes. This was indeed a proud claim at that time, for the more usual procedures were far more time-consuming. A practical point made in this paper is that when cutting tissues with a saw in this manner, the teeth should be directed back towards the operator, and this does seem helpful in practice. The instrument is shown in Figure 60.

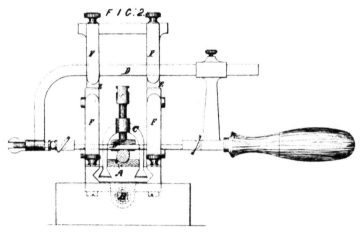

Figure 60. Hoggan's section-cutter, 1874 (from the original paper, note 96 in text).
The frame guiding the saw is the main part of the instrument, and the original paper gives important details of choice of saw and treatment of specimen, of essentially very hard nature.

Seiler's microtome of 1879[97] is strongly reminiscent of the Curtis design, but was intended to allow the blade to follow a slightly curved path to facilitate making larger sections: the author claimed that material 2 in. by $\frac{3}{4}$ in. could be sectioned without difficulty, no mean feat at that time (see Figure 61).

The Swift instrument,[98] marketed in 1882, was the outcome of a different approach to the design, relying on a large cast-iron pillar/clamp to support a Stirling-type cylinder, and to provide rails carrying a brass knife-frame. As will be seen from the illustration in Figure 62, the instrument could be used for freezing as well as embedding: the whole design combined rigidity with cheapness and convenient size.

An entirely different approach was adopted by Roy,[99] who used a

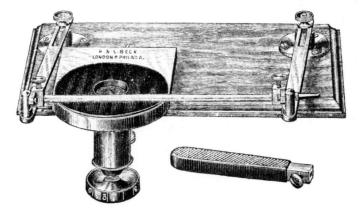

Figure 61. Seiler's microtome, 1879 (from the original description, note 97 in text).
A simple instrument fitted to a baseboard carrying knife guides which minimized lifting of the blade on contacting the specimen.

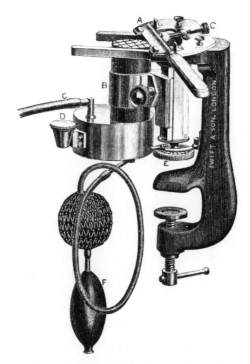

Figure 62. Swift's microtome, 1882 (from the original paper, note 98 in text).
Seen here in its freezing mode, this good design used a heavy casting as a frame to carry guide-rails for the knife and form a rigid support for the whole.

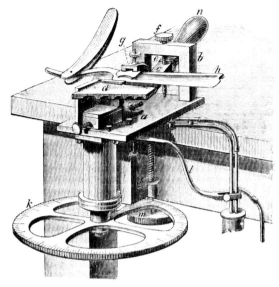

Figure 63. Roy's microtome, 1881 (from the original paper, note 99 in text).
Shown in ether-freezing form, the features of the design are the large-
diameter micrometer wheel, and the pivot (*f*) about which the blade moved
to make a controlled section. The angle of the knife could be controlled by
leather packing (*g*).

knife which pivoted about a vertical axis to cut either frozen or
embedded objects. The very large diameter micrometer head is seen in
Figure 63, but the design did not hold the knife very rigidly, as
inspection of an example has shown. In the same year, Boecker
described an instrument which carried the knife in a slide which was
itself carried on a further slide at right angles. This arrangement was
intended to give to the knife the same movement as that of the hand in
freehand sectioning, this being thought advantageous.[100] Whatever the
merit of this belief, there is no doubt that the blade was very well
supported, as seen in Figure 64.

Occasionally an instrument was designed by a group rather than an
individual. This was so with the Providence microtome, called after its
place of design. It featured a heavy knife carrier, of brass, but with only a
very simple cylinder below. Attention had been given to the design of
the knife, and as will be seen from Figure 65, there would have been
little likelihood of its springing in use. The claim that the instrument
would cut sections of uniform 10 μ thickness was probably justified.[101]
From this design, one of the original group, King, developed an
improved model, by addition of a larger head to the raising screw. The
knife carriage was made of iron instead of brass, and an extra knife was

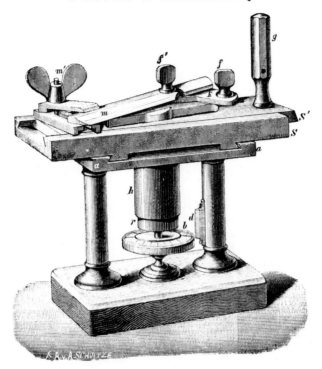

Figure 64. Boecker's microtome, 1882 (from the original paper, note 100 in text).

Evidently a Zeiss-inspired basic bench microtome with addition of a well-designed knife carrier. The two slides at right angles were intended to guide the knife in a manner reminiscent of a freehand slice, but much more important is the rigid clamping of the knife; this was a factor often overlooked at the time this design appeared, when it was assumed that any razor would suffice.

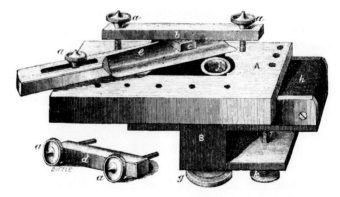

Figure 65. The Providence microtome, 1885 (from the book, note 101 in text).
This featured a heavy knife carrier, sliding on a substantial cast-iron bed.

supplied to deal with harder material.[102] The ratchet advance was arranged in one model to give a rise of 1/10000 in. per click, and five times as much in a cheaper version. The instruments were sold by King himself,[103] who had been in holy orders at the time of the Providence version, but who by 1893 was in business as a supplier of microscopical requisites (see Figure 66).

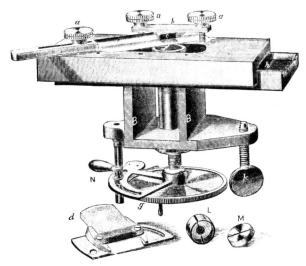

Figure 66. King's microtome, 1889 (from the original paper, note 102 in text).
A development of the Providence design, the knife carrier was of iron instead of brass, and provision had been made for finer advance of the object, with rapid-action preset lever. *d* is a stronger blade for tougher material, while *l* and *m* are alternative object-plates.

The Zeiss list for 1889 offered a microtome 'after Francotte',[104] with knife worked in a metal carrier: there seems to be no reference to this instrument in any description by Francotte himself, but it seems a serviceable design. By this date it was Zeiss policy to supply only relatively simple microtomes (see Figure 67):

> Of late years the more elaborate microtomes have undergone such manifold changes and have become so complicated, the ideal form moreover being a matter of opinion, that only workshops which devote themselves to this special branch under the guidance of experts can hope to manufacture them with any success ... therefore for the present we shall only make out two simple microtomes ... [page 92].

An entirely different approach was adopted by Beck for their design of 1892,[105] in that strong parallel rails were provided to carry the knife at both ends, in newly designed clamps which allowed the tilt of the blade to be adjusted. The object holder was mounted on a ball head, to allow of orientation, and the whole was raised by a micrometer screw at one end, to give a very serviceable and accurate instrument for its day (see Figure 68).

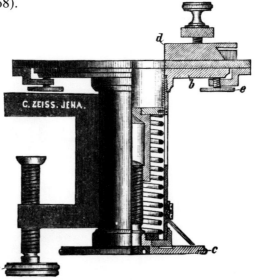

Figure 67. Francotte's microtome, 1889 (from Zeiss catalogue No. 28, 1889). Essentially a simple microtome with attached metal knife carrier, this was one of only two designs offered by Zeiss at the time, as they felt that specialist firms were needed to make more advanced instruments.

Figure 68. Beck microtome, 1892 (from the original description, note 105 in text).
Much attention had been given to properly supporting the knife in this good design. The object-clamp was on a ball-head, and the design seems most serviceable.

The need for accurate control of the knife, even in relatively simple designs which did not incorporate automatic advance, had been felt on the Continent also, and the Fromme design of 1891[106] relied on a long pivoted arm to carry the curved blade in an arc; Schaffer stressed the difficulty involved in making perfectly regular guide-ways for other patterns such as the Thoma, but that the Fromme system achieved perfectly regular sections. An interesting fully-orienting object-holder is also described in this paper, and there is no doubt that the design was a good one based on full rethinking of the mechanical principles involved (see Figure 69).

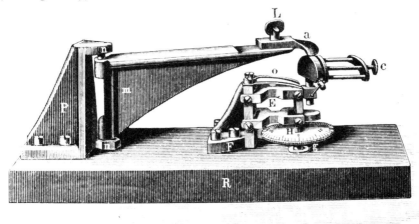

Figure 69. Fromme's instrument, 1891 (from the original paper, note 106 in text).
The long pivoted arm carrying the knife obviated the difficulty of accurately machining long grooves, but did require much accuracy in making the knife. There was a more advanced object-clamp available also.

Another original design was that of Yankawer,[107] who adopted an arc motion but in a different manner in a much smaller machine. As will be seen from Figure 70, the instrument might owe something to the earlier Swift tripod knife carrier used on the Williams freezer, but with larger spread and thus finer adjustment. It ran on a glass plate, but with its very basic object-holder could not have been intended for any other than elementary work.

The final development of the arc motion of the knife was described by Thate in 1900;[108] he had worked on the design since 1888. The straight blade was moved in a circle, carried on a pivot moving on two steel knobs to achieve an essentially tripod arrangement, with the central leg a ball-and-socket. The instrument was designed for cutting under water

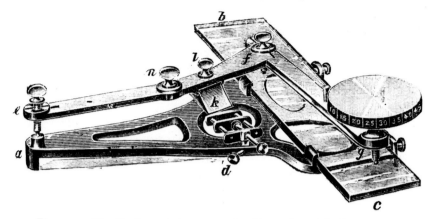

Figure 70. The Yankawer microtome, 1897 (from the original paper, note 107 in text).

A tripod blade-holder, with one leg threaded to lower it towards the object, and sliding with another across a glass plate, offered an interesting approach to microtome design; the specimen clamp, however, was so rudimentary that it could have been intended only for the least demanding work.

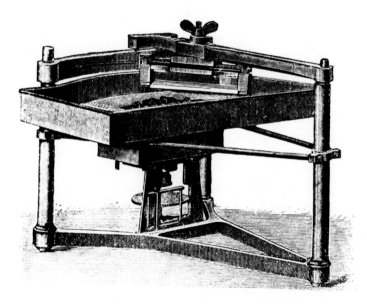

Figure 71. Thate's instrument, 1900 (from the original description, note 108 in text).

The design had been developed from 1888, and the figure shows an earlier version. It is essentially a simple microtome working in a semicircular spirit dish, which had to be raised by screw: this must have been a considerable drawback, causing much wear on the thread and much reduced accuracy. Later models obviated this difficulty.

or spirit, but the earlier form at least could not have been very accurate in its advancement of the specimen, as the screw had to raise the entire dish of liquid as well: a later form obviated this major drawback (see Figure 71).

No other models of the simple microtome with attached knife need be discussed, for all the basic designs which appeared have been mentioned, and the instrument in this form was not regarded by serious workers as meriting their attention. We may now pass to a small but important third subgroup, those instruments which had a screw rise for the object, without automatic advance, but with a knife working in a groove.

This subgroup was to achieve much use in histology laboratories. The design was first described by Körting[109] in 1880, and was marketed by Zeiss. Körting stated that it was suggested by Professor Lichteim as a desirable improvement on the Leiser form (see page 241 below). The design was thoroughly serviceable, and the relatively wide separation between clamp and slide, which may seem to introduce more possibility

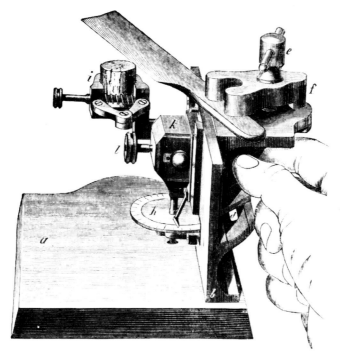

Figure 72. Korting microtome, 1881 (from the Zeiss catalogue of 1881). This is the original design of a microtome having the screw-thread advance coupled with a knife moving in a V-groove. The instrument would be the basis of many advanced models during the following twenty years.

of play to modern eyes, was deliberate, to allow a vessel to be put beneath to catch drops without wetting any other part—noticed as being especially desirable for those who worked in the house! Figure 72 shows a catalogue illustration and photographs of two interesting examples are shown in plates 37 and 38.

Three other versions of the basic design are interesting. The first is a simplified pattern for student use, marketed by Bausch & Lomb in 1885.[110] This had relatively limited adjustments, but was made from one casting for rigidity, with knife carriage moving on five points and being sprung against a flange for extra steadiness in action. Attachments were provided for ether freezing, and for using other knives, and the design was certainly meritorious, as shown by Figure 73.

Crank-drive was incorporated in the Leitz 'support' design of 1889,[111] to give smooth and rapid operation (see Figure 74). Use of chain drives became increasingly popular in the 1890s, as instruments became larger and capable of cutting sections at a high rate. An example of a very large instrument is that described by Schultze in 1893,[112] for making sections of whole organs or regions of the human body. The knife-slide was 80 cm long, and the knife itself measured 53 cm by 9 cm.

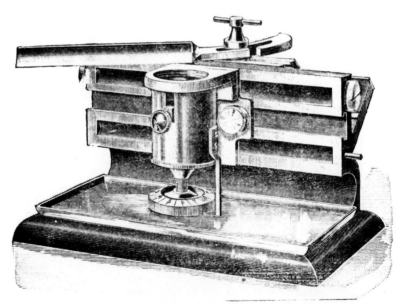

Figure 73. Bausch & Lomb student microtome, 1885 (from the original description, note 110 in text).
A simplified version of the Korting design for student use, this is a well-conceived design based on use of one casting. The knife carriage moved on five points (after Thoma), and was sprung for extra steadiness.

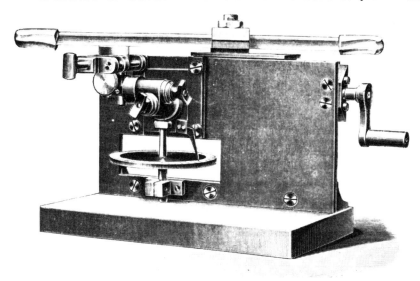

Figure 74. Leitz 'support' microtome, 1889 (from the original description, note 111 in text).
An early example of crank drive to the knife, but without an automatic advance of the specimen.

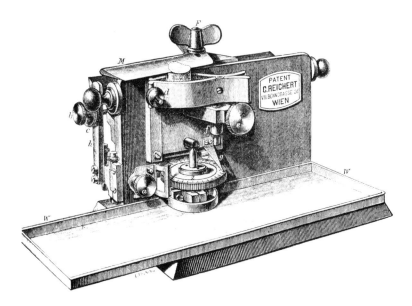

Figure 75. Reichert automatic microtome, 1884 (from the Reichert catalogue, 1884).
This was the first design based on the Korting pattern to incorporate automatic advance of the specimen on the backward stroke of the knife.

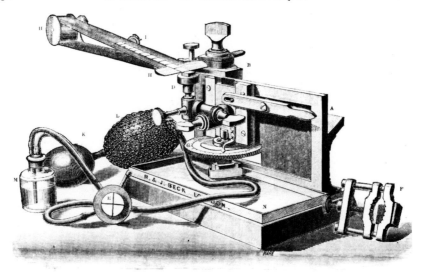

Figure 76. Beck universal microtome, 1885 (from the original description, note 114 in text).
An excellent design, shown here in its ether-freezing mode. The orientable object-clamp and ribbon to remove the sections in order were optional extras. A crank could be fitted to work the ribbon in synchrony with the knife.

A whole brain could be sectioned at 50 μ, and the machine required two operators—one to move the knife and the other to manipulate the section. It also took two strong men to move it.

We may now consider the last subsection, instruments incorporating automatic raising of the specimen. This was a very popular design which seems to have been initiated by Reichert in 1884[113] (see Figure 75), but their lead was soon followed by other makers, such as Beck, who incorporated a revolving band to catch the ribbon as it came from the knife[114] (see Figure 76). The instrument could be used for freezing, embedding, and for unembedded objects: the holder was fully orienting, and a crank handle could be supplied to provide continuous motion to knife and band. This design was well thought out, and it is interesting that an English firm supplied so advanced an instrument; when another decade had passed, the initiative would lie with the Germans.

Other firms offered similar patterns, such as that of Bulloch in Chicago,[115] another apparently serviceable design and one of the most advanced on offer in the United States. The maker did admit that he had taken the best points of German and French instruments in making his own!

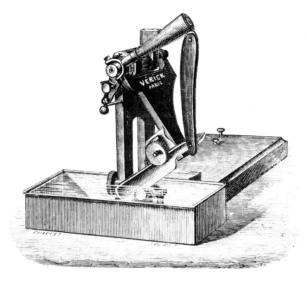

Figure 77. Malassez-Roy microtome, 1884 (from the original paper, note 116 in text).
Shown here arranged for cutting under spirit, simply achieved by tilting the instrument through a right angle, a very convenient method. The knife moved down towards the fixed specimen. Examples of this design have proved very smooth and accurate in operation.

Figure 78. Schiefferdecker's microtome, 1886 (from the original paper, note 117 in text).
Of massive construction, crank-operated for speed of sectioning. The design is similar in principle to the original Reichert instrument of this type, but has many detail improvements—the secret of efficient and repeatable operation.

The Roy microtome received attention from the French maker Vérick in 1884/5, resulting in its being called the Malassez instrument.[116] The rotation of the knife was retained, as was its descent towards the fixed specimen, and the whole instrument had the advantage that it could be placed at right-angles to its normal position to allow for cutting under spirit. This was a very neat arrangement, much less cumbersome than the usual approach to cutting under spirit (see Figure 77). Two examples of this design have been used, and their ease of operation and precision of movement were impressive; the only bad point is the use of an ordinary razor, which must allow flexure.

No flexure was likely in the Schiefferdecker design of 1886,[117] illustrated in Figures 78 and 79. This was a heavy stand incorporating glass plates as the slideways, and an object holder with controlled movement in two planes. The machine could also be provided with an attachment for cutting under spirit, and it was crank-operated for speed.

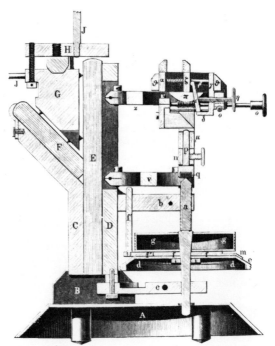

Figure 79. Schiefferdecker's microtome in section, 1886 (from the original paper, note 117 in text).
As an example of attention to detail, this sectional drawing is very revealing. The knife carrier (G) moves on five points, against glass slabs (at E and F) for complete smoothness. Generous size of parts would ensure long and accurate life.

A different mode of action was used by Miehe,[118] whereby the knife was worked by a powerful lever action: the knife was substantial and held in a strong clamp, and the instrument would cut even tough tissues with ease (see Figure 80). Many of these instruments had the defect that the specimen could be raised by only a short distance before it had to be reset, with attendant possibility of loss of one or more sections. Thoma produced a design[119] which overcame this defect, and allowed a lift of 3 cm without resetting. Other workers produced other modifications, one of the most unusual, and possibly most misguided, being that of Strasser, who went to enormous trouble to make the sections adhere to the underside of a paper roll pretreated with a gum arabic mixture:[120] the equipment was otherwise unremarkable (see Figure 81). A further, larger, version was described by him in 1892, which allowed sections of objects 10 cm broad by 15 cm long to be cut serially to a depth of 6 cm[121]: this was a considerable technical achievement.

Two microtomes by Jung were described in 1892,[122] using a swivelling bar to carry the knife (see Figure 82), and automatically raising the screw. This was a serviceable instrument, many examples of which survive to attest its wide sales.

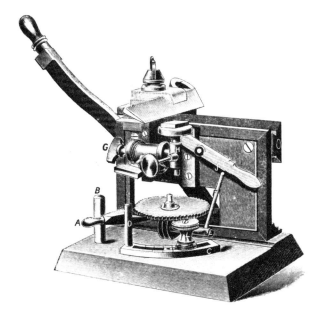

Figure 80. Miehe microtome, 1889 (from the Miehe catalogue, note 118 in text).
A strong design, operated by the prominent lever, and capable of cutting difficult connective tissues rapidly and accurately.

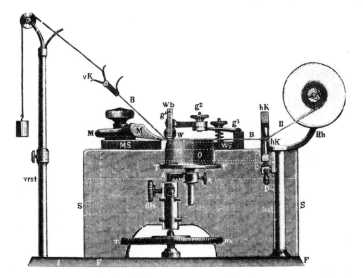

Figure 81. Strasser's paper-ribbon microtome, 1890 (from the original paper, note 120 in text).

Basically an automatic microtome of the Reichert pattern, but with attention lavished on making the sections adhere to a specially-treated paper roll (shown on the right of the illustration), which served as a temporary slide.

Figure 82. Jung student instrument, 1892 (from the original paper, note 122 in text).

Using a simple pivoting knife carrier which automatically advanced the specimen, the design was cheap to make and stood up to hard and inexpert usage. A freezing stage could be substituted for the simple paraffin stage shown in the picture. Six examples of this design have been examined, and all were still very serviceable, as nothing can get out of order.

The basic design of Reichert was modified by the same firm in 1893,[123] to make an instrument of large size capable of cutting whole brains under water. The slide was 50 cm long to allow a knife of edge 38 cm to be used, cutting at least 300 sections without adjustment for height. Other makers offered their own variations of the original Reichert design of 1884, further attesting its popularity: one such example is that of Koritska, of Milan, who by 1894 offered manual and automatic versions, as well as freezing instruments and a variety of accessory equipment.[124]

The approach adopted by Bruce, for cutting brains under spirit, was a little different: he concentrated on supporting the blade at each end while allowing its angle to be altered without altering its elevation.[125] The central well (see Figure 83) was designed to economize in spirit, for in it worked the 12 in. long knife on the wedge-slide which moved in a groove. The instrument was supplied by Frazer, and it was already in use in four universities at the time of the description.

Support of the knife was also a preoccupation of Beck, who described a new system, the Beck-Becker, in 1897.[126] This held the blade on two parallel arms operated by a lever motion. The object clamp was universally adjustable in three planes (see Figure 84), and the motion of the knife across its glass plate must have been very smooth and certain.

Figure 83. Bruce's brain microtome, 1895 (from the original paper, note 125 in text).
A large screw-operated microtome working in a shallow dish of spirit, with the knife supported at each end. The treadle-action advanced the specimen after the knife had passed.

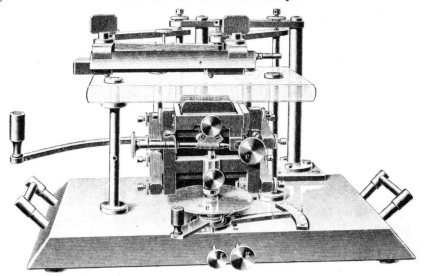

Figure 84. Beck-Becker instrument, 1897 (from the original paper, note 126 in text).

A specially-designed knife-carrier operated by a lever, and moving on plate glass was the basis of the design. This is an advanced instrument, and there is no reason to doubt its authors claims for fine accuracy.

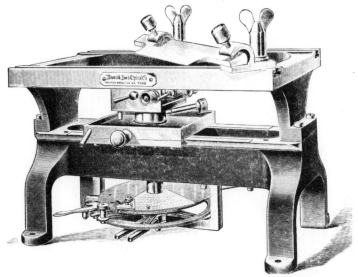

Figure 85. Minot sliding microtome, 1901 (from the original paper, note 127 in text).

It is puzzling to understand why Minot should have designed this instrument when his automatic microtome was well established. The sliding design is certainly strong, and would cut larger sections than the rotary machine, but there is little evidence that Bausch & Lomb, who manufactured the microtome, made very many of them.

Similar concern with the secure attachment of the knife was evident in the Minot design of 1901.[127] This was solidly made, incorporating a Naples object-clamp and very solid frames, and suitable for use with either paraffin or celloidin (see Figure 85).

A design in which both the knife and the object moved during cutting was that by Fiori, marketed by Koritska in 1901.[128] As will be seen from Figure 86, as the knife-spindle revolved, so did that carrying the object: they met in a manner similar to that achieved on the sliding microtome, but in a lighter machine at much less cost.

Another different design was that of Krefft in 1903.[129] This adopted a blade of unusual semicircular design (see Figure 87), which rotated eccentrically to produce the effect of a slicing motion. The blade was held the full length of its back to produce great rigidity, and while the edge completed its circle the automatic advance acted. With an easy action from its crank operation, the whole machine is a good example of original thinking applied to the problems of sectioning, especially in securing rigidity of the blade.

A chain-driven model was introduced by Leitz in 1905,[130] and it has been possible to inspect an example in good working order. It can be

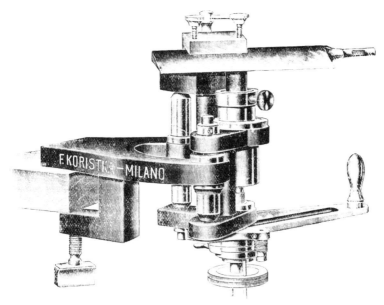

Figure 86. Fiori's microtome, 1900 (from the original paper, note 128 in text).
Another example of original thought applied to microtome design. Both object-holder and knife moved in arcs to meet each other in a slicing action, achieved by a machine that was light in weight but rigid in use.

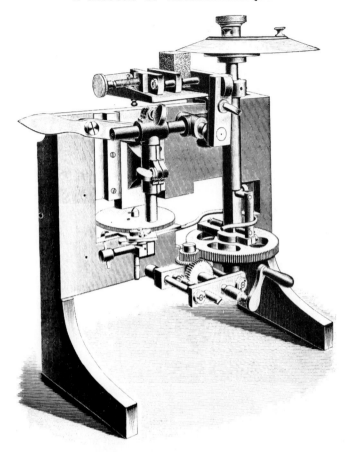

Figure 87. Krefft's rotating-blade microtome, 1903 (from the original paper, note 129 in text).
The special semi-circular blade rotated about an eccentric axis, to give a slicing action of great rigidity. This is an object which it was probably pointless to arrange for, as there is nothing proved about superiority of such an action over the ordinary guillotine-like motion of the usual machine.

operated by either hand, and was available with bed lengths of either 32 cm or 42 cm. The knife-holder incorporated two screws, to allow any degree of obliquity to be set, and a dropping apparatus was also included to keep the knife wet with spirit. The machine would not be out of place in a modern histology laboratory.

A different method of lever operation was introduced by Broek in 1907,[131] in a manner reminiscent of a modern potato-chip slicer. The knife was made to move horizontally against a screw-advanced object, with a band to remove the ribbon so formed (see Figure 88). Two other lever-operated designs should be mentioned in this subsection, both

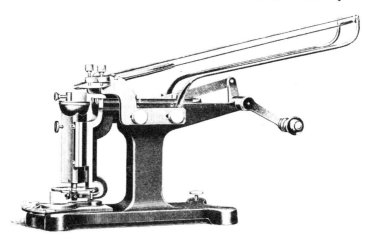

Figure 88. Broek microtome, 1907 (from the original paper, note 131 in text). In this design the lever moved the knife forward against the fixed, screw-raised, object-holder, in a rigid and powerful stroke which it was claimed would section the hardest ordinary tissues regularly at only 5 μ. The rapidly-produced sections were collected by the band, also driven by the lever action.

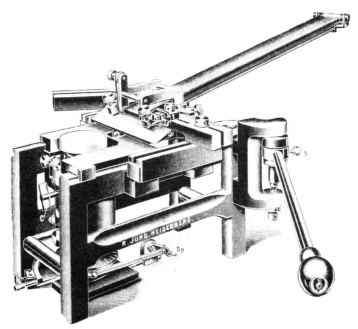

Figure 89. Jung lever-operated sledge microtome, 1910 (from the original paper, note 132 in text).

This instrument, called also the 'Tetrander', was massive, and capable of making large sections of difficult tissues, rapidly and regularly. It was very expensive, as one might expect.

Figure 90. Cambridge lever-operated sledge microtome, 1910 (from the
catalogue, note 133 in text).
First designed in 1909, this machine would cut sections up to 40 cm² in
area, of the toughest tissues such as decalcified bone. Its design has some
similarity to the Jung model, but is even more massively built.

from 1910. The first was a very substantial model by Jung,[132] having a
solid frame carrying an equally solid knife support (see Figure 89). The
object-holder was fully orienting, and an endless band could be
attached. Remarkably similar was the design of the Cambridge
Scientific Instrument Company,[133] which was even more massive and
would cut sections up to 120 × 160 mm. All wear was automatically
compensated, and an example of this machine proved to be a most
excellent piece of work on inspection, fully able to withstand
competition from any other maker (see Figure 90).

These two models conclude the survey of microtomes of class one,
based on the oldest section-cutters but developed to a very high level of
precision.

Microtomes 1870–1910 : (b) movable knife with inclined feed

We saw on page 131 above that the basis of this design, first adopted by
Capanema in 1843, was really the wooden model of Rivet, described in
1868. This was modified into a metal model by Brandt in 1870 (see
Figure 91), and described by him in the following year.[134] He acted on
the suggestion of Leuckhart in this, and got the university mechanic at

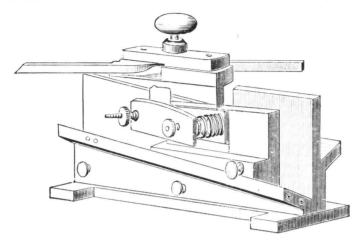

Figure 91. Leyser-Brandt microtome, 1870 (from the original paper, note 134 in text).
Brandt substituted metal for the wood of the original Rivet instrument, having it made by Leyser, the mechanic of the university of Leipzig. The quite unsatisfactory design of the original object-clamp is retained.

Leipzig to make the instrument, hence its later name of the Rivet–Leiser microtome: the name is actually spelt 'Leyser' on the instruments and in the original paper. Two examples of this instrument have been inspected, and they are quite smooth and rigid, albeit small in size. When they were first available, they must have made a very good impression compared with the Stirling machines which were their only real competitor.

Improvements followed to this basic design, an important one being that by Reichenbach in 1878, who introduced a stop to prevent the knife carrier from overshooting the end and falling out—with disastrous results when one reached out to catch it! He also described ways in which the 1 cm rise of the object could be increased, and made the machine larger (to measure 19 cm overall). His most important modification, however, was to introduce an orienting clamp to allow the specimen to be altered in position without having to remove it from the holder.[135] This is a good practical paper describing material improvements to the instrument.

In 1879 Spengel[136] described improvements which had been made in the version used for some years at Naples. This was longer to allow finer control of section thickness, and had an orienting clamp: this moved in two directions separately, and was a decided improvement (see Figure 92). As they pointed out, several clamps could be used with one instrument, economical in a busy laboratory. They also moved the

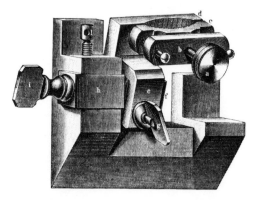

Figure 92. Rivet-Leiser modified carrier, 1880 (from the original description, note 136 in text).
An early modification to the inclined-rise microtome, offering much-needed specimen orientation and rigidity.

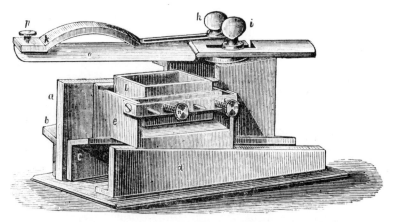

Figure 93. Windler's microtome, 1880 (from the original description, note 137 in text).
This early modification offered much-needed improvements to the original design—a better object-holder and a clamp to control springing of the blade. This was an early recognition of the difficulties caused by using a thin knife, which flexes as it hits the specimen.

object-holder along the incline by means of a continuous leadscrew: this seems an expensive modification, and one not calculated to maintain accuracy in section thickness, as this would come to depend on the accuracy of the screw-thread once more.

Some decided improvements were put forward by Windler in 1880,[137] who substituted a better specimen holder, and introduced a clamp to minimize springing of the knife, as shown in Figure 93. The principle of the inclined rise was adopted in a very different way by

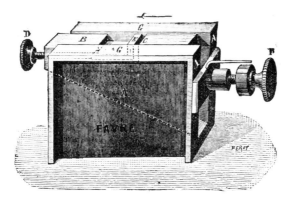

Figure 94. Lelong's instrument, 1883 (from the book by Latteux, note 138 in text).

In this little-known design the screw D advanced B to grip the object at G. The whole of this assembly was moved up and down slope A, under the control of screw F. A strong blade was pushed along the top plate to cut the sections.

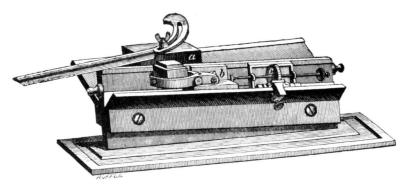

Figure 95. Thoma's microtome, 1881 (from the original paper, note 139 in text).

The important innovation in this model was that the carriers slid, not with all their surfaces in contact with the grooves, but with only five points of contact. This is the minimum number needed to maintain position, and this precision removed a major source of inaccuracy in the inclined-plane instrument. In later models the points would be made of ivory or even agate, obviating need for lubrication of the ways.

Lelong for his design of 1883,[138] pictured in Figure 94: it consisted of vertical plates containing a slide, along which the object clamp moved.

It was Thoma who really improved the original design, however. In 1881 he published a brief description of his instrument,[139] and amplified this in 1883,[140] by which time the design had achieved wide popularity. The machine is shown in Figure 95, and was made of cast

Figure 96. Becker's microtome, 1885 (from the original paper, note 141 in text).
This full-sized instrument was intended to provide the finest work very rapidly, by means of its crank drive. The orienting object-holder is of advanced design.

iron. It marked a great step forward, as it was designed to allow the carriers to touch the slideways on five points only, with geometrical accuracy to the sliding motion. The instruments were made by Jung, and could cut sections of area $4 \, cm^2$ regularly to a thickness of $10 \, \mu$: smaller pieces could be cut to $5 \, \mu$. Thoma made clear that the limitation on thinness was in the preparation of the material, and not in the machine.

By 1885 development had given the Becker instrument,[141] which was of large size and crank operated (see Figure 96). Attention had been given to the object clamp to allow accurate orientation, and the whole machine was a good example of the benefits of a specialist design from a specialist manufacturer. By this time also, both Jung and Thoma had developed the object-holder to allow a continuous ribbon of sections to be cut with proper manipulation.

Vinassa used the principle in a very substantial instrument also

Figure 97. Vinassa microtome for pharmacognosy, 1885 (from the original paper, note 142 in text).
This large instrument was intended to cut the toughest plant materials, and carried the knife on a double slide.

described in 1885;[142] this was a large machine intended to deal with the toughest materials found in pharmacognosy (as shown in Figure 97). By contrast, the design of Hildebrand was a simplified version,[143] which in spite of its low cost was claimed to equal in performance any before described. Obviously the same in principle as the much more complicated instruments (see Figure 98), this one was perhaps a corrective to the ambitious (and possibly unjustified) refinements of some specialist manufacturers. Another fairly simple version was that put out by Nachet in 1886,[144] which had the advanced feature of agate points and bearing plate for the knife carrier, but was fitted with a curiously elementary object-holder, as seen in Figure 99. The raising

Figure 98. Hildebrand's microtome, 1885 (from the original paper, note 143 in text).
A simplified version of the Rivet-based designs, intended to equal the performance of the more expensive models.

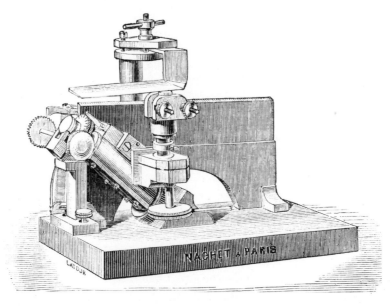

Figure 99. Nachet's design, 1886 (from the original description, note 144 in text).
On the one hand, this instrument featured agate points running on agate plates under the knife carriage, with a modified microscope fine adjustment to advance the specimen : on the other the object-holder was elementary in design, unsuitable for advanced work.

mechanism was the slow motion from a microscope, and this would give fine control to the specimen thickness—if the object carrier was sufficiently rigid. For work under spirit a tray with india-rubber bag was fitted round the object-holder stem; the special cranked shape of the knife then allowed direct use without further adaptation. This instrument was one of the few advanced microtomes available from France in the 1880s, and does show some signs of original thought.

Several variations on the original plan appeared. Some were concerned with finer control of the raising mechanism, one being the Hansen design[145] which used a 5:1 lever to reduce the speed of the micrometer screw. This gave a minimum movement of 2 μ. The design also allowed for cutting under spirit for celloidin work. Other modifications were evolved for cutting large sections; the Jung design of 1886[146] (see Figure 100) was basically just a Thoma instrument arranged so that the whole length of the 36 cm knife could be used. This was achieved by provision of a knife-support pressed on the end of the blade to minimize chatter. In the same list Jung described the instrument used at the Naples laboratory,[147] the product of a series of detailed improvements (see Figure 101). These included provision for making the blade rigid and widely adjustable, addition of a section-stretcher to minimize curling of large sections, a number of modifications to the object holder to give fine control of orientation, and better fineness of advance, accurate to 1 μ. The catalogue description of

Figure 100. Jung's microtome, 1886 (from the catalogue, note 146 in text). This was a very complete and expensive model, allowing use of the whole length of a 36 cm knife, by means of a stiffening support to minimize chatter. The arrangement to moisten the blade is shown in the picture.

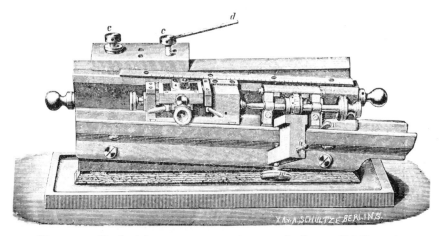

Figure 101. Naples laboratory pattern microtome, by Jung, 1886 (from the Jung catalogue, note 147 in text).
This version embodied a number of detail improvements worked out in the famous marine laboratory. Various attachments made the blade more rigid and allowed addition of a section-stretcher; the advance mechanism was especially precise.

this instrument shows how much attention to detail was required to satisfy the demands of the scientist-customers. One cannot but help feel that there was a vogue in Germany for a connoisseur's approach to the microtome, just as there was in England at that time a similar feeling about the microscope.

American manufacturers also produced basic models of this kind, such as that of Walmsley,[148] which was a simplified version (see Figure 102). So large had some designs become that makers offered them in two different sizes, the larger usually having a more complicated object holder: good examples of this trend are shown in the 1892 Reichert catalogue.[149] Although originally a manufacturer of microscopes, this firm supplied a number of advanced microtomes of its own make. Of the Thoma-like models, the larger sold for 170 marks and the smaller for 90. These prices may be compared with a cost of 110 marks for the small patent instrument, 180 for the large, and 300 for the extra-large with 36 cm knife. In 1900 this same company introduced a different design for cutting under water;[150] this was made of cast iron and incorporated a new design of knife to cut faultless sections up to 11 cm in size, as shown in Figure 103.

This represents a final stage in the design of instruments based on the Rivet plan, and was as great an advance in sectioning as had been achieved in the microscope itself in the same time.

Figure 102. Walmsley's microtome, 1887 (from the book by James, note 148 in text).
As a contrast with the complicated German models of its time, this American instrument was simplified for student use.

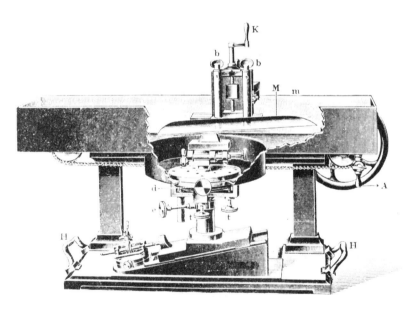

Figure 103. Reichert instrument for cutting under water, 1900 (from the original paper, note 150 in text).
The large instrument would cut objects up to 11 cm in size, for which water lubrication was necessary. The machine was of cast iron, cost twice as much as any other Reichert microtome, and incorporated a newly-designed knife.

Microtomes 1870–1910 : (c) with fixed vertical knife

The two main kinds of instrument so far considered relied on old-established principles, but the remaining two groups were new conceptions when they arrived on the scene a century after the first section cutters were devised. In these types it is the object which moves to-and-fro, rather than the knife. At first sight this might seem to reduce the likelihood of adequate control of section thickness, but it does allow the important advantage of automatic production of serial sections quite easily, as they can be taken from the stationary knife in a ribbon without difficulty.

The most famous microtome of this kind, and possibly of any kind, is the Cambridge rocking microtome, known all over the world as the Cambridge Rocker. It was designed by Horace Darwin, younger son of the famous Charles; he bought a partnership in the Cambridge Instrument Company when it was formed in 1880. The instrument was marketed in 1885, and a number of examples of about this date still exist (see plate 39). An interesting account of the kinematic basis of the design was given by Barron,[151] who showed that some instruments had been in use for sixty years without need for overhaul or repair, as any wear was automatically taken up and the bearings stayed properly seated without play. It was possible to take from store the raw castings, put them together with the springs, and cut sections on a completely unfinished instrument. This is the explanation of the comparatively low price and high accuracy and reliability of the design. It would cut sections to 0·002 mm; in 1930 when the ultra-violet microscope demanded thinner sections, it needed only slight modification to cut to 0·0002 mm; and when electron microscopy required thinner sections still, further modifications gave 0·000025 mm—a 200 × increase in its sensitivity, and a remarkable tribute to the original design.

In the 1885 description[152] the simplicity of the instrument was stressed, as was the fact that the ribbon of sections fell by its own weight onto the bench, obviating need for the elaborate silk bands required by other designs (see Figure 104). The price was one-sixth of the earlier Cambridge automatic microtome (see below, page 264), less skill was needed to operate it, the razor was fixed at the optimum angle, the block could be moved to the edge very rapidly, and no parts could get out of order. Subsequently a number of critics complained that the instrument actually cut sections which were very slightly cylindrical, and this cry was taken up from time to time as being a total bar to its use. One can only suppose that some at least of these detractors were pushing

Figure 104. Cambridge rocking microtome, 1885 (from the original description, note 152 in text).
This shows the original design of this famous instrument, a completely new approach which offered excellent serial sectioning easily and at low cost.

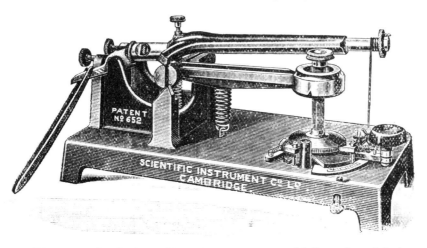

Figure 105. Cambridge rocking microtome, 1900 model (from the original description, note 153 in text).
This version was made more rigid, with better webs on the base. It also provided better clearance between block and knife on the return stroke, but the basics needed no improvement.

rival instruments, for there is no visible difference between sections cut on a rocker and those cut on any other, and dispassionate inspection would certainly have shown this to be so (see page 265 below).

The least satisfactory part of the original design was its lack of orienting object holder, but this was soon remedied. In 1900 a new pattern was introduced,[153] more rigid and obviating any tendency for the knife to touch the block on the return stroke (see Figure 105). By 1903 over 900[154] of these instruments had been sold: it would be interesting to know how many rival machines were purchased in the same period. Although one may suspect that it would be relatively few, it is impossible to establish this as the records of Jung and Reichert are not now available.

Other designs based closely on the Cambridge pattern were offered, one of the best being that made by Van der Stad in Amsterdam in 1909.[155] This was an improved version, but it was much more costly (see Figure 106). The original Cambridge pattern cost £5 0s. 0d. in 1900, had fallen to £4 10s. in 1909, and was still only £6 18s. in 1932. It is no exaggeration to suggest that it was this microtome which revolutionized section-cutting in the biological world, by making the process widely available at modest cost.

Figure 106. Van der Stad rocking microtome, 1909 (from the original paper, note 155 in text).
Although of more finished appearance, and much more expensive, this model is based closely on the Cambridge rocker, with a means of adjusting the angle of the knife.

The other type of automatic microtome in this group to gain wide acceptance was that which came to be known as the Minot model. This was designed by him in 1886, but was preceded by a similarly-conceived instrument first used about 1885. A modern paper which attempted to make clear the relative facts unfortunately succeeded in clouding them,[156] as the date of issue of the journal concerned was not checked.

In fact, the Pfeifer design, developed by the instrument maker of Johns Hopkins University, was in use from the autumn of 1885. This was stated as plain fact by the anonymous author of the paper which described the instrument,[157] when it had been in use for one year. This design (see Figure 107) was not marketed and very few were ever made, thus allowing the slightly later design by Minot completely to overshadow it. The Pfeifer instrument was certainly another novel approach to the construction of an automatic microtome. A sliding carriage held the knife, which moved only by the amount required for the thickness of the section being cut. The object was carried on a revolving wheel, also coupled to the knife carrier. As the block struck the knife a section was cut, and while the block completed its rotation, the knife was advanced, the minimum being 1/5000 in.

The Minot design, first described in 1888,[158] but certainly in definite commercial production in the year before, was a different approach again. The object was moved vertically by turning a handle, the specimen being advanced at top dead centre, by a minimum of 0·0033 mm (see Figure 108). An orienting holder was supplied, and a

Figure 107. Pfeifer's automatic microtome, 1886 (from the original paper, note 157 in text).
This large instrument was of completely original design. The knife K moved the thickness of the section on carriage C, while the object holder on the wheel E revolved a full circle between sections.

Figure 108. Minot's microtome, 1887 (from the original paper, note 158 in text).
With stationary knife and block moving up and down, advancing at the top of each stroke, this was another original automatic microtome. The star wheel E determined the number of notches the wheel Z would advance the block.

Figure 109. Minot design of 1892, from behind (from the original paper, note 159 in text).
This modification was provided with more accurately-cut grooves for the slides, and an additional toothed wheel to allow the minimum advance to be reduced to 0·001 mm.

number of early examples examined have been very smooth running. The instrument was developed to provide thinner sections by 1892,[159] with interpolation of an extra toothed wheel to give sections down to 1 μ (see Figure 109). Scheifferdecker, in this paper, was very emphatic that the sections produced were better than those from the Rocker, but had to admit that the slideways on the Minot were not as accurate. This defect was remedied in the 1901 version,[160] by use of more accurate planing tools. By 1908 Zimmermann made what was by then called the Zimmermann-Minot microtome, in refined form capable of much accuracy[161] (see Figure 110).

Figure 110. Zimmermann-Minot microtome, 1908 (from the original paper, note 161 in text).
This represents the developed form of the instrument, with detail improvements.

Other suggestions were made for improving the design, an example being in the Minot-Blake instrument of 1899,[162] which provided the moving parts with more precise guidance and support: three bearing points were kept in contact with the guiding surface by a spring. The feed-wheel was of very large diameter, giving an accurate 1 μ advance, at which it was claimed to cut sections with no gaps in the series (see Figure 111 and plate 40).

The same basic design proved capable of adaptation for large sections, still a preoccupation in the 1890s. Zimmermann brought out his machine for cutting entire cerebral hemispheres in 1898,[163] and in

Figure 111. Minot-Blake design, 1899 (from the original description, note
162 in text).
Seen from behind, with the large diameter feed-wheel giving an accurate
minimum advance of 0·001 mm. This is a better design than that
interpolating an extra wheel, as one lot of bearing backlash is thus avoided.

spite of a total weight of 33 kg the working was stated to be delightfully
easy and regular (see Figure 112). A special feature was the ability to
reduce the long travel of the block when only small specimens had to be
sectioned. Full orientation of the block was provided, as was a rubber
tube running along the back edge of the knife to convey liquids to
control its temperature—an interesting innovation. The original paper
contains a plate showing the sections as they came from the knife, and
they are certainly good in quality.

An instrument similar in basic design, albeit slightly rearranged, is
that of Reinhold-Giltay,[164] incorporating several useful features. The
operating wheel was large in size and worked quickly, to raise the block
in vertical slides. The whole equipment was supported on a special base,
and could be treadle-operated to produce sections as large as 4 × 4 cm
and as thin as 0·5 μ (see Figure 113). The author went into much detail

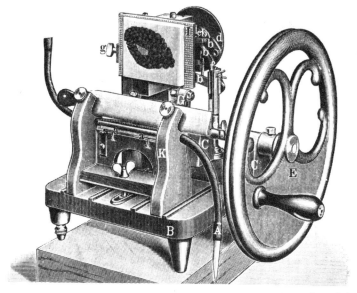

Figure 112. Large brain microtome, Minot pattern, 1898 (from the original paper, note 163 in text).
This version had longer travel for the block, giving enough clearance to accommodate an entire brain hemisphere. The machine weighed 33 kg, but was said to work easily and regularly. The tube carrying liquid at controlled temperature is seen at the back of the blade.

Figure 113. Reinhold-Giltay microtome, 1892 (from the original description, note 164 in text).
Shown here minus its frame and operating treadle. The rotating handle moved the follower h on the cam t, thus oscillating the object in the vertical slides m. At each revolution the automatic advance acted at the top of the stroke.

as to proper sharpening of the knife, including the abrasives available, in an interesting aside.

The Fromme microtome[165] used different arrangements to move the block, in the form of a massive frame to which was attached a parallelogram of arms carrying the block and advancing it (see Figure 114). A weak point of the design is lack of adequate support for the blade, a surprising feature by 1896. A design bearing a superficial resemblance to this was marketed by Reichert in the next year,[166] and is a much more satisfactory instrument (see Figure 115). The arm is very solid, and the blade made the advance in a fully-supporting carriage. An extra piece of equipment allowed the instrument to be used as its own

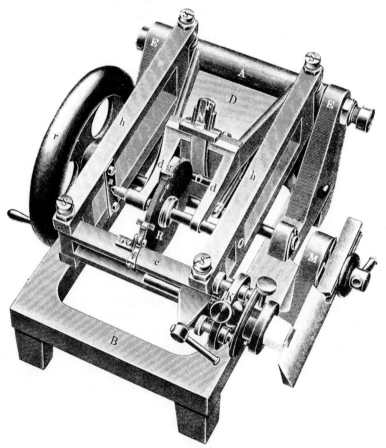

Figure 114. Fromme's microtome, 1896 (from the original paper, note 165 in text).

A different approach, using a pair of arms to carry the block towards the knife, in a parallelogram action. The blade is a weak point in the design, being supported at only one end.

Figure 115. Reichert microtome, 1897 (from the original paper, note 166 in text).
Superficially similar to the Fromme design, it was in fact rather different in principle. The substantial arms carrying the block did not also advance it: the knife moved to accomplish this. The blade was adequately supported, and the whole effect very solid.

block trimmer (see Figure 116), with the ends of the blade which could not be used for sectioning in any case. This is a very good design, and is still impressive for its smooth and accurate operation with large blocks.

Criticisms made of the Cambridge Rocker regarding its cutting of slightly cylindrical sections, and the supposed deleterious effects of this, have been noted, and resulted in attempts to make a model which cut truly plane sections. Leake[167] produced his machine to cut sections up to 45 mm square, to be marketed by Pye in 1901 (Figure 117), two years after Cambridge had also made their own design. This[168] was based on the use of the same advance mechanism as the rocker itself, the advance being arranged at right angles to the original (Figure 118). This instrument would cut sections up to 30 mm in diameter; several examples have survived, and remain impressive with their easy working and obviously long life and accuracy.

Several other original designs based on the same principles should be

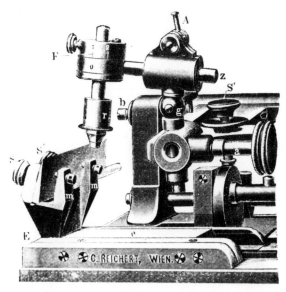

Figure 116. Block trimmer from the Reichert instrument (from the original paper, note 166 in text).

Showing the ingenious use made of the microtome, to trim blocks by means of simple attachment, and using that end of the blade which was never used for sectioning.

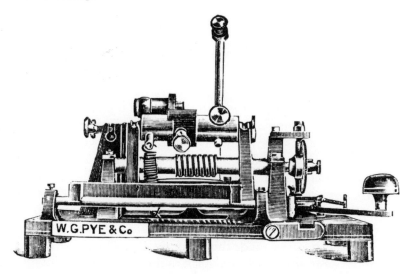

Figure 117. Leake microtome, 1901 (from the original description, note 167 in text).

Brought out two years after the Cambridge flat-cutting machine, and in any case designed as a result of confused thinking about the need to have sections cut in a perfectly flat plane. Apparently the only model marketed by Pye.

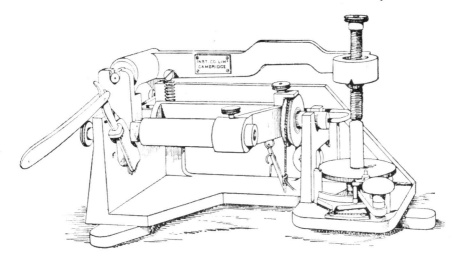

Figure 118. Cambridge flat-cutting microtome, 1899 (from the original description, note 168 in text).

Again the result of sensitivity about flat sections, but much more expensive than the original rocking model on which it was based, and never a good seller.

Figure 119. Ryder's microtome, 1887 (from the original description, note 169 in text).

Designed as an inexpensive light-weight instrument, to make large numbers of sections for students as quickly as possible. The handle was moved up and down to produce the sections.

considered, in spite of the fact that the original models which appear to have commanded wide sales were the original Cambridge and the Minot. The design by Ryder in 1887[169] was a small machine carrying the specimen on a vibrating, as opposed to rotating, lever, which advanced at each stroke to cut 100 sections per minute (see Figure 119). In contrast to most other automatic instruments, this one was light in weight and quite inexpensive, at $25.00. Two examples of this design

Figure 120. Improved Ryder design, 1895 (from the original description, note 170 in text).
Detail changes were made to give better accuracy over longer periods. This accuracy did not approach that of larger and dearer designs, but was adequate for its purpose, as trials on two examples have confirmed.

have been inspected, and were quite satisfactory. An improved version appeared in 1895,[170] with better bearings and other detail modifications (see Figure 120). The example inspected was very serviceable, while being less accurate than other designs. It was, however, offered as an inexpensive model for making large numbers of sections for student use, and for this purpose was quite satisfactory.

Another original design was that by Radais,[171] usable for both paraffin and celloidin. He had realized that unequal lubrication on the slides caused uneven thickness of the sections, and overcame this defect by using balance wheels suspended between conical steel points. The instrument would cut sections from 1 to 50 µ thick onto a paper strip for ease of handling, and the whole design bears the hallmarks of careful thought, as shown in Figure 121.

Only one other design in this section need be considered, that of Triepel.[172] This used a vertical cylinder containing a second one, raised by rotation of the first. The knife was supported on a particularly massive prism, and no lubrication was needed for the vulcanite teeth; this also overcame the effects of oil layers of varying thickness, and it is interesting to compare the two designs made with this common aim (see Figure 122).

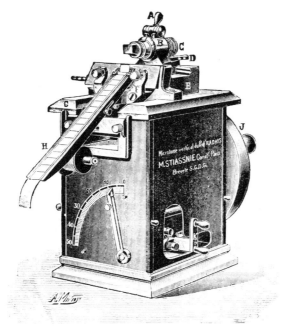

Figure 121. The Radais design, 1904 (from the original description, note 171 in text).
Designed to overcome inaccuracies resulting from differing thicknesses of oil films on slideways, this instrument used balance wheels suspended between conical steel points to cut sections down to 0·001 mm, delivered on to a paper strip.

Microtomes 1870–1910 : (d) with fixed horizontal knife

This division contains only one instrument, and one based closely on it, but the first of them has a very important place in the history of microtomy, for it was the first ever automatic microtome, inspired by the realization that ribbons of serial sections could be produced given the right instrument. Caldwell's automatic microtome was described in 1884,[173] and Threlfall gave a first-hand account of the events leading up to its appearance in a much later paper.[174] This is very interesting in its recollections of personal involvement in the momentous event of the realization of the possibility of ribbon formation (see above, page 81). From the original discussion between Caldwell and Threlfall as to the possibility of making an automatic microtome, it was in fact Threlfall who designed and had made the original instrument at his own expense. The machine was in use by Easter 1883 in the Cavendish laboratory (Threlfall was a physicist), and was then taken over to the Comparative

Figure 122. Triepel's microtome, 1905 (from the original paper, note 172 in text).
This design also sought to overcome the difficulties of oil films of differing thicknesses. The vertical cylinder had vulcanite teeth needing no lubrication, driving an inner cylinder to advance the specimen.

Figure 123. Caldwell's microtome, 1885 (from the original description, note 173 in text).
Made by the Cambridge Instrument Company to Threlfall's design as modified by Horace Darwin, this is the commercial version of the world's first automatic microtome to produce a ribbon of sections.

Anatomy laboratory where it remained in use, driven by a water motor, until after 1898. The only publication at the time was the addition of an extra page to the second edition of Foster & Balfour's textbook of embryology. Horace Darwin, who was shortly to produce the Rocker, made some similar instruments, as shown in Figure 123 and plate 41.

While abroad as an undergraduate, Threlfall told of his machine, and was surprised to find continental imitations appearing for sale : one such is the De Groot[175] (see Figure 124), its description published with full acknowledgement to Threlfall. When he went to Australia a workman who had been engaged on the original model appeared and made a duplicate, shown in plate 42.

In a very interesting footnote to his paper, Threlfall stated that Sedgwick asked MacBride to investigate in a rigorous manner in 1895 if there was any perceptible difference between sections cut with the Rocker and with an instrument producing perfectly plane sections. No difference was to be seen at any magnification.

Of the Caldwell machine, the version made by the Cambridge company would deliver 100 sections per minute as a ribbon, or double that number if a motor was used to drive it. Rotation of the handle made the horizontal carriage move backwards and forwards against the fixed horizontal knife. The object holder was fully orienting in a very definite manner (a notable advance over other instruments of its day), but the greatest novelty of the design was the provision of an endless band to receive the sections as they came from the razor.

In addition to the instruments of varying importance already described, allowing rapid and economical production of sections of known constants, a few rather bizarre suggestions were also put forward from time to time. Two examples will be mentioned, to illustrate the contrast between the approach of the professional, and that of the amateur who was still numerous and might make useful contributions to other branches of microscopy. Hart's microscope microtome[176] might at first sight appear to owe something to Hensen's design of twenty years before, but in fact it did not. What had been done was to fix a razor-blade to the slide carrier on the microscope stage, and insert the embedded object down the main tube of the instrument until it touched the blade, when it was jammed into position with the drawtube(!) (see Figure 125). The fine adjustment was used to control the thickness of the sections, and the sliding stage movements to make the cut. It is not the misplaced ingenuity of the man who made it that is to be wondered at, but rather the misplaced sense of purpose of the editors of the journal in giving space to the freak, albeit five years after it was originally described.

In somewhat similar vein one must suppose that the editors saw fit to describe the use of a lathe to cut sections in their search for completeness of reporting, again a year in arrears.[177] The object was held in the chuck, the slide-rest held the razor and gave the feed, and the ordinary

Figure 124. De Groot microtome, 1887 (from the original paper, note 175 in text).

Obviously based on a modification of the Threllfall-Caldwell design, of simplified construction but with less satisfactory object-holder.

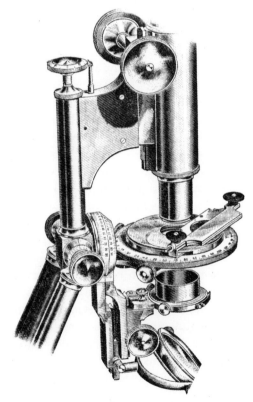

Figure 125. Hart's microscope microtome, 1885 (from the original description, note 176 in text).

This is included as an example of unproductive amateur experiments in section-cutting, this arrangement would have had the main effect of putting strains on the microscope which it had not been designed to resist.

Figure 126. Lathe as microtome, 1905 (from the original description, note 177 in text).

The only adaptation required was to give the tool-holder an adapter to grip the razor. It is certainly possible to cut sections in such a manner, as a trial has shown, but it is slow and laborious, with little hope of adequate quality.

longways movement gave the cut (see Figure 126). In a rather delightful last sentence, the report says that the device is suitable where requirements are not too exacting, for it cannot be used for the finest work or ribbands.

This survey of the evolution of the main forms of the microtome has shown that from being a largely disregarded instrument, it rapidly became one which rivalled the microscope itself in complexity and cost, once the value of serial sectioning was established. By the mid-1880s it was as fashionable to describe a new microtome as it was a new stain. The freezing microtome enjoyed a vogue, but was soon relegated to the pathologist, as there was no way to obtain serial sections without proper infiltration processes. For level-headed discussion on the merits of the many designs, the various editions of Lee's vade-mecum have chapters which are still useful, in contrast to the notably partisan discussions in the periodical literature.

Ancillary embedding equipment

Almost as vast a number of innovations, adaptations, and modifications of equipment ancillary to the microtome was proposed as for that instrument itself; in this account we must confine ourselves to noting

Figure 127. Original Leuckart embedding moulds, 1881 (from the original paper, note 178 in text).
Made of type-metal, and used on a sheet of glass to give the size of block required.

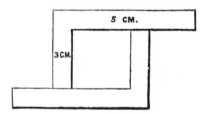

Figure 128. Dimmock embedding moulds, 1886 (from the original description, note 179 in text).
Made from brass, and of the form still in use today.

some small items of lasting value, together with embedding ovens and apparatus for making large numbers of slides in a short time.

Small items of lasting value include the L-shaped pieces of metal put on a base to provide a mould for the wax: by moving the pieces a larger or smaller mould is made with impressive simplicity. The idea seems to be due to Leuckart,[178] who used the shapes made of type-metal as a much more convenient replacement for the made-up paper boxes recommended before then (see Figure 127). These shapes were a little more complicated than those used today, which are of the form first recommended by Dimmock,[179] and made of brass (see Figure 128).

Another small item still in use is the Coplin jar,[180] introduced in 1897. This was a first approach to dealing with more than one section at

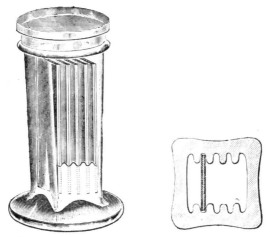

Figure 129. Coplin jar, 1897 (from the original description, note 180 in text).
Still in use today in unchanged form, although with many plastic-moulded rivals, the grooves hold slides while economically-small amounts of the processing liquids only are required.

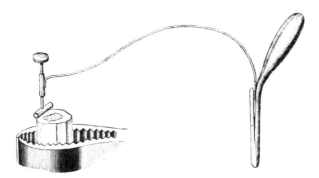

Figure 130. Schulze section-stretcher, 1883 (from the original paper, note 181 in text).
The clip on the right held the watchspring carrying a weighted bob in a sleeve, so arranged as to keep the end of the section flat while the other end was still attached to the knife edge.

once, and consisted of a jar with internal grooves to hold eight slides in a minimum of liquid, so that all could be taken through the various media rapidly and easily; it is shown in Figure 129.

Much attention was given to the tendency of sections to curl as they left the knife, and many section-stretchers were suggested to combat this nuisance: some have already been described with the microtomes they accompanied. Two other designs may be mentioned here; the first, of 1883,[181] was one of the earliest, and consisted of a small weight suspended from a watchspring in such a way that it rested lightly on the front of the block. As the knife cut the weight slid on the blade to hold the section at one end while it remained attached to the edge of the knife at the other (see Figure 130). A later version was that of Kornauth,[182] designed for Cathcart microtomes when used for paraffin sections, and typical of other designs of its time (see Figure 131). The rounded wire smoothed the section as it came from the knife, allowing it to lie on the wide blade.

Figure 131. Kornauth section stretcher, 1896 (from the original paper, note 182 in text).
Designed for the Cathcart microtome in paraffin mode, and typical of the designs of its date which usually incorporated a rounded wire to decurl the paraffin as it came from the knife edge.

Liquid-droppers, to keep the blade moist with water or spirit during the cut, were often supplied with larger microtomes, and have been mentioned as appropriate. Bernhard's dropper of 1891 is typical of most,[183] as the only real variable was the length of nozzle to correspond with the position of the knife (see Figure 132). The other device which attracted attention in the 1880s and 90s was the block trimmer, for it was apparent that accurate ribbons could only be obtained from accurately-trimmed blocks. The design by Schaffer[184] was adequate while being less complicated than some. The block was trimmed *in situ* on its carrier,

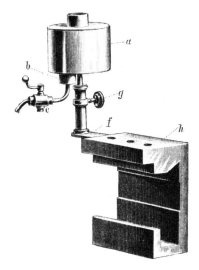

Figure 132. Bernhard's dropper, 1891 (from the original paper, note 183 in text).

Attached to the knife carrier, and thus moving with it, was this simple reservoir and tap on an adjustable stem.

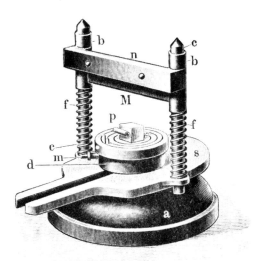

Figure 133. Schaffer block trimmer, 1899 (from the original paper, note 184 in text).

The block on its carrier was put into the slot at d, and rotated to the required position, when the blade M was pressed to make an accurate edge.

rotated under the blade to any required angle, as shown in Figure 133.

There was much preoccupation with arrangements for heating the embedding bath, as too high a temperature has a decidedly adverse

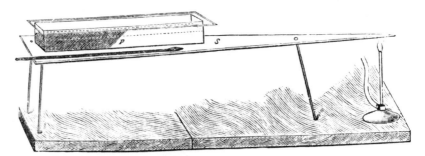

Figure 134. Ryder's embedding apparatus, 1887 (from the original paper, note 186 in text).
An entirely original approach to maintaining just the right temperature of wax, at very low cost into the bargain: the distance of the wax from the flame regulated its temperature precisely.

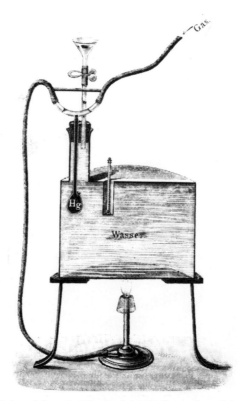

Figure 135. Sehrwald's embedding bath, 1888 (from the original paper, note 187 in text).
There is an interesting contrast between the practical and cheap Ryder design and this commercial approach which is both complicated and expensive.

effect on the material. As early as 1883 the Naples laboratory used a water bath heated by spirit lamp or gas jet, with holes and depressions in the lid and a relatively large volume of water allowing the temperature to be kept quite constant even in the absence of a thermostat.[185] A rather different and most ingenious approach was used by Ryder in 1887,[186] who stood the very long paraffin bath on a sheet of copper, warmed at its remote end by a gas heater (see Figure 134). The heat travelled along the copper and gave a zone in the middle of the bath where the paraffin was just molten: too hot closer to the flame, solid further away. Merely putting the specimen in the right part of the bath ensured its correct temperature. Unfortunately there was nothing in this for a manufacturer, as the whole apparatus cost only $2.00 to make, and in spite of its excellence, nothing more was heard of it.

The difference between the eminently practical Ryder and the German approach of the same time is seen clearly by comparison of the simple apparatus just described and that of Sehrwald,[187] who used a large and expensive water bath and mercury thermostat to achieve just the same end for much greater expenditure of time and effort (see

Figure 136. Kollosow embedding equipment, 1894 (from the original paper, note 188 in text).
Another large and expensive design, with provision for seemingly endless subdivision, and much more elaborate than would be likely to be used.

Figure 135). An even more complicated and expensive device was described by Kolossow in 1894,[188] which included a large oven with a number of drawers apparently infinitely subdivisible. Three distinct temperatures, separated by about eight degrees, were given by this equipment, all of which was in nickelled copper (see Figures 136 and 137).

A less complicated and more serviceable embedding bath was

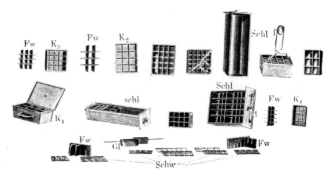

Figure 137. Kollosow ancillary equipment, 1894 (from the original paper, note 188 in text).
Showing the drawers and some of the possibilities of subdivision.

Figure 138. Cambridge embedding bath, 1901 (from the catalogue, note 189 in text).
A large bath of good design, with lids for the pots: not needlessly complicated, and relatively cheap.

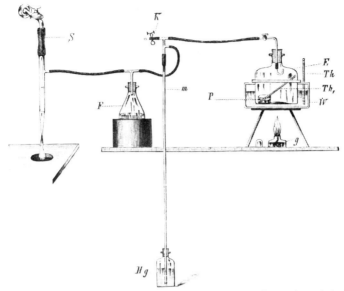

Figure 139. Hoffman vacuum embedding apparatus, 1884 (from the original paper, note 192 in text).
An assemblage of standard laboratory equipment, designed to keep the wax at the correct temperature while reducing the pressure to speed infiltration.

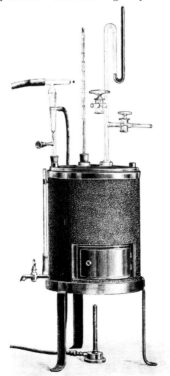

Figure 140. Fuhrmann vacuum embedding oven, 1904 (from the original paper, note 194 in text).
The lower door is an ordinary oven, while the manometer communicates

introduced by the Cambridge company in 1901,[189] as a water-jacket round a number of jars, heated by gas and controlled by thermostat (see Figure 138). In their 1909 list the same company offered two small baths, identical except for heating (one for gas and one for paraffin oil), and a large one as just described. The small one, for gas heating, cost only 55/-d., and the large one was £9 10s.

A bath heated by electricity was described in 1900,[190] it being claimed that it was much lighter and easier to operate than other forms. It had the interesting innovation that objects for embedding were suspended in the wax in small wire baskets, allowing easy transfer.

From such ordinary baths we must now turn to apparatus which allowed embedding under reduced pressure. As early as 1873 Needham[191] had seen the advantage of withdrawing the air from specimens being embedded, although at that date the process was a simple one of encasing. This seems to be the first mention of the process,

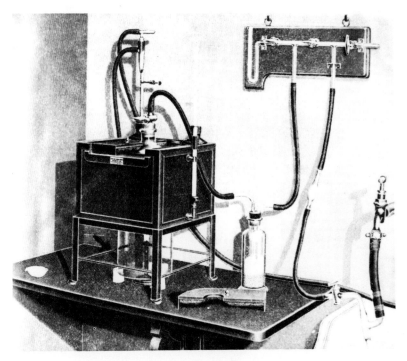

Figure 141. Materna vacuum embedding apparatus, 1908 (from the original paper, note 195 in text).
The illustration shows all the apparatus assembled, with filter pump and manometer on the right, and the oven itself on the left. The guard round the gas burner minimized the effect of draughts when the flame was low.

and he specially recommended it for such organs as lung and cochlea—cavernous structures—to ensure that the tissue was as well supported within as without. The organ was put in the molten wax and exhausted until bubbles ceased to appear: this was a very advanced technique for 1873.

In 1884 Hoffman put forward his suggestion for vacuum embedding, as we would now call it.[192] He used an ordinary filter-pump connected to other ordinary items of equipment, as shown in Figure 139. When the manometer moved no further, embedding was considered complete. This was a simple and doubtless effective arrangement, the longest stay in the paraffin being only about twenty minutes. The amateur Pringle described virtually the same apparatus and process again in 1892,[193] apparently in ignorance of the earlier paper.

Two other vacuum embedding procedures should be mentioned. Fuhrmann in 1904[194] described a more permanent arrangement, using a water-jacketed oven for both ordinary and vacuum embedding. The ordinary bath was below the vacuum bath, and the whole appearance is neat and serviceable (Figure 140). In 1908 Materna described[195] another neat apparatus for the process, leaving the equipment in essentially its modern form: see Figure 141.

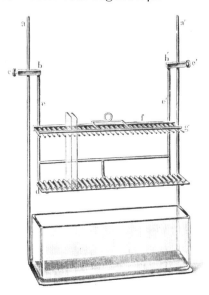

Figure 142. Borrmann's rack for multiple staining, 1894 (from the original paper, note 196 in text).
A sensible and practical arrangement. Nowadays the rack would not be fitted with handles and supporting frame, but would be put right inside the bath to be covered by a lid.

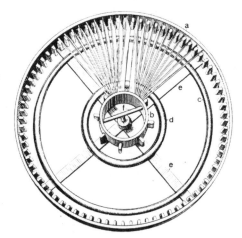

Figure 143. Neumayer circular multiple rack, 1905 (from the original paper, note 197 in text).
A cheap arrangement which would have fitted into one of the widely-found gas troughs very conveniently.

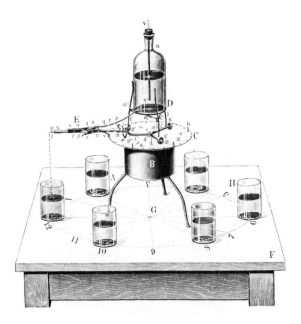

Figure 144. Sanzo's embryo fixer, 1904 (from the original paper, note 198 in text).
The device poured fixative into each dish at a preset time, to fix the embryos at the predetermined age required. This was the basis of the automatic control of other histological processes.

Attention was also being directed towards processing a number of slides together. In 1895 Borrmann described what was really a large rack, to hold 60 slides in a special bath, so that they could be moved easily and economically between the different liquids[196] (see Figure 142). Ten years later a circular rack and bath were proposed,[197] to hold 80 slides at once (see Figure 143). This was all quite straightforward, but in the meantime attention had been given to making a device which would produce embryos fixed at different ages. Sanzo's description of the result in 1904[198] showed dishes with embryos of the same initial age arranged round a clockwork device which could be set to deliver fixative into a dish at a predetermined time, thus arresting development at whatever stage was wanted without the bother of waiting for the stage to

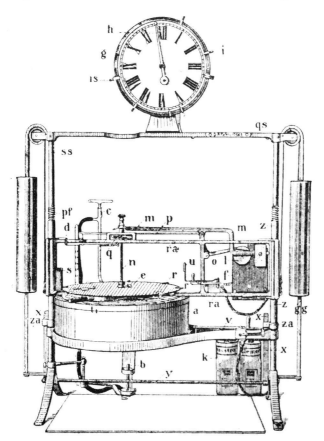

Figure 145. Arendt's automatic tissue processor, 1909 (from the original paper, note 199 in text).
The first tissue processor. The whole apparatus is seen in this illustration, with counterweights for the dishes.

be reached: Figure 144 illustrates the device.

This was obviously capable of adaptation for other purposes, and in 1909 Arendt described the first automatic processor.[199] This received pieces of fresh tissue and after due time returned them infiltrated with paraffin—a notable technical achievement! The apparatus, illustrated in Figures 145 and 146, is in essentials that still used in busy hospital laboratories today.

The summary of equipment used in microtechnique is now complete, with notice of the shape of things to come in the form of automatic processing. By 1910 all the main techniques had been worked out, and many procedures then in use are still followed in virtually unchanged form today. The importance of preparation techniques was to be emphasized with the advent of electron microscopy; for this, insistence on excellence of fixation and specialized forms of metallic shadowing were to become dominant, stimulating work on the background theory of the processes. On the other hand, as methods of interference microscopy developed to practicality, attention would also

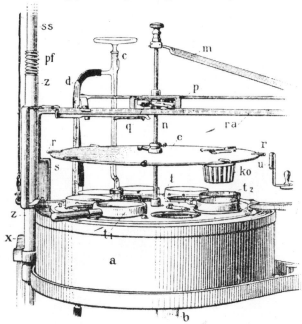

Figure 146. Arendt's processor basket, 1909 (from the original paper, note 199 in text).
The core of the machine was the basket. Pieces of the specimen were put in the basket (ko), which revolved between dishes of reagents, staying in them for longer or shorter periods as set by the clockwork control, to form the basis of modern processors.

be directed to the study of living cells, from tissue culture and using those techniques of micromanipulation just being born in 1910.[200]

It is probably true to say that the histologist of today has a better fundamental knowledge of his subject than had most of his predecessors. The same is true of commercial suppliers of microscope slides, whose market is now largely the schools following the great growth of science teaching before the 1939 war. In the next section we shall consider some commercial suppliers of the nineteenth and early twentieth centuries, to assess their contributions to histological practice.

References for Chapter 5

1.   D Cooper (1841)—Microscopical memoranda, *Mic. J.* *1 :32.*
2.   H Welcker—*Ueber Aufbewahrung mikroskopischer Objecte nebst Mittheilungen über das Mikroskop und dessen Zubehor.* (Giessen: Ricker, 1856.)
     This small book provided information for the use of members of the Giessen microscopical society, and gave German workers their first accessible notes on English methods.
3.   See the footnote to p. 32 of the *London Physiological Journal 1 (1843).*
4.   Exhibition of the Works of Industry of All Nations, 1851—*Reports by the juries on the subjects in the thirty classes into which the exhibition was divided.* (London: W Clowes for the Royal Commission, 1852.) See p. 533.
5.   A Ross (1837)—An adjusting object-glass, *Trans. Soc. Arts* *53 :99–107.*
6.   D Cooper (1842)—Hints to microscopists, *Mic. J.* *2 :78.*
7.   J Wilson (1702)—The description and manner of using a late invented set of small pocket-microscopes, *Phil. Trans. R.S.* *23 :1241–1247.*
8.   J Quekett—*A practical treatise on the use of the microscope.* (London: Bailliere, 1848.) See p. 296.
9.   Quekett, op. cit. note 8 above, 2nd edition 1852. See p. 289.
10. Quekett, op. cit. note 8 above, 3rd edition 1855. See p. 312.
11. G G Valentin (1839)—Die Fortschritte der Physiologie im Jahre 1838, *Valentin, Repert.* *4 :1–358.* See p. 30.
12. See, for example, the statement on p. 54 of *M.M.J. 2 (1869)*, that the editor of that journal (Henry Lawson) always used it for sectioning.
13. Quekett, op. cit. note 8 above. See pp. 371–372.
14. R L Maddox (1869)—Triple-bladed section-knife, *M.M.J.* *1 :55–56.*
15. On p. 77 of *Priced and illustrated catalogue of microscopes and accessories, magnifying glasses, stereoscopes, graphoscopes, etc.* (Philadelphia: James W Queen & Co, 1890), item 3748 is a Valentine knife, at $6.50. This was, however, a somewhat old-fashioned factoring company.
16. P Harting—*Das Mikroskop.* (Braunschweig: Vieweg, 1866.) See vol. 2, p. 62 for a figure of the Strauss-Durckheim instrument.
17. C Chevalier—*Des microscopes et de leur usage.* (Paris: Crochard, 1839), see p. 192.
18. J A Ryder (1884)—On some points in microtomy, *Amer. M. Mic. J.* *5 :190–191.*
     . . . a new art has been developed which we may call microtomy.
19. The instrument is shown in table form but otherwise largely unmodified in the 1872 edition of the catalogue of Nachet, Paris. On p. 27 some equipment is crossed through, marked 'supprime', being replaced by an improved instrument which is actually a modified Adams design.

20. (1836)—Letter addressed to R H Solly, Esq., from H Bowerbank, Esq., respecting his method of obtaining specimens of wood for exhibition and examination in the microscope, *Trans. Soc. Arts* *50 :227–228.*

21. A D Michael (1895)—The history of the Royal Microscopical Society, *J.R.M.S. 1895 :1–20.* See p. 16.

22. Quekett, op. cit. note 8 above. See his p. 309.

23. A Oschatz (1843)—Ueber Herstellung und Aufbewahrung microscopischer Praparate, *Froriep, Notizen 28 :20–23.*

24. P Harting—*Das Mikroskop.* (Braunschweig: Vieweg, 1866.) See Vol. III, p. 409.

25. G T Fisher—*Microscopic manipulation . . .* (London: T & R Willats, 1846.) See p. 48.

26. G F de Capanema (1848)—Beschreibung eines Mikrotoms zu botanisch-anatomischen Untersuchungen, *Flora 31 :465–469.*

27. H Schacht—*The microscope, and its application to vegetable anatomy and physiology,* translated by F Currey. (London: Highley, 1853.) See pp. 5–6.

28. H Welcker—*Ueber Aufbewahrung mikroskopischer Objecte nebst Mittheilungen über das Mikroskop und dessen Zubehor.* (Giessen: Ricker, 1856.) See pp. 33–39, and his plate.

29. W S Gibbons (1856)—Microscopical investigation, *Q.J.M.S. 4 :299–302.*

30. J Smith (1860)—On a section and a mounting instrument, *Trans. R.M.S. 8 :1–3.* The paper was read on 29th June 1859.

31. W L Notcutt—*A handbook of the microscope and microscopic objects.* (London: Lumley, 1859.) See the plate facing p. 152.

32. H D Schmidt (1859)—On the minute structure of the hepatic lobule . . ., *Amer. J. Med. Scis. 37 :13–40.* See pp. 36–38.

33. R Beck—*A treatise on the construction, proper use, and capabilities of Smith, Beck, and Beck's achromatic microscopes.* (London: van Voorst, 1865.) See pp. 117–118.

34. A Chevalier—*L'étudiant micrographe.* (Paris: Delahaye, 1865.) See p. 236.

35. M Mouchet (1870)—On a new instrument for cutting thin sections of wood, *M.M.J. 3 :75.*

36. V v Hensen (1866)—Ueber ein Instrument für microscopische Praparation, *Arch. mikr. Anat. 2 :46–55.*

37. W His (1870)—Beschreibung eines Mikrotoms, *Arch. mikr. Anat. 6 :229–232.*

38. J Luys (1868)—Instrument pour les coupes microscopiques, *Bull. Soc. Anat. Paris 13 :448–449.*

39. G Rivet (1868)—Sur un nouveau microtome, *Bull. Soc. Bot. France 15 :31–32.*

40. J G Hunt (1869)—A new section instrument, *M.M.J. 1 :370.*

41. T Hawksley (1870)—New section machine, *Q.J.M.S. 18 :190–191.*

42. A B Stirling (1870)—Description of a section-cutter for microscopical purposes, *J. Anat. Physiol. 4 :230–234.*

43. F V Raspail (1825)—Développement de la fécule dans les organes de la fructification des céréales, et analyse microscopique de la fécule, suivie d'experiences propres à en expliquer la conversion en gomme, *Ann. Sci. Nat. 6 :224–239.*

44. A good account of Stilling and his work is given by E Clarke & C D O'Malley *The human brain and spinal cord.* (Berkeley: Univ. California Press, 1968.) See pp. 270–275, 833–835. The 1859 book mentioned is: B Stilling—*Neue Untersuchungen über den Bau des Ruckenmarks.* (Cassel: Hotop, 1859.)

45. W Rutherford (1871)—On some improvements in the mode of making sections of tissues for microscopical observation, *J. Anat. Physiol. 5 :324–328.*

46. W Rutherford (1873)—A new freezing microtome, *M.M.J. 10 :185–189.*

47. J Needham (1873)—On cutting sections of animal tissues for microscopical examination, *M.M.J. 9 :258–267.*

48. W J Fleming (1875)—A modification of Dr. Rutherford's freezing microtome, *M.M.J.* *14:79–81*.

49. L Tait (1875)—On the freezing process for section-cutting: and on various methods of staining and mounting sections, *J. Anat. Physiol.* *9:249–258*.

50. W Rutherford (1875)—On the freezing microtome. A reply to Mr. Lawson Tait. *J. Anat. Physiol.* *10:178–185*.

51. R Hughes (1876)—An improved freezing microtome, *J. Anat. Physiol.* *10:614–6*.

52. W B Lewis (1877)—On a new freezing microtome for the preparation of sections of the brain and spinal cord, *J. Anat. Physiol.* *11:537–539*.

53. W B Lewis (1878)—Application of freezing methods to the microscopic examination of the brain, *Brain* *1:348–359*.

54. Anon (1887)—Rutherford's combined ice and ether-spray microtome, *J.R.M.S.* *7:508–509*.

55. U Pritchard (1875)—Freezing tissues for sectioning, *Lancet 1875:833*.

56. J Williams (1881)—Williams' freezing microtome, *J.R.M.S.* *1:697–699*.

57. J W Groves (1881)—Williams' microtome adapted for use with ether as the freezing agent, *J.Q.M.C.* *6:293–295*.

58. J W Groves (1882)—A further improvement in the Groves-Williams ether freezing microtome, *J.R.M.S.* *2:755–756*.

59. Anon (1883)—Fearnley's modification of the Groves-Williams ether freezing microtome, *J.R.M.S.* *3:913–914*.

60. Anon (1882)—Taylor's freezing microtome, *Amer. Mon. Mic. J.* *3:168–169*.

61. Anon (1882)—Satterthwaite and Hunt's freezing section-cutter, *J.R.M.S.* *2:710–711*.

62. C W Cathcart (1883)—New form of ether microtome, *J. Anat. Physiol.* *17:401–403*.

63. Anon (1888)—Cathcart improved microtome, *J.R.M.S.* *1888:1047–1048*.

64. Anon (1906)—Cathcart-Darlaston microtome, *J.R.M.S.* *1906:734–735*.

65. Anon—*Catalogue of Watson microscopes*. (London: Watson, 19th edn 1909.)

66. M Gray (1884)—Gray's ether freezing microtome, *J.R.M.S.* *4:981–982*.

67. W E Boecker (1884)—Ueber ein neues Mikrotom mit Gefriereinrichtung, automatischer Messerfuhrung und selbstthätiger Hebung des Objectes, *Zeits. Instrumentenk.* *4:125–127*.

68. Editorial note—Jacob's freezing microtome, *J.R.M.S.* *1885:899–900*.

69. Editorial note—Hayes' ether freezing microtome, *J.R.M.S.* *1887:1051–1052*.

70. Editorial note—Jung's freezing microtome, *J.R.M.S.* *1887:331–332*.

71. Editorial note—New Delepine microtome, *J.R.M.S.* *1900:128–131*.

72. Personal communication from G L'E Turner.

73. Editorial note—Delepine improved microtome, *J.R.M.S.* *1901:592*.

74. British Patent Specification No. 9900 (1888), in the name of H F Dale.

75. A Bruce (1888)—Microtome for cutting whole sections of the brain and other organs, *J.R.M.S.* *1888:837–839*.

76. C R Bardeen (1901)—New freezing microtome for use with carbon dioxide, *J. App. Micr.* *4:1320–1323*.

77. Editorial note (1904)—Preparation of frozen sections by means of anaesthol, *J.R.M.S.* *1904:474–475*.

78. W J V Osterhout (1904)—Contributions to cytological technique; 1. A simple freezing microtome, *Univ. Calif. Pubs. Botany* *2:73–77*.

79. R Krause (1908)—Eine neuer Gefrier- und Kühlvorrichtung für das Mikrotom, *Z.w.M.* *25:289–300*.

80. G A Piersol (1888), discussed in *J.R.M.S.* *1889:462*.

81. C H G Bird (1875)—Imbedding in elder pith, for cutting sections,

Q.J.M.S. *23:23–27.* The paper was read at a meeting of the Medical Microscopical Society on 19th June 1874.

82. P Schiefferdecker (1876)—Ueber ein neues Mikrotom, nebst Bemerkungen über einige neuere Instrumente dieser Art, *Arch. mikr. Anat.* *12:91–102.*

83. A Fiori (1899)—Nuovo microtomo a mano con morsetta tubulare, *Malpighia 13:193–199.*

84. H F Hailes (1877)—On a new form of section-cutting machine, *J.Q.M.C. 4:243–245.*

85. C S Roy (1879)—A new microtome, *J. Physiol.* *2:19–23.*

86. W Hailes (1880)—Hailes' polymicrotome, *J.R.M.S.* *1880:1034–1035.*

87. W Teesdale (1880)—The Pearson-Teesdale microtome, *J.R.M.S.* *1880:1034–1035.*

88. Illustrated in the *Zeiss catalogue No. 25* (1881), and for many subsequent editions.

89. B A von Gudden (1875)—Ueber ein neues Microtom, *Arch. f. Psychiatrie 5:229–234.*

90. W Betz (1873)—Die Untersuchungsmethode des Centralnervensystems des Menschen, *Schultze, Archiv* *9:101–117.*

91. Editorial note (1882)—Katsch's large microtome, *J.R.M.S.* *1882:126.*

92. W Deecke (1882)—Cutting and mounting sections through the entire human brain, *Proc. Amer. Mic. Soc.* *5:275–277, 279–280.*

93. Reported in *J.R.M.S.* *1891:145–146.*

94. Reported in *J.R.M.S.* *1905:766–769.*

95. E Curtis (1871)—An apparatus for cutting microscopic sections of eyes, *Trans. Amer. Ophthal. Soc.* *8:60–68.*

96. G Hoggan (1874)—On a new form of section-cutting machine for the microscope, *J.Q.M.C.* *3:266–278.*

97. Editorial note—Seiler's mechanical microtome, *J.R.M.S.* *1879:328–329.*

98. S Marsh (1882)—Swift & Son's improved microtome, *J.R.M.S.* *2:432–434.*

99. C S Roy (1881)—Neues Schnellgefrier-Microtom, *Arch. mikr. Anat.* *19:137–143.*

100. E Boecker (1882)—Ein neues Mikrotom mit automatischer Messerführung, *Zeits. Instrumenttenk.* *2:209–212.*

101. J W Behrens—*A guide for the microscopical investigation of vegetable substances.* Translated and edited by A B Hervey & R H Ward. (Boston: Cassino, 1885.) See pp. 188–191 for the Providence microtome.

102. J D King (1889)—The King microtome—numbers 1 and 2, *Microscope* *9:76–77.*

103. See, for example, advertisement page iii of *The Microscope 13 (1893).*

104. See p. 93 of *Microscopes and microscopical accessories No. 28.* (Jena: Carl Zeiss, 1889.)

105. Editorial note—Beck's double slide microtome, *J.R.M.S.* *1892:894–896.*

106. J Schaffer (1891)—Fromme's Patent-Mikrotom ohne Schlittenführung und eine neue Präparatenklammer, *Z.w.M.* *8:298–302.*

107. S Yankawer (1897)—A new microtome, *The Microscope* *5:145–148.*

108. Editorial note—Microtome with arc-movement of knife for section-cutting under water, alcohol, &c., *J.R.M.S.* *1900:645–647.*

109. Editorial note—Zeiss' microtome, *J.R.M.S.* *1881:699–700.* The original description, much overshadowed by the Zeiss catalogue eulogy, was given in *Jena Zeitschr. Naturwiss.* *14:193–199.*

110. Editorial note—Bausch and Lomb Optical Co's laboratory and student's microtome, *J.R.M.S.* *1885:1089–1091.*

111. Editorial note—Leitz's 'support' microtome, *J.R.M.S.* *1889:304.*

112. O Schultze (1892)—Demonstration eines neuen Schneidapparates für grosse

Schnitte, *Würzb. Phys. Med. Sber.* *1892 :116–117.*
113. Editorial note—Reichert's microtomes, *J.R.M.S.* *1884 :823–825.*
114. Editorial note—Beck's universal microtome, *J.R.M.S.* *1885 :344–346.*
115. Editorial note—Bulloch's combination microtome, *Amer. Mon. Mic. J.* 6 :45–46 *(1885).*
116. L C Malassez (1884)—Microtome de Roy perfectionné, *Arch. de Physiol.* *4 :348–363.*
117. P Schiefferdecker (1886)—Ueber ein neues Mikrotom, *Z.w.M.* 3 *:151–164.*
118. *Preis-Verzeichniss von G Miehe 1889.* (Hannover: Hildesheim.)
119. R Thoma (1890)—Ueber eine Verbesserung des Schlittenmikrotoms, *Z.w.M.* 7 *:161–164.*
120. H Strasser (1890)—Das Schnitt-Aufklebe-Mikrotom, *Z.w.M.* 7 *:289–304.*
121. H Strasser (1892)—Weitere Mittheilungen über das Schnitt-Aufklebe-Mikrotom und über die Nachbehandlung der Paraffinschnitte auf Papierunterlagen, *Z.w.M.* 9 *:1–14.*
122. P Schiefferdecker (1892)—Ueber zwei von R Jung gebaute Mikrotome, *Z.w.M.* 9 *:168–175.*
123. J Pal (1893)—Ueber ein neues grosses Mikrotom für Gehirnschnitte von C Reichert in Wien, nebst einschlägigen technischen Notizen, *Z.w.M.* 10 *:300– 304.*
124. See F Koritska—*Catalogo illustrato descrittivo, N.7.* (Milan, 1894.)
125. A Bruce (1895)—Microtome for cutting sections under spirit, *J.R.M.S.* *1895 :487–490.*
126. A Beck (1897)—Ein neues Mikrotom (System Beck-Becker), *Z.w.M.* 14 *:324– 331.*
127. C S Minot (1901)—Improved automatic microtomes, *J. App. Mic.* 4 *:1317– 1320.*
128. A Fiori (1900)—Nuovo microtomo automatico a doppia rotazione, *Malpighia* 14 *:411–424.*
129. P Krefft (1903)—Rotations-Mikrotom 'Herzberge', *Z.w.M.* 20 *:7–11.*
130. A Henneberg (1905)—Neues Mikrotom von Leitz, *Z.w.M.* 22 *:125–130.*
131. A J P von Broek (1907)—Ein einfaches Mikrotom für Serienschnitte, *Z.w.M.* 24 *:268–274.*
132. P Mayer (1910)—Ein neues Mikrotom: das Tetrander, *Z.w.M.* 27 *:52–62.*
133. See the *List No. 57A.* (Cambridge: Scientific Instrument Co, 1910.) Pp. 8–10.
134. A Brandt (1871)—Ueber ein Mikrotom, *Arch. mikr. Anat.* 7 *:175–179.*
135. H Reichenbach (1878)—Ueber einige Verbesserungen am Rivet-Leiser'schen Mikrotom, *Arch. mikr. Anat.* 15 *:134–136.*
136. J H Spengel (1880)—Further improvements in the Rivet-Leiser microtome, *J.R.M.S.(O)* 3 *:334–337.*
137. Editorial note—Windler's microtome, *J.R.M.S.* *1882 :711–712.*
138. P Latteux—*Manuel de technique microscopique.* (Paris: Coccoz, 2nd edn 1883.)
139. R F K Thoma (1881)—Ueber ein Mikrotom, *Virchow, Archiv* 84 *:189–191.*
140. Editorial note—Thoma's sliding microtome, *J.R.M.S.* *1883 :298–307.*
141. J W Spengel (1885)—August Becker's Schlittenmikrotom, *Z.w.M.* 2 *:453–459.*
142. E Vinassa (1885)—Beitrage zur pharmakognostischen Mikroskopie, *Z.w.M.* 2 *:309–325.*
143. H E Hildebrand (1885)—Ein vereinfachtes Mikrotom von grosser Leistungs-fähigkeit, *Z.w.M.* 2 *:343–345.*
144. Editorial note (1886)—Nachet's microtome, *J.R.M.S.* 6 *:1082–1084.*
145. P Groult (1886)—Le nouveau microtome à levier, *Naturaliste* 8 *:241–243.*
146. *Preis-Verzeichniss über Instrumente und Apparate.* (Heidelburg: R Jung, 1886.) See p. 24.

147. ibid., p. 18.
148. F L James—*Elementary microscopical technology*. (St Louis: Medical & Surgical Co, 1887.) See p. 37.
149. *Mikroskope No. 18*. (Wien: C Reichert, 1892.)
150. J Starlinger (1900)—Das neue Reichert'sche Schlittenmikrotom zum Schneiden unter Wasser, *Z.w.M.* *17:435–440*.
151. S L Barron (1954)—Kinematic design applied to instruments, *Trans. Soc. Inst. Technol.* *6:66–82*. See pp. 80–82.
152. Editorial note—Cambridge rocking microtome, *J.R.M.S.* *1885:549–552*.
153. Editorial note—Cambridge rocking microtome, 1900 pattern, *J.R.M.S.* *1900:131*.
154. See the advertisement in *J.R.M.S.* *1903* end-papers for any month.
155. J Boeker (1909)—Über ein verbessertes 'Rocking-Microtome', *Z.w.M.* *26:242–255*.
156. R P Cowles & O W Richards (1947)—The Pfeifer and Minot automatic rotary microtomes, *Trans. Amer. Micr. Soc.* *66:379–382*.
157. Anon (1886)—Revolving automatic microtome, *Studs. Biol. lab. Johns Hopkins Univ.* *3:477–479*. Part 8, in which this paper occurs, is dated October 1886.
158. P Scheifferdecker (1888)—Mittheilungen von den Ausstellungen wissenschaftlicher Apparate auf der Anatomen-Versammlung zu Wurzburg und der 61. Verslammung Deutscher Naturforscher und Aertze in Koln im Jahre 1888, *Z.w.M.* *5:471–481*.
159. P Schiefferdecker (1892)—Ueber das von E. Zimmermann genaute Minot'sche Mikrotom, *Z.w.M.* *9:176–179*.
160. Editorial note—Minot automatic rotary microtome, *J.R.M.S.* *1901:597*.
161. M Wolff (1908)—Über Gerfiermethoden und Gefriermikrotome im allgemeinen, sowie über einen neuen Gefriertisch fur die Zimmermannschen Mikrotome und über die Behandlung freier Schnitte, *Z.w.M.* *25:169–184*.
162. Editorial note—Minot-Blake microtome, *J.R.M.S.* *1901:91*.
163. G C von Walsem (1898)—Ueber ein neues von E. Zimmermann in Leipzig gebautes grosses Mikrotom, *Z.w.M.* *15:145–154*.
164. J W Moll (1892)—Das Mikrotom Reinhold-Giltay, *Z.w.M.* *9:445–465*.
165. J Schaffer (1896)—Neue Mikrotome aus der Werkstätte der Gebrüder Fromme in Wien, *Z.w.M.* *13:1–9*.
166. J Nowak (1897)—Ein neues von der Firma C. Reichert construirtes Mikrotom, *Z.w.M.* *14:317–324*.
167. Editorial note—Leake's microtome, *J.R.M.S.* *1901:92–93, 106*.
168. Editorial note—Rocking microtome to cut flat sections, *J.R.M.S.* *1899:345*.
169. Editorial note—Ryder's automatic microtome, *J.R.M.S.* *1887:682–685*.
170. Editorial note—Automatic microtome, *J.R.M.S.* *1896:132–133*.
171. Editorial note—Radais' microtome with vertical slideless carrier, *J.R.M.S.* *J.R.M.S.* *1904:720–722*.
172. H Triepel (1905)—Ein Zylinder-Rotations-Mikrotom, *Z.w.M.* *22:118–125*.
173. Editorial note—Caldwell's automatic microtome, *J.R.M.S.* *1885:150–153*.
174. R Threlfall (1930)—The origin of the automatic microtome, *Biol. Reviews* *5:357–361*.
It is worth quoting from the only contemporary notice: M Foster & F M Balfour— *The elements of embryology*. Revised by A Sedgwick & W Heape. (London: Macmillan, 2nd edn 1883.)
> The microtome we are most accustomed to is a 'sliding microtome' made by Jung of Heidelberg; it gives excellent results. Recently however Messrs Caldwell and Threlfall have designed an automatic microtome which has been used with success at the Cambridge Morphological Laboratory and promises

to effect a great saving of time and trouble in cutting sections (vide p. 471 and Proceedings of the Cambridge Phil. Soc. 1883) ... [page 434].

Since writing the account of section-cutting on p. 434, we have obtained more experience as to the practical working of Messrs. Caldwell and Threlfall's microtome there mentioned. We find that it cuts more accurately and better than any other microtome with which we are acquainted, and can confidently recommend it to investigators and teachers with large classes ... [page 471].

175. J G de Groot (1887)—Ueber ein automatisches Mikrotom, *Z.w.M.* *4 :145–148.*
176. Editorial note—Hart's microtome-microscope, *J.R.M.S.* *1890 :504–506.*
177. Editorial note—Using a lathe as a microtome, *J.R.M.S.* *1906 :106–107.*
178. P Mayer (1881)—Ueber die in der Zoologischen Station zu Neapel gebrauchlichen Methoden zur mikroskopischen Untersuchung, *Mitt. Zool. Stat. Neapel* *2 :1–27.*
179. Editorial note—Imbedding-box, *J.R.M.S.* *6 :165 (1886).*
180. Editorial note—Laboratory dish, *J.R.M.S.* *1898 :237.*
181. F E Schulze (1883)—Schnittstrecker fur Paraffinschnitte, *Zool. Anzeiger* *6 :100–103.*
182. K Kornauth (1896)—Schnittstrecker fur Paraffinschnitte, mit dem 'the Cathcart improved Microtome', *Z.w.M.* *13 :160–163.*
183. W Bernhard (1891)—Kleiner Tropfapparat fur Mikrotome, *Z.w.M.* *8 :305–310.*
184. Schaffer (1899)—Eine Zuschneide-Vorrichtung fur Paraffinblocke, *Z.w.M.* *Z.w.M.* *16 :417–421.*
185. Editorial note—Water-bath and moulds for embedding, *J.R.M.S.* *1883 :913.*
186. J A Ryder (1887)—Paraffin imbedding apparatus, *Amer. Nat.* *21 :597–600.*
187. E Sehrwald (1888)—Einfache Vorrichtung, die Temperatur im Paraffin-schmelzofen constant zu halten, *Z.w.M.* *5 :331–334.*
188. A Kolossow (1894)—Ein neuer Apparat für Paraffineinbettung der Objecte, *Z.w.M.* *11 :154–162.*
189. Cambridge Instrument Company List for 1901, p. 12.
190. C Regaud & R Fouilliand (1900)—Bain de paraffine à chauffage electrique, *Robin, J. Anat.* *36 :574–579.*
191. J Needham (1873)—On cutting sections of animal tissue for microscopical examination, *M.M.J.* *9 :258–267.* See pp. 264–265.
192. F W Hoffmann (1884)—Einfacher Einbettungsapparat, *Zool. Anzeiger* *7 :230–232.*
193. A Pringle (1892)—Paraffin infiltration by exhaustion, *J. Path. Bact.* *1 :117–119.*
194. F Fuhrmann (1904)—Uber einen Universal-Paraffineinbettungsthermostaten, *Z.w.M.* *21 :462–467.*
195. L Materna (1908)—Ein neuer Vakuum-Paraffinofen, *Z.w.M.* *25 :439–445.*
196. R Borrmann (1894)—Ein neuer Apparat zur bequemen, schnellen, und gleichmassen Färbung und Weiterbehandlung von Serienschnitten, *Z.w.M.* *11 :459–464.*
197. L Neumayer (1905)—Objekttragergestell zur Massenfärbung von aufgeklebten Paraffinschnitten, *Z.w.M.* *22 :181–185.*
196. L Sanzo (1904)—Apparecchio utile in embriologia per la fissazione automatica a tempi voluti di embrioni in via di sviluppo, *Z.w.M.* *21 :449–457.*
199. G Arendt (1909)—Apparat zur selssttätigen Fixierung und Einbettung mikro-skopischer Präparate, *Munchen. med. Wochschr.* *56 :2226–2227.*
200. It may be helpful to offer some literature references to techniques outside our period, in some disciplines which were to become important:
     Histochemistry
     A Prenant (1910)—Methodes et resultats de la microchemie, *J. de l'Anat. et de la Physiol.* *46 :343–404.*

M Parat (1927)—A review of recent developments in histochemistry, *Biol. Rev.* 2 :*285–297*.

Micromanipulation

R Chambers—The nature of the living cell as revealed by micromanipulation, *in* J Alexander (Ed.)—*Colloid chemistry theoretical and applied*. (New York: Van Nostrand, vol. 2, 1928.) See pp. 467–486.

Tissue culture

R G Harrison (1912)—The cultivation of tissues in extraneous media as a method of morphogenetic study, *Anat. Rec.* 6 :*181–193*.

R G Harrison (1928)—On the status and significance of tissue culture, *Archiv. exper. Zellforschung* 6 :*4–27*.

6. Notes on Commercial Mounters 1800–1910

Introductory notes and the earliest mounters

Unarguably, the Victorian era was the golden age of microscopy. The development of the instrument itself was paralleled by equally far-reaching advances in techniques of specimen preparation, as we have seen; and in addition there was an amateur interest in the microscope which was important commercially, quite apart from any contributions which it made to scientific progress.

It would have to be admitted that such contributions were few, at least in histology. Certain amateurs of private means were able to achieve great things in microscopy as such, and the name of E M Nelson is an obvious example of this essentially English tradition. Most amateurs of the microscope were much less dedicated and able, however, preferring to give their time to mere inspection of nature's wonders. To satisfy their need for specimens a large number of commercial mounters flourished, especially in England but also in other countries. Very few of these commercial workers contributed to histological progress, but some of the countless thousands of their mounts can still provide some idea of techniques successfully applied in the nineteenth century. For this reason some outline of their work must find a place in this book, although a mere outline will have to suffice. Therefore attention will be directed in greater detail to mounters who provided histological specimens, rather than to the rest who provided other specialist preparations such as diatoms. Many of the smaller suppliers will have to be omitted or treated very briefly.

We have already seen in Chapter 2 above that commercial mounts in the form of sliders were sold in the eighteenth century, in large numbers but of very stereotyped specimens. This was the result of gross limitations of preparative technique, dry mounts alone being feasible. Such sliders were only rarely marked with the name of the supplier, and such remained the case in the first thirty years of the nineteenth century. Probably the first supplier to add his name to a preparation of the glass type which was starting to succeed the bone slider by 1830 was J West, as shown in plate 43. Another supplier of the 1820s and 30s was Andrew

Pritchard, a selection of whose mounts is also shown in plate 43. He occasionally included his name on the reverse of the label naming the specimen, and he was the main supplier of preparations in the 1830s. His famous 'List'[1] of 1835 serves as his catalogue. It contains a few more than two thousand objects, and it is likely that Pritchard would not have been able to supply them all! Many insects are featured, in whole or in part: for example, 94 eyes of insects and other arthropods. Sections of about 250 different trees and shrubs are also listed, plus over 120 different crystals. His price list (from 18 Picket-Street, Strand, London) includes all manner of optical instruments, and items 22 to 32 are preparations. Objects selected from the lists cost from 2/6d. to 18/–d. per dozen, and the best objects from 1/–d. to 5/–d. each. Thin sections of fossil woods were from 2/–d. to 7/–d. each. It will be obvious that the relative cost of the specimens was high, and it may have been that this gave opportunity for competitors to appear.

C M Topping started work as a mounter in 1839 or 1840,[2] and rapidly acquired great skill. He had a prolific output, and his preparations are numerous and still sought after by collectors. His minute injections were some of the most successful ever to have been produced, and he was responsible for preparing the proboscis of the blowfly, in 1842, in a manner which became a favourite preparation in the succeeding years.[3] His preparations were highly esteemed from the first,[4] but no lists have been found until that of Baker[5] of 1862, where Topping preparations are mentioned but not distinguished in the text, so that no guide can be given to the range of preparations beyond that offered by the surviving objects. As will be seen from plate 43, they were far-ranging and of uniformly excellent presentation. His son Amos succeeded his father as a mounter, some time before Topping senior died in 1874. This commercial preparer is probably the most famous ever known, and his mounts can be taken as typical of the best techniques of the 1840s and 50s.

It is difficult to know to what extent other mounters worked before 1850, for Pritchard's and Topping's mounts were factored, and although the list of the Willats brothers of 1846[6] mentions various preparations, it is probable that they themselves did not make them. Their prices are high, tooth sections being 8/–d. each, and balsam mounted objects costing 2/–d. each: 2/–d. in 1846 was about the equivalent of rather more than five pounds nowadays!

Up to the middle of the century, therefore, there were relatively very few professional mounters, and their products commanded high prices. As we shall see, competition suddenly grew apace.

Mounters at work in and from the middle years of the century

The second edition of Quekett[7] has several pages of advertisements at the end, as does the third edition of 1855. These remain invaluable as a guide to commercial mounters of the 1850s. Those mentioned in 1852 were C Poulton of Reading, J B Dancer of Manchester, J T Norman of London, Smith & Beck of London, and Hett of London. In the 1855 edition Poulton and Norman are omitted, but M Pillischer, S Stevens, W Ladd, and F L West, all of London, are added. Something should be said of all of these, with the addition of Bourgogne of Paris, who was also at work in the 1850s.

Of Poulton, nothing else seems to be known: a few preparations with the name C S Poulton have survived, but are of no special merit.

Norman, on the other hand, went on to prepare objects for many years. In the 1852 advertisement he offered (from 10 Fountain Place, City Road), recent and fossil infusoria, spicules and test objects, urinary deposits and polariscope objects, injections of various animals, and vegetable and animal tissues. He was at 178 City Road from 1862 to 1893, and received a prize medal in the 1862 International Exhibition for excellence of manufacture. His mounts were very distinctively labelled (see plate 44), and his 1872 catalogue[8] filled 32 pages. The cover gives the date of his establishment as 1846, and he claimed that his objects are equal to any others offered to the public anywhere. He numbered all his preparations for the catalogue, and a breakdown of them is quite typical of the time. Numbers 1–258 were diatoms and desmids, 259–267 were Algae, 268–401 various plant parts, 402–613 timber sections, 614–684 seeds and petals, 685–785 insect eggs, 786–816 insect larvae and pupae, 817–1005 insect mounts of whole animals, 1006–1551 mounts of insect parts, 1552–1655 sets of insect parts by species, 1656–1761 spicules of sponge etc., 1762–1827 various shells and spines, 1828–1875 bones, spines and scales, 1876–1893 teeth and cartilage, 1894–1951 hairs, 1952–1968 feathers and wool, 1969–2084 opaque injections of a wide range of chordate organs, 2085–2157 transparent injections of a range of mammalian organs, 2158–2211 urinary preparations, 2212–2219 'entozoa', 2220–2366 crystals, 2367–2492 minerals and fossils, 2493–2527 polariscope preparations, 2528–2578 microphotographs, and 2579–2584 microscopic writing.

Of the preparations listed, therefore, fewer than 10% could be described as histological in the widest possible meaning of the word. On the other hand, no less than 40% were insect preparations of one kind or another. His preparations, of whatever kind, were of the best quality, and many have survived to the present day.

Dancer is a well-known name in microscopy and in other scientific fields also. He offered reasonably-priced microscopes and microscopic objects of every description in the Quekett advertisements, and a selection is illustrated in plate 44. He remains best known for his devotion to microphotography, the production of tiny pictures for viewing through the microscope. This process seemed to many scientists of the time to be trivial in the extreme, but it was adopted by Dagron to send messages to besieged Paris in 1870, by pigeon post. Since then the science of microphotography has developed apace, not only in document copying but for the very demanding microcircuitry of modern electronics: some indication of the scope of the process started by Dancer is given by Stevens,[9] who outlines a little of the history. The standard source on Dancer is the pamphlet by Ardern,[10] which amplified the earlier papers by Stirling.[11] Dancer died in poverty in 1887, and in 1896 his sisters sold his negatives to Richard Suter, a well-known mounter in London, for fifty pounds, the sum to include instruction in their use. In 1960 the negatives, never used by Suter, were passed to Arthur Barron following Suter's death (at the age of 95) the previous year.[12] Although Suter seems not to have used the negatives, he did offer some of Dancer's slides for sale in his 1900 list.[13] In spite of a lack of chronological sequence, it may be mentioned here that Suter started to supply mounts about 1875, and many of his sections were of high quality, as will be seen from plate 44. Dancer's microphotographs were factored by many dealers, and are today collector's items in much demand.

Alexander Hett became M.R.C.S. in 1829, and in the 1852 Quekett advertised his excellent microscopic injections, for medical gentlemen and others engaged in researches. He claimed to prepare every variety of microscopic object required to illustrate minute anatomy and physiology, and such pathological subjects as it was possible to mount as permanent preparations. He had been awarded a prize medal at the 1851 Great Exhibition for these mounts. He was at 24 Bridge Street, Southwark in the 1852 edition, and at 8 Eastbourne Terrace Paddington, with the degree of M.D. as well as his M.R.C.S., by 1855. In the 1880 edition of Beale he advertised from 4 Albion Grove, Islington, but neither the Medical Register nor the Medical Directory has any mention of him, and the R.C.S. took him off the books in 1893 when he could no longer be traced. His preparations are deep-cell fluid mounts, and are usually signed and occasionally dated: the earliest seen had the date 1847. All include a small air bubble, and it is interesting that this was apparently intentionally enclosed, as Suffolk told

Rousselet.[14] The cell was sealed with a coat of gold size once the specimen had been immersed in its dilute spirit, and then covered with a final coat of lamp-black and shellac as a finish. Hett preparations are diamond marked with the name of the specimen, and are of superb quality, being unsurpassed by any other fluid-mounted opaque specimens from any source. Of the considerable numbers which survive today, more than two thirds are in good order, a considerable tribute to the permanency of usually impermanent fluid mounts in the hands of an expert.

Beck preparations also had a long life, although the firm acted mainly as factors as they primarily manufactured microscopes only. It is possible to date some of their mounts: the firm was called Smith & Beck, at 6 Coleman Street, London, from 1847 to 1857: the name then became Smith, Beck & Beck as in 1864 they removed to 31 Cornhill; and then became R & J Beck,[15] following Smith's retirement in 1860. Later they moved to 68 Cornhill, where they remained until the 1930s. They offered a series of slides in the 1850s of very general type, but acted as agents for a fine set of transparent injected preparations, apparently imported from Germany.[16] A significant number of these have survived (see plate 19) characterized by being mounted on slides of irregular size and unnecessarily large dimensions for the section. The range of subjects was wide, and the quality high, for the date. By the 1880s they supplied first a physiological and then a pathological series, followed later by a bacteriological series. Their mounts, even in the 1850s, were hardly ever paper covered; the slides have a decidedly modern appearance compared with their contemporaries—usually there is no ringing to the cover slip, and only one relatively plain label.

Samuel Stevens was a well-known agent for natural history supplies in the 1850s, and his list of 144 different preparations in the 1855 Quekett is a glorious jumble of specimens in no order of any kind, consisting only of objects of the most general interest for the amateur's cabinet: nothing more need be said of them here.

Similarly, F L West, advertising from 39 Southampton Street, Strand, was obviously a dealer offering a number of instruments as well as preparations. W Ladd, of 29 Penton Place, Walworth, was a definite manufacturer of stands, having an honourable mention in the 1851 Exhibition for them; one must assume that any preparations he sold were offered only as factored objects to accommodate his customers.

Pillischer, of 88 New Bond Street, was in a bigger way of business so far as can be judged from his catalogue of 1865,[17] but was essentially a factor of slides and not a mounter. He offered a wide range of injections and sections, with one or two oddities such as blood discs from *Lepidosyren* (at 2/–d., this was twice the price of ordinary blood mounts), but generally the catalogue contains an ordinary amateur assortment, and we may assume that this would also have been the case in the 1850s.

The beginnings of what would become an important amateur pursuit, as well as a commercial activity of considerable financial worth, also took place in the 1850s, when Shadbolt first mentioned making slides of diatoms in an arranged manner.[18] It seems likely that Thomas Comber started to make mounts of selected diatoms about 1858, but did not at that time arrange them in patterns. This seems to have occurred in the early 1860s, as other mounts of diatoms of the 1850s also lack symmetrical arrangement.[19] The Norman catalogue of 1872, already mentioned above, did not offer arrangements of diatoms, but the 1876 list of Wheeler[20] mentions that most of his recent and fossil diatoms are in symmetrical groups. In the 1890s several catalogues list specially beautiful groups for demonstration at soirees, etc.; several such slides have been examined and are works of art, but of course without scientific worth.

A series of preparations by Bourgogne of Paris has been preserved, the first dated July 1854, by J Bourgogne, of a marine alga. This is a standard 3 × 1 slide (see plate 45), but an earlier slide of TS human cranium is much smaller, 58 × 15 mm, and although signed is not dated. J Bourgogne seems rapidly to have abandoned use of paper covering, for a slide with his name and addition of a note about receipt of a 1st class medal in 1855, a strewn mount of diatoms, has only a plain balsam-attached round cover and two labels in the modern style. Others in this seemingly-large family of mounters include H Bourgogne, with LS human canine dated March 1859; Charles; Eugène with two slides after 1873 and after 1878; and finally the Bourgogne brothers with a very adequate section of sheep's kidney. The 1866 catalogue of Baker[21] mentions that preparations by Bourgogne are supplied, but unfortunately does not distinguish between mounters in the lists. In addition to the foregoing mounters, slides by Bourgogne and Alliot of Paris have also been seen, but it has not been possible to date them.

More information has survived from the 1860s than for the previous decades. Baker catalogues of preparations exist for 1862 as well as for 1866, as we have already seen (page 290 above). The 1862 list has 23 pages, and that of 1866 has a further four, but much the same kind of preparation is included. Opaque injected preparations were 1/9d. each, as were transparent injected slides; other general objects, including those by Topping were 1/–d. each. Many test objects were listed, including the Möller probe platte with list at 10/–d. (By 1868 this had been developed into the famous Typen-Platte, containing 400 types, at 60/–d.[22]) Many different diatom slides were offered, as well as a wide variety of plant parts such as cuticles, petals, and pollen. Sections of wood, coal, and various plant material were plentiful, and there were very many insect mounts, as well as miscellaneous matter such as hairs, spicules, polariscope preparations, and shells. Anatomical slides included the usual urinary series, bone preparations from many species, teeth of various kinds, striped and unstriped muscle fibres from a variety of chordates, and a few skins and spermatozoa, in addition to the usual large number of opaque and transparent injections. Three pages of microphotographs are also included. This company continued to maintain an interest in the supply of preparations, in addition to their manufacture of microscopes, and their acting as agents for such famous foreign firms as Grubler, Zeiss, and Reichert. Their 1887 list does not give details of their stocks, but shows that they kept preparations by Cole, Norman, Topping, Möller, Boecker, Thum, and other eminent preparers. Typical prices at this time were 1/–d. each for general mounts, 1/6d. for physiological and pathological mounts by Cole, and 2/–d. for test objects in special mountants.

In 1892 they offered a whole series of lantern slides of various histological specimens, doubtless to cater for the increasing amount of elementary science teaching going on at that time. They offered one of the most complete sets of bacteriological preparations obtainable anywhere, and in 1898 started to offer a lending library of slides on very favourable terms. Their mounted slides included 156 specimens illustrating plant histology, 144 mounts (15 of which included more than one section) of slides for physiology and pathology, 84 slides for zoology, and 24 specimens for teaching bacteriology. This was a very favourable collection for the histologist, and was even further developed over the next few years, until by 1909 Bakers claimed that over 69 000 slides had been borrowed, quite apart from those purchased. In addition to the basic physiology set of 72, there was a supplementary group of 35 well chosen mounts, and a further set of 164 pathological mounts, making this one of the most complete lists ever offered. Tropical diseases were covered as well as bacteria in great profusion, besides all the various purely amateur preparations. Some examples from this admirable range are shown in plate 46.

Other companies beginning to supply mounts in the 1860s included Amadio, whose 1864 catalogue[23] contained 56 pages, no fewer than 40 of which were devoted to preparations. These were sold at from 12/–d. to 18/–d. per dozen, and were mostly numbered for ease of ordering. Numbers 1–152 were a mixture of items, some of them test objects, 153–277 polariscope objects, 278–398 diatoms, 399–424 spicules, 425–455 sections of shells, 456–632 insect mounts and dissections, 633–669 various parasitic insects, 670–716 opaque mounts, 717–899 vegetable preparations, 900–1026 miscellaneous objects, 1027–1045 sections of hairs, 1046–1064 sections of fossil teeth and bone, 1065–1084 blood discs and spermatozoa, 1085–1135 sections of bones and teeth from a wide range of animals, 1136–1160 various simple histological sections, 1161–1319 a large selection of injected anatomical preparations. The similarity between this list and that already quoted from Norman is notable. So far as surviving specimens labelled with the name Amadio would show, all of them were brought in from various professional mounters, had a new label applied, and were re-sold. It seems unlikely that Amadio ever mounted his own preparations. In addition to the above specimens, there was also a section on microphotographs, numbering about 60, but with no indication of their source. Here, however, the catalogue states that Mr Amadio reduces ordinary photographic portraits to microscopic size, at a fee of 21/–d. He may or may not have done this for himself, of course.

Other dealers in preparations in the 1860s included Crouch,[24] who merely mentioned in a few lines that he supplied objects including entomological, physiological, botanical, and injected specimens, at prices ranging from 9d. to 2/3d. each.

A very specialized type of mount was that engraved with a diamond point on glass, either as a test object or as a curiosity. The first category is synonymous with the name of F A Nobert of Barth in Pomerania, who made a ruling engine (which still survives in the Smithsonian Institution in Washington) suitable for making stage micrometers. This concept was developed into an ever finer series of test rulings, for as soon as the finest band of a series was resolved by an expert microscopist, he ruled another band with finer spacings. The ability of an instrument and an individual to resolve these bands was a source of stimulation to both opticians and microscopists in the nineteenth century, and progress in microscopy owes much to such use of test objects. The earlier natural objects proposed by Goring exhibit enough individual variation to be unreliable when used for critical work, and in spite of the continuing use of certain species of diatoms throughout the century, the real test was the ruled line. It has been shown that the finest bands of Nobert's rulings could never have been resolved by the light microscope,[25] as their separation lay beyond even its theoretical limits. There is a particularly complete collection of Nobert rulings in the Optisches Museum of the Carl-Zeiss-Stiftung in Jena, but other examples are widely found for they were sold in fairly large numbers, albeit at high prices. For example, Nobert is known to have made a series of seven test plates between 1845 and 1873, and the 1880 edition of Wheeler's catalogue[26] states that Nobert rulings were received direct from Nobert to be sold at prices of £8 for the 19-band (to a fineness of 112000 to the inch), and £15 for the 20-band (up to 200000 lines per inch).

In the same list Wheeler also offered his own rulings, after the style of Nobert. These did not go more than a mere 20000 to the inch (at 10/–d.), and were obviously not in the same class at all. After Nobert's death in 1881 the rulings naturally became increasingly scarce, and the only adequate successor to Nobert was Grayson. He was a Yorkshireman who settled in Melbourne and developed a ruling machine, which was gradually perfected until it could make 120000 lines per inch—the best ever ruled. He offered three test-plates: 10 groups for 1000 to 10000 per inch; 12 groups 5000 to 60000 per inch; and 12 groups 10000 to 120000 per inch. These did not become available until 1894, when they were inspected by Nelson,[27] who reported on them most favourably, saying that the irregularities of spacing so often seen in Nobert rulings were absent in the Grayson ones. As others appeared, Nelson and Merlin appraised them in favourable terms. Bale summarized Grayson's work in his obituary,[28] and more recently Reed has added a few details.[29] He ruled diffraction gratings and mounted diatoms in addition to his micrometric work, and used realgar as mountant for his later slides.

In a very different category were the rulings made by W Webb on glass. In 1880 these included various epigrams and religious texts, made in microscopic writing: the 2nd chapter of St John, all 2070 letters, fitted into so small an area that the Bible would have gone into one square inch, cost 42/–d.; but the Lord's Prayer on the same scale was only 10/–d. The lines were much more coarse than those of Nobert (with whose productions Webb tried in vain to make them comparable), and the whole idea was that of a novelty of essentially the same kind as mounts of butterfly scales.

Overseas, a range of commercial slides similar to those available in England was offered. As examples we may quote the so-called Essex Institute preparations,[30] a few of which have survived for inspection, including wood and cereals, and a few histological slides of rather mediocre worth. The French suppliers Nachet[31] and Chevalier[32] mentioned preparations only very briefly. Nachet offers general objects for 1 f. to 1 f. 25, while anatomical injections cost from 2 to 4 f. Chevalier, whose list included many different optical goods as well as microscopes, listed ordinary preparations at 75c, entomological ones at 1 f. 50, teeth and bone sections at 2 f. 50, and timber preparations from 1 f. 25 to 3 f.

In 1875 a letter to the editor of the *M.M.J.* compared English and foreign preparers. Cole, Enock, Norman, and Amos Topping were acknowledged to be of the highest class, with the various members of the Bourgogne family their equal in their own specialities. Other specialists were Rodig and Möller, although the latter's prices and authenticity in naming were criticized. Cole and Son were stated to have recently branched out from preparing only diatoms to making also physiological and pathological specimens, many of which were excellent.[33]

We have already mentioned the Cole series of preparations and their accompanying descriptions (see pages 36–38 above), and we saw there that the support such a venture would have needed for success was not forthcoming. Nonetheless, a large number of preparations was supplied, and are all of first-rate quality. There is evidence that Cole carried out work for medical men's original research (see plate 46), as well as making his own slides, and there is no doubt that his publications as well as his preparations gave him an authority in histology never before or since enjoyed by any commercial mounter. We saw (page 40 above) that when Cole crystallized his methods in his book of 1895, they were not those of the research worker of the time, but rather those of the amateur tinged with a dash of commercial expediency, for example in his use of the freezing microtome. His son Martin was actually responsible for many of the preparations, and Wells gives a vivid account of the trade in 1888:[34]

> ... and I made my sketches under the Bloomsbury Dome and enlarged them as diagrams in a small laboratory Jennings shared with a microscopist named Martin Cole in 27 Chancery Lane. Cole, at the window, prepared, stained and mounted the microscope slides he sold, while I sprawled on a table behind him and worked at my diagram painting. Cole's slides were sold chiefly to medical students and, neatly arranged upon his shelves were innumerable bottles containing scraps of human lung, liver, kidney and so forth, diseased or healthy, obtained more or less surreptitiously from post mortems and similar occasions.

This volume of Wells' autobiography contains many reminiscences of biology and its teaching under T H Huxley, who was primarily responsible for its acceptance in the latter part of the nineteenth century in England.

Many other suppliers offered slides in the 1870s, among them diatom specialists such as Firth who started in an amateur way in 1877 and developed into a supplier of first-rate preparations for over thirty years, at remarkably cheap prices. A much more important supplier was Edmund Wheeler, who had a catalogue of 24 pages by 1876;[35] he had been awarded a prize medal as early as 1866, and by 1876 had a stock of 20000 slides on offer. He supplied mounting materials as well as preparations, and in addition to two pages of anatomical specimens, had 20 different slides of the human eye—a rarity for that date. They included sections of retina and choroid, plus macerated retina for Jacob's rods and ultimate termini of nerve fibres, with a section of the whole eye, adult or foetal which must have been a very difficult slide to make. At 3/–d., these were at least half as dear again as the usual preparations. The remainder of his slides included a great many diatoms, opaque mounts, insects, and microphotographs, in addition to the Nobert tests already referred to. He was also sole proprietor of the Webb engravings. The catalogue is interesting in having specimens of the special papers for covering slides stuck on as samples, the plain yellow or red covers for the backs being 6d. per 100, and the gilt ones, crimson or green, for the fronts being 1/–d. per 100. The front of this particular catalogue is interesting, for it shows that Wheeler had bought up the residue of the materials and stock of the late C M Topping. Thus Wheeler was successor to Topping. An eighth edition of the catalogue was issued in 1880, with much the same contents except that the total stock was now 40000, and the testimonial from Dr W B Carpenter reprinted from the earlier edition in more prominent display. The next edition is dated 1884, but bears the name of W Watson & Sons, in spite of its being called the ninth edition. Wheeler retired in the middle of 1884 due to ill-health, and transferred all his stock to Watsons, who thus stood in direct line of descent from C M Topping, and had some of his slides left. Apparently four first-class medals and diplomas had been awarded to Wheeler, in the United States as well as this country and Germany, and the only change in the catalogue was to name Watson as the supplier and add details of a few microscopes.

A tenth edition was issued in 1885, by which time a stock of 50000 slides was claimed. The histological series had not been extended, and the same preparations of the eye were still listed, but attention had been given to more mounts of the amateur decorative class, with a number listed for use at soirees etc., and the first mention of artistic groups of scales and hairs. New slides of bacteria, mounted with the greatest care as Specimens, were noted: they included anthrax, tuberculosis, cocci, etc. The twelfth edition offered a range of pathology slides also, and the 17th edition of 1905 advertised depots in Glasgow and Australia as well as branches in Edinburgh and Birmingham: these would be for the principal business of microscope supply and not primarily for specimens. Watsons also now had a circulating library of slides, similar in scope to that already mentioned as operated by Baker. Over thirty pages of closely-set type are given over to preparations, and the range of human histology is quite wide, with most organs represented. The eye is still represented more or less in the same terms as in the Wheeler list of thirty years before, but slides such as muscle end-plates, pacinian corpuscles, cochlea, and cerebrum are also included. A number of such preparations have survived for inspection, and compare well with any others of their date, although by the best modern standards they are deficient: thick sections and a rather diffuse stain do not allow the finest details to be made out. Sets of slides for students of physiology are offered, at 35/–d. for two dozen complete in box, and the range of pathological and bacteriological preparations had been extended. The usual range of amateur preparations was still offered, with a page of special slides for soirees, including cornucopias of flowers and pictures of a hen with chickens made from butterfly scales. The 20th edition of the catalogue was issued in 1910, and the contents were much the same as before, but with more bacteria and pathological preparations. The firm of Watson always had a strongly traditional approach to its customers, keeping to hand a wide range of accessories and specimens which one would have thought had long since become unobtainable. When the company was merged with another in the 1960s, during a visit to their Barnet works and stockroom one could almost have been back in Edwardian times, so little had been altered. This may go a long way towards explaining their strongly amateur bias in both instruments (always a deference towards the connoisseur) and preparations.

At the time Wheeler was transferring his interests to Watson, in the 1880s, much competition was evident, in spite of Pelletan's strictures in 1879,[36] where he said that most commercial objects had small value in spite of being pretty. Histological preparations he thought had the least value of any, the preparers having neither knowledge nor technique. In this context it may be worth noting that Pelletan himself had started a journal[37] in 1877, devoted to histology and related subjects; in, for example, the 1883 numbers he advertised himself as directing a laboratory for general micrography, able to supply a range of preparations! Others who advertised that year included Boecker of Wetzlar, Wheeler, A C Cole, and Thomas Bolton. The advertisement pages of the *J.R.M.S.* for the same year carried notes from Enock, How (for rock sections), A C Cole, Wheeler, Smith (of Carmarthen, for botanical preparations), Russell (rock sections), Hunter & Sands, Joshua (of Cirencester, for freshwater algae), and Boecker; while Beck and Ross included mention of preparations in their optical advertisements.

Of the above, Enock deserves more attention here, in spite of his not offering histological slides. It is likely that Fred Enock was the first to mount insects for the microscope without pressure, probably for the Loan Collection of Scientific Apparatus at South Kensington in 1876. Some of these slides were deep cells, preserving the form of the living animal. He went on to become famous for his mounts of this kind, including admirable dissections of insects (see plate 22). Enock was very secretive about his methods, apparently divulging it to only one person, who described it after Enock's death.[38] Two other claims to fame belong to this mounter: the educational merits of supplying his mounts with descriptions illustrated with figures of the chief points were undoubted, as largely increasing their value. None of these figures seem to have survived, unfortunately. The other merit is a scientific matter, for Enock devoted himself to studying the tiny 'fairy-flies' or Mymaridae. At his death in 1916 he had almost completed a full monograph on the family, noted in entomological circles as egg-parasites and most difficult to handle: Enock supplied slides of them, as flat balsam mounts. In a modern issue of the Quekett journal, Enock's work on the group has been investigated and extended,[39] and the nature of his liquid mountants more fully established.[40]

Two mounters from overseas should be noted from the 1880s, both dealing largely with marine life. The Naples laboratory started to offer preparations commercially from 1880, Fritz Meyer being largely responsible. With their first price-list, including over 400 preparations, the point was made that slides sold before that date had only occasionally had a scientific value, as to make useful preparations technical skill had to be combined with scientific understanding if the important points of a specimen were to be brought out. The Naples slides inspected do indeed combine both requirements, offering some marine larvae which are now very scarce, well fixed and stained more than eighty years after they were mounted.[41]

The other marine-life preparers lived in Jersey. Sinel offered embryological slides in 1883,[42] mounts of developing ova and larvae in fluid, a number of which have survived for inspection. They have probably never been surpassed for quality, and are still usable for demonstration to undergraduates. Sinel joined forces with Hornell in the 1880s before Hornell himself supplied a very good range of specimens from about 1894.[43] In 1901 he published a volume[44] of articles put out in the late 1890s, dealing with life histories of a series of coelenterates, annelids, echinoderms, and other groups. They are well written and illustrated, and the half-tones give an indication of the large size of his marine biological station. From this he supplied museum mounts of skeletons, as well as his series of preparations. Some of these dealt with elementary botany and zoology, but the most remarkable were his two sets of 14 slides on marine zoology: these included not only various well prepared larvae, but also difficult material such as sections of the eye of the cuttlefish. A number of such slides inspected have demonstrated their lasting excellence. The two series of slides, plus a bound volume of letterpress cost 42/–d. in 1901.

Specimens for amateur mounting

An interesting development in the 1880s was the arrangement made to supply unmounted objects, labelled and selected as suitable, for the amateur to make into his own preparations. Doubtless this had a deep psychological satisfaction for the amateur, who could justly claim to have made the mount instead of merely buying it. Ward of Manchester, who also supplied a variety of ready-made preparations, and was noted especially for his superb micro-crystals for the polariscope,[45] offered a wide range of such objects. In 1883 he offered (from 249 Oxford Street, Manchester) no fewer than 34 series, most of them containing 24 different unmounted objects. Some sets have been inspected, and are all similarly made up. A packet with the series label contains folds of paper, rubber-stamped with the serial number of the specimen, and with a somewhat ill-duplicated sheet of instructions on top. The specimens ranged through seeds, zoophytes, spicules, molluscan palates, starches, pollens, and hairs, and most of them appear to be quite accurately identified. They were well dried, of course, and the usual amateur treatment of the 1880s and 90s would hardly be calculated to make first-quality mounts from them, but they were certainly a starting point:

> The above series of objects were introduced to the Microscopical world some few years since, and have met with universal approbation.
>   To the possessor of a Microscope they offer a cheap and ready method of obtaining a cabinet of interesting objects in various classes of Natural

History, and with the concise instructions accompanying each packet, even the beginner experiences but little difficulty in making good Microscopic slides.

The Series is very varied, and includes in the list above, upwards of 700 specimens carefully and accurately named, and every one different, obtained from almost every portion of the Globe, and from the Animal, Vegetable and Mineral Kingdoms.[46]

The sets stayed popular for many years, being sold widely by dealers without Ward's name, although always having his same advertisement. Different sets were introduced as time went by, but the same very brief instructions were included; they have proved entirely adequate when mounting some of the specimens.

Rather different specimens were provided by Thomas Bolton, who specialized in sending out tubes of living organisms. He started, on a subscription basis, in 1878, with a tube of *Lacinularia socialis* on 13th September. At weekly intervals thereafter he undertook to send out further specimens, and seems to have done so, for by 28th December 1883 the list of species supplied had continued without a break. Viewed in retrospect, this was a somewhat heroic undertaking, to say the least. To accompany the specimens he produced, from August 1879, a portfolio of drawings,[47] from 17 Ann Street, Birmingham. He had moved to 57 Newhall Street, Birmingham, by February 1881, by which time he had supplied his specimens to many different science schools and individuals. The drawings and their descriptions were often rather crudely done, with handwritten text and very sketchy drawings poorly duplicated, but by the time the tenth issue was put out (February 1884), the front page at least was quite grand! He was a thoroughly practical observer of living objects under the microscope, as his little leaflet[48] testifies, and this was in marked distinction to much work of perhaps a more scientific character going on at that date: one microscopist at least had not gone over totally to the microtome.

One other venture of the kind might be noted, and that is the marketing of ready-cut sections of various tissues, supplied in tubes ready for further processing, by both Watson and Baker in the first decade of this century. Baker also supplied blocks of embedded material from their series of slides on human physiology; each block was sufficient for about 200 slides, and cost only 2/6d. in their 1899 list. There is no doubt that for the student of histology these would represent a useful purchase, providing plenty of material on which to practise staining without the considerable labour of fixing and infiltrating.

Mounters working from the beginning of the twentieth century

Three firms require mention here. The first offered preparations in fluid mounts, largely for the amateur but of excellent quality. Clarke & Page advertised for the first time about 1905,[49] but the partners had advertised separately concerning microscopical goods for the previous two years. They supplied their still-admired slides until 1923, some of them being ordinary balsam mounts of flattened insects and the like, but almost all of them as realistic whole mounts of small arthropods in liquid. Some examples are shown in plates 22 and 49.

The second mounter in this section was Abraham Flatters, whose book was discussed above on page 41. Flatters worked in Longsight from the beginning of the century, probably 1901. After a few years he was joined by two partners, and the firm became Flatters, Milborne & McKecknie, still at 20 Church Road, Longsight in Manchester. They offered a wide selection of slides of all kinds, including histological mounts of good quality, and also started their own journal[50] in 1910. The purpose of the venture was simple, to help the amateur to mount his own objects, but a purely commercial motive was disavowed in the preface to the first issue of 1st July 1910. The contents were highly practical, with photographs showing the various procedures in detail, plenty of diagrams, and actual photomicrographs of the results which could be obtained: in fact, the recipe used in Flatters' earlier book, but covering both plant and animal preparations. The specimens mentioned in the text were available for sending to those who wished to mount them, or prepared slides were offered as an alternative to those who did not. Some of these preparations have survived, and are good of their kind; the modest cost was from 7d. to 1/6d. each. The journal continued until 1916, and the company ceased only in the 1960s, after passing through a number of stages outlined from the slide labels shown in plate 49. The 1930s catalogues of Flatters & Garnett, as it was by then called, offered all manner of things for the microscopist. The list of preparations covered 87 pages in the 1933 edition, many of them being first-class histological mounts of the widest possible variety of tissues and organs. Many of these slides are still in active use in school and college laboratories.

The final commercial mounter we should consider is a rather special case, that of a highly-qualified scientist offering sets of histological slides with accompanying descriptive illustrated guidance notes. Some examples of the preparations, which were of excellent quality, are shown in plate 49, and a few sets of the descriptive matter have survived.

Although Sigmund had offered other histological slides before the start of this venture, it was his advertisements just before the first world war which brought him a great deal of success.[51] The material offered was on the physiological histology of man and mammalian animals, ten parts to be issued, one every three months, to cost 10/–d. per part including ten slides. There are many similarities between this venture and that of Cole thirty years before, but Sigmund's descriptive matter was never nearly as ambitious. The organ systems covered included skin, muscle & bone, central nervous system, reproductive organs, respiratory & urinary organs, the eye, hearing, smell & touch, and the circulatory system. The slides of nervous preparations are still of much interest and use, being of the highest commercial quality.

This brief survey of commercial mounters has been written with the emphasis on any contributions made to histological technique by such people. It will have become obvious that contributions to knowledge were minimal when seen from the point of view of direct original work. However, it should perhaps be remembered that their work stimulated interest in the microscope, and that they had a part to play in the educational system insofar as they provided students with preparations for their courses in the life sciences at different levels. It may be that with the increasing emphasis on work in biology in the schools in England towards the end of the nineteenth century, this contribution was more important than might be suspected.

References for Chapter 6

1. A Pritchard—*A list of two thousand microscopic objects*. (London: Whittaker, 1835.) As mentioned in the text, Pritchard offered sections of fossil woods, and he was the first commercial mounter to do so. Interest in the subject was only just developing in the early 1830s, largely as a result of the work of Witham. His small book of only 84 pages contained 16 plates, and may be regarded as the beginnings of the subject: H T M Witham—*The internal structure of fossil vegetables found in the Carboniferous and Oolitic deposits of Great Britain, described and illustrated.* (Edinburgh: Blackwood, 1833.)

2. C Brooke (1875)—The President's address, *M.M.J.* 13:97–107. See p. 107.

3. See the note on p. 132 of *J.Q.M.C.* 4 (1875).

4. [D Cooper] (1841)—Microscopical memoranda, *Mic. J.* 1:15–16.

5. *A catalogue of microscopic preparations, by C. M. Topping, and the most esteemed makers.* (London: C Baker, 1862.)

6. *Plain and achromatic microscopes, magnifying glasses, microscopic objects, lenses, &c.* (London: T & R Willats, 1846.)

7. J Quekett—*A practical treatise on the use of the microscope.* London: Bailliere, 2nd edition 1852.)

8. *Classified list of first class microscopic objects, with many unique, rare & interesting specimens.* (London: J T Norman, 1872.)

9. G W W Stevens—*Microphotography.* (London: Chapman & Hall, 2nd edn, 1968.)

10. L L Ardern—*John Benjamin Dancer*. (London: Library Association, 1960.)
11. J F Stirling—A forgotten genius. John Benjamin Dancer, microscopist, optician and instrument-maker, *Watson Microscope Record*   *44 :3–6, 45 :12–16 (1938), 46 :8–10, 47 :12–14 (1939)*.
12. A L E Barron (1960)—On J B Dancer & the discovery of his microphotographic negatives, *The Microscope   10 :234–238*.
13. *Revised catalogue of microscopical slides etc.* (London: R Suter, 1900.) (He lists 512 different microphotographs, all at 1/–d.)
14. C F Rousselet (1898)— On some micro-cements for fluid cells, *J.Q.M.C   13 :93–97*.
15. G L'E Turner (1969)—Hugh Powell, James Smith and Andrew Ross: makers of microscopes, *in* J North (Ed.) *Mid-nineteenth-century scientists*. (Oxford: Pergamon Press, 1969.) See pp. 104–138.
16. Editorial note (1861)—Catalogue of transparent injected preparations, sold by Smith and Beck, London, *Q.J.M.S.   9 :129–131, 217*.
17. *Catalogue of achromatic microscopes etc.* (London: M Pillischer, 1865.)
18. G Shadbolt (1852)—On the structure of the siliceous loricae of the genus *Arachnodiscus, Trans. M.S.L.(O)   3 :49–54*.
19. D S Spence (1958)—Thomas Comber, *The Microscope   11 :294–295*.
20. *A classified list of first-class microscopic objects*. (London: E Wheeler, 7th edition 1876.)
21. *A catalogue of microscopic preparations by Bourgoyne, Moller, Norman, Topping, and other esteemed English and foreign preparers*. (London: C Baker, 1866.)
22. See the note in *J.Q.M.C.   1 :141 (1868)*.
23. *A catalogue of achromatic microscopes and other optical, philosophical, and mathematical instruments*. (London: F & J Amadio, 1864.)
24. *A catalogue of achromatic microscopes etc.*, (London: H Crouch, 1866.)
25. G L'E Turner & S Bradbury (1966)—An electron microscopical examination of Nobert's finest test-plate of twenty bands, *J.R.M.S.   85 :435–447*.
26. *A classified list of first-class microscopic objects,* (London: E Wheeler, 8th edn, 1880).
27. See the note in *J.R.M.S.   1895 :134–135*.
28. W M Bale (1919)—Obituary. H J Grayson, *J.R.M.S.   1919 :20–22*.
29. F Reed (1957)—H J Grayson, *The Microscope   11 :123–124*.
30. See the rather cautious note in *M.M.J.   2 :55 (1869)*.
31. *Maison Nachet & Fils fabrique d'instruments de micrographie*. (Paris: Martinet, 1872.)
32. *Catalogue illustré des microscopes de précision*. (Paris: Chevalier, 1885.)
33. See the letters to the editor, *M.M.J.   14 :209–210 (1875)*.
34. H G Wells—*Experiment in autobiography : discoveries and conclusions of a very ordinary brain, since 1866*. (London: Gollancz, 1934.) See pp. 313–314.
35. *A classified list of first-class microscopic objects*. (London: E Wheeler, 7th edn, 1876.)
36. See the report in *J.R.M.S.(O)   2 :943–944 (1879)*.
37. *Journal de Micrographie*. Paris, vol. 1, 1877, editor Dr J Pelletan.
38. J S Pratt (1921)—Mr Fred Enock's method of mounting heads of insects without pressure, *J.R.M.S.   1921 :141–146*.
39. C H Ison (1958)—Fairy-flies: their capture, mounting, and identification, *J.Q.M.C   28 :59–72*.
40. E D Evens (1958)—On fluid mounts by F Enock, *J.Q.M.C.   28 :51–53*.
41. See the notes in *J.R.M.S.   1880 :700–701, 1031–1032*.
42. See the note in *J.R.M.S.   3 :147 (1883)*.
43. See the advertisement in the end-papers of *J.Q.M.C.   12 (1895)*.
44. J Hornell—*Microscopical studies in marine zoology*. (Jersey: The Biological Station, 1901.)

45. E Ward—*Microscopical mounting, and micro-crystallization.* (Manchester: Ward, 1883.)
46. From an advertisement inside the wrapper of a series of Ward's unmounted objects.
47. T Bolton—*Portfolio of drawings and descriptions of living organisms (animal and vegetable) illustrative of freshwater and marine life.* (Birmingham: T Bolton, 1879.)
48. T Bolton—*Hints on the preservation of living objects and their examination under the microscope.* (Birmingham: Bolton, 1880.) Bolton died in 1887.
49. See the advertisement in the end-papers of *J.Q.M.C. vol 9 (Nov. 1905)*.
50. *The Micrologist* was published by the company from its offices.
51. See the advertisement in *J.Q.M.C. (April 1913)*.

7. Microscopy, Microtomy, and Histology in the Nineteenth Century

Earlier in this book there has been frequent reference to general histological matters while surveying progress in specimen preparation. These references were necessarily concerned with matters of detail: we should now paint a broader canvas, putting advances in microscopy and microtomy in the wider perspective of medical education and research.

Little attention has been given to the history of histology. The name itself appears to have been introduced in 1819,[1] in a small booklet of only 40 pages, but with sufficient importance to be translated into French within a year. Its new division of human tissues did not, in fact, differ very much from Bichat's arrangement which will be mentioned below.

Papers on the history of histology tend to be short and rather elementary, and need not be quoted here, with the possible exception of O'Rahilly's paper,[2] which is interesting as an example limited to a listing of significant dates derived from secondary sources, with the notable errors and omissions which that unfortunately implies in this subject. In a very different category are the five papers by Baker,[3] which go into much detail and present a most stimulating series of ideas on the development of theories of the cell, and thus incidentally of histology. The book by Hughes[4] also offers a well researched synthesis, in an enjoyable and informative format, valuable for giving insights into a variety of approaches to the questions of histological and cytological evolution. In more specialized areas, Clarke & O'Malley[5] provide a superb synthesis of neurological history vividly supported by numerous extracts from original works ably translated.

The general history of medical education is provided in the very readable book by Newman;[6] specialized information is given in Flexner;[7] the German scene in the 1870s is ably summarized by Billroth;[8] and useful additional material is given by O'Malley.[9] In addition to these surveys, a general knowledge of the history of medicine is required, and the standard medical history of Garrison,[10] with the books by Major[11] and Singer & Underwood[12] provide adequate coverage.

Specialized books on certain aspects of medical history are also helpful. The history of anatomy is ably described by Keen;[13] more modern works include that by Craigie,[14] which condenses a great deal of information into a relatively short article; Corner[15] gives another terse summary, and Hunter[16] is also still useful. The special aspect of embryology, owing so much to the microscope, is ably dealt with by Needham's excellent account,[17] and by Meyer.[18]

Works on pathology and bacteriology to be recommended include those by Long,[19] Krumbhaar,[20] Bulloch,[21] Ford,[22] and Grainger.[23] Many other sources are suggested by these works, and selected references will be given as appropriate.

The earliest years of histology

The history of histology begins for all practical purposes with Bichat's work.[24] He put forward a science of anatomy and pathology based on a classification of tissues in the body, with their distribution in organs and their special diseases. This was a monumental effort, and nothing of much consequence for animal tissues followed it for many years. In this respect the date 1830, selected as a watershed in the history of microtechnique, is also a watershed in histology. Before that date very little was done to shed further light on Bichat's work; afterwards there were ten years of hard work in microanatomy succeeded by thirty years spent developing the concepts of functional histology and pathology, before a virtual explosion of original work in all branches of microscopical work on tissues.

Bichat did not use a microscope in his work, actually mistrusting the device. Later, when good instruments were available, this kind of prejudice was removed from most of the scientific world; however, it would still be necessary to urge the employment of the instrument in medical studies for many years after the famous paper by Hodgkin & Lister was published in 1827.[25] If Bichat had laid the foundations of modern histology, then this paper secured the cornerstone. For the first time an accurate description of tissues was published, but this was not immediately apparent. In particular, one of their findings, that human red blood cells have no nucleus, was taken to mean that the microscope was still unreliable, as most earlier observers had claimed to find one.[26] It required Weber[27] to make the comparison between human and other vertebrate corpuscles to clear up the confusion: this was not done until fourteen years later, which seems an astonishing lapse of time for not too difficult a problem. Even more remarkable, as late as 1869[28] Savory asserted that everyone had been wrong and that it was only the method of preparation which made observers think that non-mammalian erythrocytes had a nucleus, as none was present until post-mortem changes occurred!

These events offer a revealing glimpse of histological progress in the middle part of the nineteenth century, with isolated steps forward but much confusion. In retrospect a lot was achieved, but one must doubt if findings became widely known very quickly, and to what extent they influenced the work of others. In this context the account of Addison's work by Rather[29] is also very revealing of the ebb, flow and eddy of biological research in this period.

One important way in which the gospel of the mic oscope was spread was by the formation of a school of workers, and Purkinje provides an excellent example of such a school. In 1832 he acquired a Plössl microscope of adequate performance, and shortly afterwards gave classes in his own house in Breslau.[30] The 1830s were years of intense activity in Germany, with release from the constraints of Naturphilosophie coinciding with the availability of good microscopes, and impressive results soon accumulated in spite of a lack of any sophisticated methods for preparing specimens for inspection. The example of Purkinje influenced Müller in Berlin to use the microscope, and he in turn influenced his own pupils, including Henle and Schwann. It is probable that only a few men were responsible for the great impetus given to scientific microscopy in Germany at this time, while other countries achieved only sporadic results. Their instruments had only the same order of magnification and resolution as the best of Leeuwenhoek's simple microscopes but they were much easier to use and freely available commercially. The first flush of enthusiasm inevitably led to conflict between new and old observers, but much was discovered.

The rise of histomorphology

We have seen that preparative techniques in the 1830s were limited to teasing and/or squashing fresh material: the low apertures of the instruments of that time were entirely suited to the resulting observations. During only the first five years of the 1830s, for example, Wendt investigated sweat-gland ducts,[31] Purkinje and Valentin described ciliated tissues,[32] Wagner published on muscle fibres,[33] and

Gurlt dealt with sebaceous glands.[34] This is not a comprehensive list, but a selection to show the range of work undertaken. In 1836 some British results were available in Boyd's notes on stomach glands,[35] but it was in Germany that the main work went ahead with Remak on nerve fibres,[36] Purkinje on brain-cells,[37] and Valentin on nerve endings,[38] as a few selected examples. Before the end of the decade further work on the nervous system was carried out by Purkinje and Remak, while others dealt with muscle, epithelia, gastric glands, and general cell theory, all in Germany. In 1839 Wagner wrote his book on physiology,[39] which was much in advance of its time, being based on microscopic functions.

By the beginning of the 1840s interest was beginning to shift towards the functional aspect of histology, also in countries other than Germany. Bennett's little summary[40] was well reviewed,[41] and a quotation from the review is interesting:

> The time has at length arrived, when it has been deemed expedient to institute a Class in Edinburgh, for instructing the rising members of the profession in the manipulation of the Microscope. This really becomes daily the more necessary, seeing that scientific practitioners throughout Europe, are resorting at length to this means, with such acknowledged advantage and success. ... It will be a matter of surprise to us, if the London Medical Schools do not appoint Professors for the same purpose; and this appears really the more looked for and demanded, seeing that many sources of error which creep into the experiments made by the tyro, are too apt to cause him to hastily publish results, and thus render more confusion in the science than is necessary.

The review goes on to agree with Bennett that some histologists have gained their knowledge from books rather than from preparations. The twenty-seven pages of the pamphlet are thoroughly scientific, but this is likely to be the result of Bennett's working abroad rather than a reflection of British attitudes and knowledge at that time.

It would be incorrect to minimize the English interest in histology at that time. Grube's paper on the macula[42] was reprinted in translation; the usual techniques of slight compression and very patient observation were used. Bowman's work on muscle[43] and on the kidney[44] stand comparison with any: in neither case were his results bettered for a long time, again tending to confirm the impression that it was not the ability which was lacking in this country, but perhaps the inclination. The recently-instituted Microscopical Society of London, first in the world, displayed little interest in histological matters in spite of the fact that its senior members and officers were largely medical men. In the volumes of its papers read before 1850 only fourteen titles out of a total of fifty could possibly be called histological, and the greater part of these are by John Quekett. One of them is a good example of the use of the microscope being applied to identification of a tissue: a fragment of material from a church door was proved to be of human skin, from the back of a fair-haired person, thus supporting the tradition that a captured Danish pirate was flayed and had his skin nailed to the door 900 years before.[45] This is all very interesting and amusing, but it was not of the same order of importance as the work being done in Germany at that time, and one cannot but feel that the Society lost a great opportunity when it failed to stimulate the application of the microscope to important matters in the life sciences.

There had, of course, been some interest in histology on an organized basis expressed in England by that date, and an excellent example is the fifty-one page booklet by Paget.[46] This is a very impressive account of histological knowledge not bettered anywhere in the world at that date. The work is thoroughly documented and any student of the time who read it would have been fully in command of the then current histological situation. As an example, the section on nerves may be quoted. Paget himself remarks that a great deal had been discovered in the past four years, and quotes all authors who had contributed to current views, in nearly 100 references. No mention is made of the retina, or, of course, the inner ear, and nothing is recorded of nerve endings or the nature of nerves. Paget says in the preface:

> The design of the present Report ... is to bring together, in the briefest possible space, the principal conclusions regarding the structure and functions of the several tissues of the human body which have been rendered certain or most probable by microscopic investigation. .... In no department of medical science has there been so great an addition of facts in the last ten years as in minute anatomy; and in none has the access to knowledge been more difficult ...

Shortly afterwards Bennett gave some lectures in Edinburgh,[47] in which he included a short account of the history of histology. A quotation regarding the relative status of the subject in England and Germany is instructive:

> On the present position of histology I need not dwell long. The microscope is now generally acknowledged to be an instrument absolutely essential to the anatomist and physiologist. If a practitioner who had received his education only eight years ago, were to look into any of the modern works on these subjects, he would be surprised to see the textures described, not as he was formerly accustomed to consider them, not as they present themselves to his sight, but as they appear under the microscope ...

It must be acknowledged that the Germans have taken the lead in histological studies, and the names of Ehrenberg, Muller, Schwann, Schulz, Wagner, Weber, and Valentin, whose reputations principally depend on the discoveries they have made by means of the microscope, must be well known to you. The activity, also, the Germans now display in the examination of healthy and morbid structures, is extraordinary, and can only be appreciated by an actual observer. Much of this, undoubtedly, depends on the systematic manner in which histology is studied, and the ready support afforded those who dedicate themselves to its cultivation.

... this study has made such progress in Germany, where private courses of lectures have been given upon it for some years, and where, in consequence, the young anatomist commences his career with a knowledge which would have astonished the older observers. By means of the instruction thus received he soon learns, in a science so unexplored as that of histology, to rival his teachers ...

This important summary of the position rightly emphasizes the importance of the German system of schools of microscopical anatomy. Müller had written his book on glands as early as 1830,[48] but even more important was his personal influence on his pupils at Bonn and Berlin: there were no parallels outside Germany at the time.

Histophysiology and medical education

By the 1840s there was a clear basis of histological knowledge. Henle's book on the epithelia[49] must have been widely circulated, and his slightly later book classifying tissues on a sound histological basis[50] was a key work. Later in the decade came the first English textbook-proper of microscopic anatomy,[51] including the first description of the corpuscles of the thymus. The word 'protoplasm' had been used for the first time by Purkinje in 1839, as recorded in the inaugural dissertation of one of his pupils;[52] and we have already seen that the first textbook of microtechnique was written by the English histologist John Quekett in 1848. There were thus available both theoretical and practical guides to stimulate work in teaching and research.

The course of lectures given by Bennett in Edinburgh had a parallel in the work done by Acland in Oxford, as the quotation at the beginning of this book so vividly shows. In 1843 Guy's Hospital offered superintended observations with the microscope,[53] and in 1854 microscopical meetings started in the Apothecaries' Hall.[54] Quekett, admittedly perhaps of unusual zeal for his time, had arranged for instruments to be available in the histological theatre of the Royal College of Surgeons by 1855,[55] and his lectures on the subject had been published in 1851.[56] To modern eyes these are a curious blend of plant

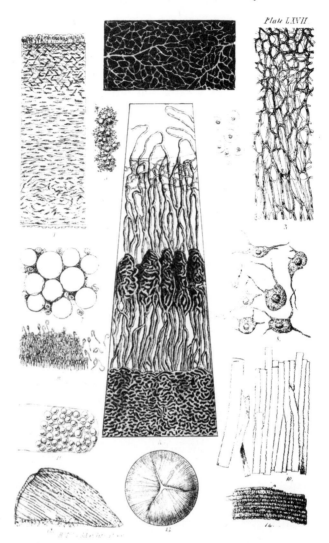

Figure 147. Retina as illustrated by Hassall, 1849 (from the book, note 51 in text: his plate LXVII).
Numbers 2 and 5–9 show retina. Number 2 is the surface view of injected capillaries (from a Quekett preparation), at the top of the plate in the centre. Number 5 is the bubble-like mass to the left below 2, showing the nuclei of the granular layer of the retina. Other numbers are supposed to show other details, but it is obvious that retinal structure was unknown.

and animal histology, omitting many of the now usual tissues. They were, however, based on his splendid series of preparations, some of which are still to be seen in the Hunterian Museum. This collection was probably the most extensive in the world by the 1850s; Kölliker[57] stated that it ranked with the Hyrtl collection of injections in Vienna, with the collections of injections and some sections in Utrecht, and with the London collections of Tomes (teeth) and Carpenter (hard tissues of the lower animals). In fact, this was probably an understatement of its importance, in spite of the fact that the interest and industry of one individual alone was responsible for its creation.

A cry from Ireland uttered in 1850[58] is interesting for two reasons. The first is the point made that no original communication in histology had been made from that country, in spite of the fact that

> ... In the schools of London and Edinburgh not a few medical men have risen to fame and eminence from the prosecution of this method of research, while amongst our continental brethren the study of the microscope has long occupied a prominent position in medicine, and proved highly conducive to the advancement of professional knowledge. ...

Lyons had the declared object of stimulating the authorities to add histology to the Irish curriculum. His twenty-one pages relied heavily on French authorities, and herein is the second point of interest. He had worked in Paris, which was largely isolated from other medical centres at that time in spite of developing the mere observation of clinical signs into a system of clinical medicine. However, as Ackerknecht has pointed out,[59] France had a number of active microscopists (including Raspail, Donné, Dujardin, Lereboullet, and Davaine) and these must have exerted their personal influence in spite of the prejudices of the clinicians. Donné taught microscopy privately from the 1830s, while Lebert taught from 1836, Gruby from 1841, and Mandl from 1846. Medical science was poorly regarded in France for many years after the rise of histological studies in Germany and England,[60] which must go a long way towards explaining the relatively few contributions coming from France. However, some individuals did receive instruction and inspiration in histology in Paris, and Lyons was among them. Only one year after his first rallying cry he had obviously started a course of instruction privately, for he gave details of it which are interesting as probably presenting a mirror image of those he had himself received abroad.[61]

His discourse contains forty-one pages, twenty more than that of the previous year, and shows that he had given a course between February and April in 1851, and specified the content of a second course to be offered in May and June for three evenings per week at 8 p.m., and being illustrated by use of a number of movable microscopes. The lectures dealt with the general employment of the instrument, its parts and their manipulation, sources of error in making observations, what we would now call basic cytology, animal fluids, capillary circulation, epithelia, cartilage, muscle, nerves and brain, and glands. Pathological work included inflammation, 'pyogenesis', and diseases of glands, epithelia, kidney, and a few miscellaneous conditions. The fee for the lectures was 21/–d. In addition he offered a class in practical microscopy, for one hour a day for a fee of 63/–d., attendance at which admitted the student to the lectures free of charge. For a few especially interested pupils he offered a two-year microscopic pupilage for a fee of fifteen guineas, Lyons undertaking to provide all details of his cases in addition to guiding the studies of his pupils and helping in the selection of subjects for original investigation. This kind of arrangement was certainly calculated to improve the application of the microscope to medical studies, and may have followed the continental pattern quite closely.

Such histological studies must have been much stimulated by the publication the next year of the first textbook of histology in the modern form by Kölliker,[62] who was among the first to apply the cell theory developed by Schwann to descriptive embryology. The book was rapidly translated into English,[63] most curiously by two men who seem to have had little or no reputation for work in histology: this is an interesting commentary on matters in the 1850s. It is still a very interesting source, not least for its numerous illustrations. Kölliker had also produced a larger textbook on microscopic anatomy, a few parts of which were incorporated into the English version, although it was not thought necessary to translate it entirely. The author's preface says:

> Medicine has reached a point, at which Microscopical Anatomy appears to constitute its foundation, quite as much as the Anatomy of the Organs and systems; and when a profound study of Physiology and Pathological Anatomy is impossible, without an accurate acquaintance also, with the most minute structural conditions.

The account of the retina was the best to have appeared to its date; it is brief and free from gross error.

In referring to the work of Schwann, Kölliker recognized the essential similarity of plant and animal cells. Although more than a decade had elapsed between the publication of Schwann's book[64] and that of Kölliker, few others had based histological details on the cell theory. It is true that Purkinje had made observations of cell structure even before Schwann, but the question had become clouded by the fact that Schleiden had pressed quite wrong views of cell formation, thus causing much confusion in the 1840s[65] and delaying acceptance of the basic theory. In fact, it was botanical investigations which finally

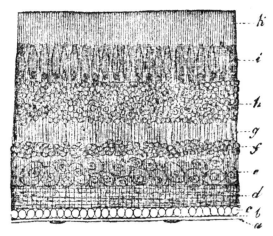

Figure 148. Retina as illustrated by Kölliker, 1854 (from the book note 57 in text: Volume 2, page 369).

The figure has been enlarged from its original size, for extra clarity. *c* Was labelled as clear globules, *d* as expansion of optic nerve, *g* as fine-granular layer, where the radiating fibres are more distinct than elsewhere, *i* as the internal division of the bacillar layer, and *k* its external division, with the prolongations of the 'cones' and the true 'rods'.

It is clear that some progress has been made in retinal histology, but much remains to be discovered.

showed up the error, in spite of some very clear illustrations of the animal nucleus in division, for example, Barry's drawing of cleavage in the rabbit,[66] and Kölliker's own work on division in *Sepia*,[67] both of which were intentionally labelled (wrongly, as was later apparent) to accord with Schleiden's ideas.

The work and influence of Virchow

It remained for Virchow to clarify the situation. In the first issue of his famous *Archiv* in 1847 he stated current views on the organization of tissues, only to reject them; since they were based on Schleiden's unjustified generalizations they did not accord with his own micro-scopical observations on pathological processes in tissues. For a young man of 26 to start his own journal was unusual, and for him to upset the beliefs of his senior colleagues in its first issue even more remarkable, but Virchow was quite unusually remarkable in many ways: Ackerknecht's excellent study of his life and work helps us appreciate the story.[68] From this it is plain that Virchow slowly came to hold his different opinions, as his paper[69] on the evolution of cancer started with Schwann-based statements and gradually progressed to less orthodox views. His famous aphorism '*omnis cellula e cellula*' was first stated in a paper of 1855[70] which is largely concerned with putting the case for wider use of the microscope in pathology. He was by no means the first to make this generalization, however. It had been stated by Raspail in 1825, Goodsir in 1845, and Remak in 1852, as shown by Ackerknecht.[71] Nevertheless, it was Virchow who made the scientific world take note of it: he seems to have arrived at it independently, and he certainly propagated it with tireless energy. A series of lectures given in the new institute of pathology at the beginning of 1858, was published later that year with virtually no revision.[72] In this book Virchow dealt with pathology at the cellular level, but this was not a new approach. Lebert's excellent work[73] had an atlas of pathological cytology as its third volume, and both Vogel[74] and Wedl[75] had treated the subject similarly. The area in which Virchow made a momentous contribution is perhaps better summarized in his 1855 paper than in the book: that all diseases are basically disturbances of cells, which all arise from each other and depend on intracellular processes to determine function, so that all abnormalities are based on interference with the normal cell structure. He decisively rejected the Schwann's ideas on cell formation, and transferred the whole basis of life and disease to the cell. His book is a survey of physiological and pathological histology based on this precept. He had previously put forward some of the synthesis, in terms of other people's ideas, but this book dealt with his own observations and ideas with a wealth of illustrations from the microscope. The book, which is not easy reading even in translation, was a decided success; it went through two more editions within three years of publication and was translated into a good English version in 1860. The work stimulated a vast amount of microscopical investigation in Germany and abroad, even tending to focus attention on anatomy to the detriment of function until almost the end of the nineteenth century. However that may be by 1860, as a direct result of the influence of Virchow and other German workers, and with help from a few English teachers, the microscope was widely employed in medical education and research, as Dale[76] noted for England.

Nerve and sense-organ histology

Even before the stimulus of Virchow, a lot had been accomplished on the histology of nerve and sense-organs; crude methods only were available until the 1870s, but some significant results were recorded in spite of this. The tissues and organs concerned are among the most difficult to prepare and the most difficult to understand, even today. A useful summary of the information which has accrued on the neuron was edited by Hydén in 1967,[77] with an account of the history of the neuron as its first chapter;[78] the book gives a picture of results obtained from use of the most sophisticated modern methods, and it is small wonder that there was confusion and controversy about such matters 130 years ago. Purkinje's work on the cerebellar cells, which still bear his name, was published in 1838,[79] and while it may be surprising that such prominent structures had not been described before then, in the absence of any description by him of the methods he employed it is likely that fresh material was observed in a compressorium: it is still not easy to make out the cells even when they are known to be there.

Pacini described his corpuscles in 1840[80] although they had been mentioned by Vater as early as 1817. In spite of the complexity of modern methods which reveal in detail the structure of these sensory endings, it is not too difficult to demonstrate them using very simple methods with stretched fresh tissues. Bowman's book on the eye[81] contained some simple retinal histology, and more on the ciliary muscle, but Lockhart Clarke's work on the spinal cord, with its popularization of clearing and mounting sections, was a decided landmark in histology as well as in microtechnique.[82] This is one of the very few earlier papers which gives an account of the methods used, but here the method itself was being reported as well as the results it gave. This paper stimulated others to investigate both the method and its results.

We have already seen that the opposite was the case with Corti's paper on the inner ear,[83] in spite of an impressive number of technical innovations and histological results on a very difficult organ. There can be little doubt that both histology and microtechnique would have advanced more quickly than was the case if his findings had not been overlooked for many years afterwards. The year 1851 produced yet another significant paper, that of Müller describing visual purple.[84] By this time, in a period of great scientific and technical endeavour in industry as well as in medicine, much of value had started to accumulate on the histology of the most mysterious organs of the body.

The eye had been given attention quite apart from the histological, for in the next year Helmholtz published his paper on the theory of colour vision,[85] still a most interesting gross approach to fine function, which must have directed attention to the organ as a considerable challenge to the microanatomist. Work on fine nerve endings was also going forward at this time, for Wagner & Meissner published their account of tactile endings in the same year.[86] Little is said in their account about preparative methods, but their results are certainly substantially correct. Virchow himself had demonstrated neuroglia in 1854,[87] in an impressive paper which applied histological knowledge to elucidating a problem: no technical details are included. In 1856, however, a book[88] by His on the histology of the cornea did offer some technical details. The cornea could be sectioned with no previous preparation at all if a very sharp knife was used to cut it at one stroke. If pyroligneous acid was used as a preservative and hardening agent very thin sections could be produced, when subsequent treatment with hydrochloric acid destroyed the intercellular substance to leave cells, vessels, and nerves defined in a few hours. He was able to prove that the cornea and sclerotic were continuous, and to describe accurately the nerves of the cornea, in addition to showing that it is nourished at least in part by the aqueous humour. His concluded that injury to the cornea caused changes which were those of the cells themselves, and that division of the nucleus is an activity of the cell which reacts when stimulated. This is an important little book which was in accord with what Virchow had stated in his 1855 paper, but which was not then widely accepted.

More work on the eye was published in 1858, by Nunneley.[89] He was firmly of the opinion that the retina could be examined only immediately after death and without addition of any substance whatsoever. Even when taken from the living eye, the rods and cones could be seen to undergo changes while being examined, and chromic acid greatly modified their structure. Making thin sections was very difficult, and if another usual method, of drying before cutting, was employed, the results were so altered as to be totally useless. He found that Goadby's fluid, diluted to 50%, was the best preservative. The plates which accompany this paper are surprisingly good in their figures of rods and cones, but much less so in their general retinal arrangement.

There was obviously a long way to go, as Nunneley himself said when he compared the results obtained by other workers on the cones of the retina: Bowman was the first to describe them, Hannover thought they did not generally exist, Hassall made no mention of them at all, and Kölliker's descriptions were at variance with all others! This was a very fair summary of the situation in 1858.

In the same year the first of three papers by Max Schultze on fine endings of nerves in different organs was published, that on the inner ear.[90] These papers mark something of an epoch in histology, for his results were quite remarkably accurate; with the 1862 work on the olfactory organ[91] and that of 1866 on the retina[92] a turning point was reached in the understanding of the function and structure of very complicated organs. The difference between the structure found by Nunneley and that described only eight years later by Schultze is quite remarkable. The eight plates are excellent in quality, and the author produced a very complete paper on the subject, including not only the structure of the organ in a variety of animals, but also its embryology. Where Schultze had the advantage over earlier workers was in his use of osmic acid to fix the material. This allows the retina to be split into laminae, which in the absence of any means of proper sectioning was excellent for the investigation. Iodized serum was used to hold small dissections of the eye, and was found excellent for the purpose. With these very few reagents, and a great deal of patience, a vast amount of new information was revealed: it is a chastening experience to repeat the substance of this work today for much can be seen which most histologists would feel obliged to demonstrate by extremely complicated techniques. The thought must occur that much later work was merely facilitated by advances in histological technique, and not actually made possible.

In the 1860s, before the widespread use of stains other than carmine and well before use of the microtome was established, other significant results were published on the histology of nerves in various organs. In 1859 a paper by Lister & Turner,[93] on the structure of nerve fibres, mentioned that they had been inspired to the work by inspection of some of Lockhart Clarke's preparations. They repeated his process to find out if the colourless area round the carmine-tinted centre was a space, or the Schwann cell which had not taken up the colour. The area proved to be the Schwann cell, and the plate with the paper is accurate in its details of sections of nerve cells; they make the point that the chemical composition of the two parts must be very different. In an appendix Lister went on to say that Stilling's view of the sheath as being essentially similar to the axial cylinder was incorrect. Lister had seen margarine under the microscope give the fibrous appearance which Stilling had believed to show a similarity between the parts of a nerve fibre. Lister also confirmed Kölliker's opinion that the axial cylinder retained its full diameter in the cord, the sheath alone being reduced. There is no doubt that these later workers were able, by use of a carmine solution not available to the great Stilling, to correct his views on the structure and arrangement of neurones—a significant step forward in the space of only four years.

In 1867 Trinchese published work on motor end-plates,[94] having demonstrated their existence in a number of animal groups, and saying

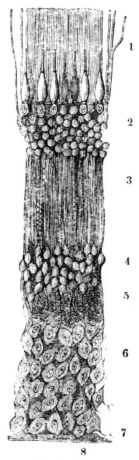

Figure 149. Retina as illustrated by Kölliker, 1860 (from A Kölliker—*A manual of human microscopic anatomy*. (London: Parker, 1860.) See page 553).

This was a version of his textbook brought up to date to compare with the 1859 German edition. There is a lot of difference to be seen in the pictures for only a six-year gap in publication, and clearly much had been achieved in the 1850s.

Layer 1 is the bacillar layer, and shows the rods and cones quite clearly. Layer 3 is the intermediate granular layer.

that only two other workers (Beale and Kölliker) had failed to recognize their existence. A review[95] of the paper quite rightly made the point that Trinchese was not only very slighting about Beale, but had neither repeated his experiments nor tried out other techniques. In spite of obvious sympathy on the part of the editors with Beale, the summary of the findings of Trinchese as to the existence and arrangement of end-plates and their neurones would not be disputed today: only the finer details of structure remained to be described, thus closing a notable dispute of the 1850s and '60s. A note about Beale's work is given below on page 326.

In the next year Hulke published a paper on the retina,[96] a prelude to a further description of the eye in general which appeared the following year.[97] This author wrote widely on the eye and its histology for some years, perhaps in consequence of his being surgeon to a London eye hospital, but his work was derivative compared with the approach of German workers of the time. He used osmic acid in the Schultze manner, but found nothing new.

It may not be altogether surprising, therefore, that German scientists were thought to be ignorant of work being done abroad, and a certain sensitivity in the criticism made of this ignorance suggests that German workers were also rather scornful of others, including the English. There are some revealing passages in the journals of the time, as in a review of Stricker's textbook of histology:[98]

> ... Again we must say that we consider that English work has been almost completely overlooked. . . . One would naturally be led to believe that Deutschland was essentially the country where the microscope flourished in the hands of the human anatomist. This is of course objectionable ... and it is an evil of very old growth.[99]

This textbook was in fact a resume of German research and teaching, and it certainly fails to consider much foreign work. The chapters, in volume II, on the brain (by Meynert), and the spinal cord (by Gerlach) are long, detailed, and of high quality. The cranial nerves are traced much further than in English works of the time, and the illustrations were of the highest standard. The prompt translation of the rather turgid German into good English[100] must have done a great deal to make the latest histological information available to a wider audience at a time when there was great need for an up-to-date treatment.

There was more preoccupation with the nerve fibre. Ranvier published a good deal of information in 1871,[101] showing that nerves did not have a continuous sheath, but that it was interrupted, in what came to be called the nodes of Ranvier. These are not easy to demonstrate even in modern preparations, and Ranvier had recourse to osmic acid and silver nitrate to demonstrate their presence. He supposed that the nodes allowed the fibres to be nourished, the sheath being impermeable. He went on to write his wellknown and ambitious book on nerve histology[102] on the basis of lectures given in the academic year 1876–7, by which time he was well-known as an histologist. There are over 700 pages in the two volumes, with excellent plates, and the

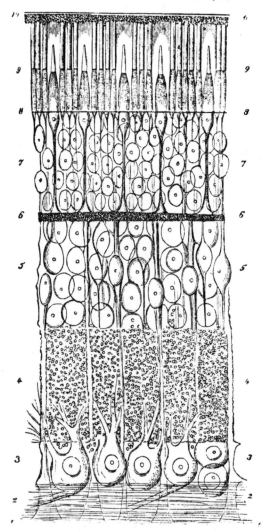

Figure 150. Retina as illustrated by Stricker, 1873 (from the book note 100 in text: volume 3, page 219).
The account was written by Max Schultze, and represents an enormous gain in histological knowledge compared with Köllicker's drawing of 13 years before. Layer 3 is the ganglion-cell layer, and 9 is the layer of rods and cones; the overlying pigment is correctly shown at 10.

histology of nerves is summarized in a thorough but somewhat idiosyncratic manner; helpful historical surveys are included in addition.

With this book on nerves and Stricker's manual dealing with the subject, plus the sense organs, in as great a degree of detail as could be imagined at the time, a good foundation had been laid for student and researcher. The eye and its appendages were described by no fewer than ten separate authors in Stricker, including Rollett on the cornea, and (the recently deceased) Max Schultze on the retina. It is quite doubtful if there were ten expert histologists concerned with the eye in any one country outside Germany at that time.

Work did come from England, however, in connection with the brain. Bevan Lewis suggested new techniques for brain sections in 1876,[103] with a plate showing better pictures of pyramidal cells than Stricker contained. He urged the most rapid possible processing and inspection to obtain results far superior to those obtained by hardening and sectioning in the normal way, and in only two days compared with eight weeks. He cut sections by hand, compressed the tissue under a coverglass, and then transferred it all to spirit before raising the cover and washing with water before applying a metallic stain. The specimen was then dried over sulphuric acid under a bell-jar, treated briefly with chloroform, and then mounted in balsam. The process does not seem to have been taken up, in spite of the excellence of his plate which bore out his claims. By the mid-1870s there was much preoccupation with making microtomes which would section a complete brain, and workers were not interested in quick methods even if they were superior, unless a whole-brain section was achieved.

Thereafter there was steady progress in the investigation of the central nervous system, as has been mentioned as appropriate when discussing reagents and apparatus in other sections of this thesis. A particular highlight was Kölliker's substantial contribution on the cerebellum in 1890,[104] in which substantially correct views on its structure were put forward, by the application of Golgi's method as outlined by him in 1886.[105] Further important work was published only two years later by Cajal,[106] who developed his own methods of metallic impregnation to produce impressive results in neuro-anatomy. In 1896 he published his classical account of the vertebrate retina,[107] using refinements of technique and displaying considerable insight into the interrelationship between structure and function. All of these advances relied totally on improved preparative techniques, far outpacing those of such as Barrett[108] only ten years before. Golgi's rapid method of impregnation was a notable achievement in specimen preparation, requiring much investigation of minutiae before it was repeatable, but yielding impressive results after practice.

Work on the retina continued to interest histologists. Bernard, in a series of papers from 1900,[109] published some searching investigations, having shown that amphibian rods were transversely divided before being filled with refractive material during development; enormous histological detail based on most careful preparative techniques is included in these papers. He disputed some of the results of Cajal, and provided a very interesting comparison of the views on retinal structure of various histologists.

A few outstanding points were cleared up at the beginning of this century, and two examples may be given. One was on the structure of the nerve endings in human muscle, and Grabower was able to show that this was essentially the same in man as in other animals.[110] Joseph clarified the nature of neuroglia in a variety of animals, showing them to have an epithelial origin and a supporting function.[111] Perhaps a final reference to the application of histological methods to the understanding of the human cerebral cortex would be appropriate; the book by Campbell[112] marked the development of the study of the architectonics of the cortex, which could never have been possible without the development of microtomy.

Following this record of unalloyed success in histological investigation, it would be proper to consider the other side of the coin, and mention a few of the cases where wrong conclusions were obtained.

Histological errors

There are, of course, many examples in the literature where poor technique or some other factor has caused wrong conclusions to be drawn, and the work of Lionel Beale on motor end-plates is a good example. Beale was well known in England for his textbooks and his university work in London, and was responsible for popularizing the use of carmine in this country. He gave much attention to nerves and their endings, especially in muscle, but in spite of great industry came up with totally wrong findings on end-plates. He stated that fibres never end in voluntary muscle, and he also thought they had no endings in the central nervous system. It was his considered opinion that a fibre left a central nerve cell and returned to it after its course round the body. A paper in 1863[113] summarized his views in the context of a report on a series of demonstrations given by Beale in response to German criticisms of his well-known paper of 1860,[114] which had stated that nerves going to striped muscle terminate in a network outside the sarcolemma. Kühne and Kölliker, separately, had denied this statement, causing Beale in 1862[115] to read another memoir to the Royal Society in which he claimed to have been able to follow the nerves further than his critics. In the lectures thirty-one specimens were inspected, and Ciaccio agreed that what Beale had stated was correct. He added that for three authorities to reach such different conclusions is not altogether surprising, for different observers are preconditioned to see different things, and different methods of preparation reveal different things. Nothing could be more correct, but what neither Beale nor Ciaccio allowed for was that carmine simply does not stain fine nerves or their terminations, so that Beale was confusing two separate nerves by failing to see the terminations of either: in such case it might well seem that the one nerve was continuous. This is a devastating condemnation of the methods of Beale, and we now know that it would not be until all workers could prepare tissues so as to present identical pictures with a variety of techniques that histology could be free from such gross error: that date was still far in the future in the 1860s.

Beale's error, then, was one of faulty preparative technique compounded by the failure to realize that one method could never reveal all possible information about a tissue. In 1879 Thin emphasized this fact in his paper on hyaline cartilage[116] and how its apparent structure depended on its method of preparation. Different methods of hardening and staining produced quite different pictures, and the author considered that serious doubts were cast on the interpretations put forward by various authors on the structure of cartilage. Taken alone, use of carmine or eosin staining was not conclusive evidence of the existence or limits of cellular protoplasm in any animal tissue. This was a useful note of caution at that date.

Another possible source of error is that resulting from ignorance of the optical principles of the microscope itself: as an example we may take the paper by Haycraft[117] which purported to explain the striated appearance of voluntary muscle fibres as an optical effect of their undulating shape. The idea had been suggested by Bowman over forty years before, but by 1881, when Abbe had started to elucidate the principles of microscopic optics, it is quite surprising that the paper should have been published by the Royal Society. It is true that the *J.R.M.S.* had for years been urging that histologists should have a grounding in the theoretical principles of their instrument, to obviate their putting forward erroneous interpretations of finer structures, as well as their missing important points. Manifestly, neither Haycraft nor the editors of the journal concerned had had such instruction, and it remains true to the present day that very few histologists have such training.

An example of information which should have been noted but

apparently was not was given by Nelson, when he looked at the stripes on muscle for himself.[118] He could make out the full pattern of the striations, and commented that Krause's membrane (the dark line on the white band) had been figured in an illustration dated 1852, but he believed that a microscope of 1841 would have resolved this line. Naturally, one must wonder if he was knowledgeable about the preparative techniques used in 1841 and 1852, for there would be a notable difference between them.

Histology and medical education in Europe

We have seen that early instruction in microscopical and histological work was undertaken by private tutors, and only slowly found a place in the formal medical curriculum in the 1840s and '50s. Collingwood's address[119] in 1859 offers a useful summary of some pertinent material. He noted that London had recently divided the medical examination into two parts, and thought it a helpful alteration as allowing the student to work at the much extended curriculum without cramming. Part of the extended syllabus was the new discipline of embryology, an important auxiliary to the study of function. It was to the microscope that they were all indebted for advances in physiological science, and it was to the microscope that the students must turn if they were to be properly educated as practitioners or researchers.

An interesting account of a student in the 1860s has come down in a diary[120] which includes references to some of the best-known scientific men of the time. Some examples of the use of the microscope are instructive:

> October 3rd ... Attended a lecture on Physiology by Lionel S. Beale, M.B. and F.R.S., an individual apparently not over-popular with the students, perhaps because he has a somewhat irritable temper and a finical way of lecturing besides ...
> March 21st ... Viva voca examination in Physiology. Had to inspect microscopical preparations and say what they represented.
> February 24th ... Physiology: Nose and Eye. Class very excitable, as well as Dr. Beale.
> November 26th ... Went round the wards with Dr. Beale, who was very gracious to me on the subject of the microscope.
> December 4th ... Tea-dinner at Dr. Beale's, after which we adjourned to his laboratory and saw various microscopical specimens under a 1/26th-inch object-glass, magnifying 2,000 diameters.
> January 29th ... Examined vomit containing sarcinae ventriculi under the microscope.

We cannot know how typical of an 1860s student this would be , but it is clear that Beale encouraged his use of the microscope, that it was used in the course and the examinations, and in the wards. It is also noteworthy that the eye and the nose could be dealt with in one one-hour lecture.

There were several routes to a medical qualification in Britain in the years before 1870, and the work described by Taylor would not have been required of all medical students, by any means. He undertook a definite three-year course, and in 1869 the General Medical Council's committee on medical education reported to suggest a logical order of subjects, including histology in the second half of the first year. After 1885 the course was increased by another year to five, with histology put in the second year. This at last meant that medical students would follow a satisfactory scheme in histology wherever they trained.

Naturally, in some colleges this had been the case for some long time. At University College in London William Sharpey had been made professor of anatomy and physiology in 1834, having trained at Edinburgh.[121] He was to remain until 1874, exerting much effect on the teaching of histology. His special table, still to be seen in the College, for using the microscope to illustrate histological discourses (he was one of the first to do so), has a central hole and a brass rail let into the top edge. A bar pivoted round the centre to allow a microscope to revolve round the rail, thus passing from student to student. A lecturer in practical histology was appointed from 1856,[122] and from 1859 the work had a special room set aside. Harley, who wrote a small textbook of practical histology, was the first lecturer, and was succeeded in 1867 by Michael Foster, who was made professor in 1869, before going to Cambridge as Praelector in Physiology the following year. It is interesting that medical students still take histology courses in the department of physiology at Cambridge, for Foster was quite outstanding a person. Burdon-Sanderson succeeded him, to a chair made suddenly much more important in 1871 by the new University requirement that all medical students must study practical physiology (which in those days was about two-fifths histology). The influence of the school thus established was remarkable, and the famous handbook[123] issued in 1873 played a lasting part in giving Britain some prestige in the subject which most associated by that time only with Germany.

By 1872 also the new regulations of the University of London, matched by those of the Royal College of Surgeons, had stimulated new courses at King's College London, where Rutherford gave a course developed over several years, some details of which he published[124] with the following introductory remarks:

At the present time there is much discussion regarding the manner in which Practical Physiology should be taught to medical students. Perceiving the difficulties which physiological teachers have to encounter in carrying out the new regulations of the College of Surgeons, the Editor of this Journal has asked me to give a resume of my course of Practical Histology. I do not regard my course as perfect, nevertheless I have found it work successfully through several years, and, after having ascertained the modes of tuition adopted elsewhere in this country and in Germany, I have come to the conclusion that on no other system does the student accomplish so much in so short a time.

The system of teaching is essentially that adopted by my old master, Professor Bennett, of Edinburgh, the gentleman who first taught Practical Histology in this country. Under Professor Bennett's direction I taught the subject in a manner differing somewhat from that which I now adopt. In Edinburgh, the students of my ordinary class merely examined the tissues. I now find no difficulty in providing every pupil with the means for making for himself a little cabinet of microscopic specimens of the various tissues. This involves little trouble, and greatly increases the interest with which the student follows the class.

A full set of instructions for operating the course is given, from preparing a set of tissues a month before the work is to start, to an exact list of exercises to be undertaken in order; drawing and mensuration were used wherever possible. A wide variety of preparative techniques was included, and general instructions were provided for making all manner of mounts. This was really a most adequate course, not anachronistic even today, when the main difference would be that a wider range of tissues would be examined, but fewer of them prepared. From a general investigation of German and other courses of the time, one must conclude that a Rutherford-trained student would have come favourably out of any comparison. In 1872, of course, the microtome was coming into use for original work in some places. Rutherford advises use of the Valentin knife or razor according to the subject, with simple embedding in carrot or wax if the material was especially difficult. Stirling's section cutter was mentioned only for those who found difficulty in using the razor.

This was by no means an unusual attitude. Atkinson reported results of work investigating the best stain for brain and spinal cord, and used a section machine only if he required uniformly thin sections for any special reason. He did, though, allow that many found hand-sectioning a difficult and rather frustrating procedure.[125]

There is other evidence that British interest in medical microscopy was growing in the 1870s, perhaps following the established German lead. A Medical Microscopical Society was formed in 1873, with Jabez

Hogg as President. Its meetings were largely concerned with practical matters—observing living tissues, sectioning animal organs, preparing brain and spinal cord—and at its monthly meetings not only was a lot of information conveyed but the use of the microscope in medical studies was obviously stimulated.[126]

At a meeting of the British Association in 1873 Rutherford acknowledged that the instrument itself had to be perfected if it was to contribute to advances in histological knowledge (the best objectives of the day had a performance about that of a present-day student $\frac{1}{6}$ in.). He was sure that the application of new methods of specimen preparation would give most future advances in this knowledge, on the basis of the improvements seen in the previous fifteen years. These included the use of new fixatives, new stains, silver nitrate and gold chloride, osmic acid, and so forth. The retina had yielded some of its secrets to osmic acid, and only some of the most profoundly difficult organs such as the brain and peripheral nerve terminations remained unknown. If the techniques applied to human tissues were to be put to comparative animal histology, no one could fail to make original discoveries: comparative embryology, in addition, was almost completely unexplored. A few of his words are revealing:

> Speaking of microscopical study ... I cannot but remember that in this country more than in any other we have a number of learned gentlemen who, as amateurs, eagerly pursue investigations in this department. I confess that I am always sorry to witness the enthusiastic perseverance with which they apply themselves to the prolonged study of markings upon diatoms, seeing that they might direct their efforts to subjects which would repay them for their labours far more gratefully. I would venture to suggest to such workers that it is now more than ever necessary to abandon all aims at haphazard discoveries, and to approach microscopy by the only legitimate method, of undergoing a thorough preliminary training in the various methods of microscopical investigation by competent teachers, of whom there are now plenty throughout the country.[127]

Regretfully, it has to be admitted that this advice fell on deaf ears, for diatom-dotting continued apace, and no organized school of histology emerged for a very long time. There was a slightly more hopeful note struck in 1875, when the Royal Irish Academy made a grant for inquiries on staining agents in microscopic anatomy. Admittedly, this was for only twenty five pounds, but nothing further was ever heard of this small step in the right direction, nor of Dr Reuben Harvey, its recipient.[128]

Some of the blame for this notable falling behind must lie with the nature of the universities in this country. These were few in number (although there were several university colleges in the larger cities), and outside London, which had its own problems in the 1870s and '80s, neither Oxford nor Cambridge were places where the claims of science were likely to be well received. In Germany, however, the situation was very different, and this must account for the more intensive work in histology undertaken there. With many universities, each vying with the rest for prestige, the stimulus for original work was enormous, and provision was made for it. Not all was straightforward even there, however. In Berlin in 1876, for example, a new chair in microscopical anatomy was established for a particular person, on the pretext that the holder of the chair in gross anatomy could not be expected to devote himself to both the gross and the microscopical aspects. It was apparently correct that the anatomist did not devote himself overmuch to the microscope, but the gap was filled, as the examinations approached, by the assistants to the professor of pathology. Seeing their income threatened, they sought the help of the professor of physiology, who had a new institute almost completed, with room and assistants enough to teach microscopical anatomy to all who could so desire. So was the tale recounted,[129] of a matter which has a curiously contemporary ring!

The situation in the United States

The account of the influence of Johns Hopkins University on American medicine[130] shows that, before that university was established in 1876, the so-called medical schools in the United States were simply low-grade largely commercial cramming institutions, with medical research pursued only occasionally by a few individuals usually not associated with any particular college. This accounts for the poor quality of the papers in such journals as gave space to microscopical matters before about 1890. Johns Hopkins was founded with the ideal of creating an institution where research and proper teaching were to go hand in hand. The medical school, not opened until 1893, was a model of its kind, and soon exerted much influence on medical education in the United States, and later throughout the world. The hospital was opened in 1889, and attracted quite outstanding men to its staff, so that by the 1890s first the university and then the medical school had started to produce adequate work in histology.

Before that date there were, of course, a few other meritorious institutions. A description by Stowell[131] of microscopy in the university of Michigan shows that in 1872 there was only one microscope in the whole university. By 1882 there were five courses available for students. The one on normal human histology had thirty lectures and thirty practical classes, in which most students made about 20 mounts while working on the structure of tissues—enough to allow him to work alone later, for professional purposes or research. The plant histology course took twenty lectures and practicals, and the student had to produce a minimum of 60 accurate drawings of structures he had investigated. A second plant histology course dealt with pharmaceutically-important plants especially for the detection of adulteration. Work in advanced normal and pathological histology, and in chick embryology and urinalysis was also available. By this time the university had 116 microscopes, and an official report recommended some centralization of resources. It is also clear that this university was much more committed to work with the microscope than most others at that date, and that it could not be taken as representative of the country as a whole.

In 1890, it was clear that the general situation was not very satisfactory, and that the university of Michigan was highly atypical, due to Stowell's personal exertions.[132] Only 66 people had attended the last annual meeting of the American Society of Microscopists, and many 'trashy' instruments were on sale, being bought by schools as well as uninformed amateurs.

In the 1890s there were pleas for the introduction of systematic instruction in microscopy in American universities,[133] advocating provision of advanced outfits to allow European work on the subject to be matched by American. The example of the Zeiss works, where academics were called on to direct its operations and manage its design departments, was held up as an example of scientific eminence devoted to the pursuit of excellence in optics. A background of theoretical knowledge was required, to be applied to the optical and mechanical parts of the instrument. Some of Cox's remarks on the microtome are pertinent:

> ... We use the instrument our teacher used without asking whether his success as a teacher was because he used it or in spite of his using it. I remember very well, and so do most of you, when students from certain great and deservedly popular schools in biology scouted all microtomes and argued that free-hand cutting was a better way of section-making than any mechanical one. Now the microtome is a complicated machine, and the most elaborate have been devised where once simplicity was a cult.

Cox was president of the American Microscopical Society, then in its sixteenth year, and well known in the United States as an academic of standing.

Latham made the point in 1896 that in the last decade medical education in the United States had been widely discussed and seen to have imperfections when compared with medical education in Europe, and also in comparison with other subjects in America itself. She suggested that a proper balance between subjects was often lacking, many professors considering only their own subjects, so that it had been known to be allowable for some students to take pathology without having done any histology, thus making the pathology quite unintelligible. Pathology had tended to become an anatomical study, and a return to the physiological aspect was now urgently required. In this call the author showed a considerable awareness of the global situation, and a notable far-sightedness.[134] That very little had been done to remedy the situation regarding proper teaching of histology, as well as many other aspects of medical education, was made devastatingly clear by Flexner in his 1910 report,[135] so all of these cries had been made in vain. It seems inescapable that both teaching and original work in histology were far worse off in America than in Germany, Britain, or France before 1910. We must now conclude with a last glance at the scene in Europe before the first world war.

Histology in Europe at the turn of the century

The Franco–Prussian war, and the defeat of the French in 1870, caused many changes in the approach to medicine, and of its associated sciences. The essentially German method of collecting many data and analysing them to help a truth emerge became the pre-eminent technique. There is a tendency to assume that any German publication of the last part of the nineteenth century must have authority, but authors other than those working in Germany wrote in that language, not always authoritatively. An example is the paper by Betz, summarizing methods of investigating the central nervous system.[136] This recommended a simplified version of the Stirling machine, and used carmine as the only stain, although one extra was the use of polarized light to trace fibres in the hemispheres. Betz was in fact Russian, wrote in very bad German, and published about the last old-fashioned paper in a reputable German journal!

On the other hand, the preoccupation with dyestuffs and microtomes, more as vehicles for possible prestigious publication than as true contributions to knowledge, is another characteristic of the German histological scene in the 1880s and '90s. It is quite true that German research and teaching went hand in hand; as Waldeyer said,[137] professors were paid by the state to train doctors, and to do that properly had to be concerned also with advancing science. Particular standpoints had to be abandoned in favour of giving the student a broad foundation, just as the original investigator had to attack a problem from the broadest possible base if he was to succeed. German laboratory organization, and the direct interest taken by the state in the educational system, plus the competition engendered among a number of universities seeking prestige for their founders and/or patrons, provided the means for histological advance, as in other disparate fields. At the end of the nineteenth century virtually all other countries were much less organized in these respects, and valued the sciences less into the bargain. Precise delimitation of histology, in curriculum or research, is impossible, for it appears in anatomy, physiology, and pathology, and continued to form part of each of these departments of knowledge. When the unaided eye reached the limits of its capacity, microtome and microscope were invoked to enable the worker to go further, in the natural context of a research institution, which was the German idea of a university. Quite certainly, from the late 1880s to the 1930s Germany (this includes Austria for our purposes) was the country to go to for postgraduate work in medicine, and its journals remain a fascinating source of original work in histology and its associated subjects. German influence naturally spread to other countries, including England and America, as the postgraduate workers brought back not only technical know-how but also ideals. By the end of the century, laboratories for various branches of medical science had been built in most hospitals of any standing in Britain, with bacteriology and pathology usually coming before physiology. By 1907, the addition of extra subjects and subdivision of existing ones, had increased the British medical course to six and a half years overall.

By the end of the nineteenth century most German institutes were also in process of rebuilding, and all possessed class-rooms for microscopical courses.[138] By 1910, for example, 2000 students were enrolled at the Munich laboratories, 500 of them attending courses in histology, and 300 working in the practical aspects. Only two professors supervised all this work, and this low level of staffing was general in Germany; for example, for 250 students at Marburg there was a professor with six assistants. One wonders where they found time for original work, but German students were much less supervised than their contemporaries in, for example, England. So far as the Germans were concerned, discipline appropriate to a school was not to be extended to the university: if a man had chosen his profession, responsibility to qualify for it rested squarely on his own shoulders. Most students attended far more than the minimum courses required to qualify.

Outside Germany the situation was much less satisfactory. The medical sciences were denied independence, and original work came from isolated individuals, working in relatively ill-equipped departments. This subservience of medical sciences to clinical medicine in both France and Britain goes far to explaining their backwardness in histology and work with the microscope generally. It was only·the repercussions of the 1914 war, starving German laboratories for many years afterwards, that allowed Britain and France to start to assert their increasing abilities, to end almost half a century of German leadership in histology.

It has already been said, and it is worth repeating, that pure histological technique alone did not make possible much of the work done: it may have facilitated it, but bearing in mind that a vast amount of knowledge had been accumulated before such techniques were developed, it cannot be said to have allowed it to happen. Naturally, some techniques such as osmic acid fixation and metallic impregnation did give results obtainable in no other way. It is also correct that when bacteriology developed towards the end of the nineteenth century, the microtome and the microscope had to be applied together to pinpoint the foci of infections, and in this respect histological technique was the key to a very dramatic advance in medicine. On the other hand, the pivotal work of Metchnikoff on inflammation and phagocytosis was carried out largely without recourse to the main techniques of serial sectioning and differential staining which were all the vogue after 1885; it may be that too much attention was given to developing the microtome and the various procedures, and too little to sheer careful observation.

References to Chapter 7

1. A F J K Mayer—*Ueber Histologie und eine neue Eintheilung der Gewebe des menschlichen Körpers bei Gelegenheit der Eröffnung seiner Vorlesungen über Anatomie.* (Bonn: Marcus, 1819.)
2. R O'Rahilly (1958)—Three and one-half centuries of histology, *Irish J. Med. Sci.* *1958:288–292.*
3. J R Baker—The cell-theory: a restatement, history, and critique:
   I—(Introduction, discovery of cells), *Q.J.M.S.* *89:103–125 (1948)*;
   II—(Discovery of protoplasm and the nucleus), *Q.J.M.S.* *90:87–108 (1950)*;
   III—The cell as a morphological unit, *Q.J.M.S.* *93:157–190 (1952)*;

IV—The multiplication of cells, *Q.J.M.S.* *94 :407–440 (1953)* ;
V—The multiplication of nuclei, *Q.J.M.S.* *96 :449–481 (1955)*.

4. A Hughes—*A history of cytology*. (London: Abelard-Schuman, 1959.)
5. E Clarke & C D O'Malley—*The human brain and spinal cord*. (Berkeley: Univ. California Press, 1968.)
6. C Newman—*The evolution of medical education in the nineteenth century*. (Oxford: University Press, 1957.)
7. A Flexner—*Medical education in Europe*. (New York: Carnegie Foundation, 1912.)
8. T Billroth—*The medical sciences in the German universities*. Translated from the German edition of 1876. (New York: Macmillan, 1924.)
9. C D O'Malley (Ed.)—*The history of medical education*. (Berkeley: Univ. California Press, 1970.)
10. F H Garrison—*An introduction to the history of medicine*. (Philadelphia: Saunders, 1929.)
11. R H Major—*A history of medicine*. (Springfield:Thomas, 1954.)
12. C Singer & E A Underwood—*A short history of medicine*. (Oxford: Clarendon Press, 2nd edn 1962.)
13. W W Keen—*A sketch of the early history of practical anatomy*. (Philadelphia: Lippincott, 1874.)
14. D Craigie—*History of anatomy* (article in Encylopaedia Britannica, revised by F G Parsons, 11th edn, 1910).
15. G W Corner—*Anatomy*. (New York: Hoeber Clio Medica, 1930.)
16. R H Hunter—*A short history of anatomy*. (London: Bale, 2nd edn, 1931.)
17. J Needham—*A history of embryology*. (Cambridge: University Press, 1934.)
18. A W Meyer—*The rise of embryology*. (Stanford: Univ. California Press, 1939.)
19. E R Long—*A history of pathology*. (Baltimore: Williams & Wilkins, 1928.)
20. E B Krumbhaar—*Pathology*. (New York: Hoeber Clio Medica, 1937.)
21. W Bulloch—*The history of bacteriology*. (Oxford: University Press, 1938.)
22. W W Ford—*Bacteriology*. (New York: Hoeber Clio Medica, 1939.)
23. T H Grainger—*A guide to the history of bacteriology*. (New York: Ronald Press, 1958.)
24. M F X Bichat—*Traité des membranes en général et diversees membranes en particulier*. (Paris: Richard, Caille & Ravier, 1800.)
    M F X Bichat—*Anatomie générale, appliquée a la physiologie et à la médecine*. (Paris: Brosson, Gabon, 1801.)
25. T Hodgkin & J J Lister (1827)—Notice of some microscopic observations of the blood and animal tissues, *Phil. Mag.* *2 :130–138.*
26. See, for example, the treatment by J Bostock—*An elementary system of physiology*. (London: Baldwin & Cradock, 3rd edn 1836.)
27. E H Weber (1841)—Ueber menschliches Blut, *Deutsch. Naturf. Verslamml. Bericht.* *1841 :93–94.*
28. W S Savory (1869)—On the structure of the red blood-corpuscles of oviparous vertebrata, *R.S. Proc.* *17 :346–350.*
29. L J Rather—*Addison and the white corpuscles*. (London: Wellcome Institute, 1972.)
30. A good account of various aspects of Purkinje's work, at Breslau and later, appears in V Kruta (Ed.)—*Jan Evangelista Purkyně 1787–1869 centenary symposium*. (Brno: Univ. J A Purkyně, 1971.) Other papers in this symposium deal with various aspects of nineteenth century histology and physiology, including the purkinje cell.
31. A Wendt (1834)—Ueber die menschliche Epidermis, *Müller, Archiv* *1834 :278–291.*

32. J E Purkinje & G G Valentin (1834)—Entdecklung continuirlicher durch Wimperhaare erzeugter Flimmerbewegungen, *Müller. Archiv.* *1834:391–400.*

33. R Wagner (1835)—Ueber die Anwendung histologischer Charactere auf die zoologische Systematik, *Müller, Archiv.* *1835:314–320.*

34. E F Gurlt (1835)—Vergleichende Untersuchungen über die Haut des Menschen und der Haus-Saugethiere, besonders in Beziehung auf die Absonderungsorgane des Haut-Talges und des Schweisses, *Müller, Archiv.* *1835:399–418.*

35. S Boyd—*Inaugural essay on the structure of the mucous membrane of the stomach.* (Edingurgh: Stark, 1836.)

36. R Remak (1836)—Vorlaufige Mittheilung microscopischer Beobachtungen über den innern Bau der Cerebrospinalnerven, *Müller, Archiv* *1836:145–161.*

37. J E Purkinje (1836)—Über Flimmerbewegungen im Gehirn, *Müller, Archiv.* *1836:289–290.*

38. G G Valentin (1836)—Ueber die Verlauf und die letzten Enden der Nerven, *Acad. Caes. Leop. Nova Acta* *18:51–240, 541–543.*

39. R Wagner—*Icones physiologicae.* (Leipzig: Voss, 1839.)

40. J H Bennett—*The employment of the microscope in medical studies; a lecture introductory to a course of histology.* (Edinburgh: Machlachlan, 1841.)

41. Anon (1841)—Bibliographical notices, *Mic. J.* *1:172–173.*

42. A E Grube (1840)—On the structure of the macula lutea of the human eye, *Mic. J.* *1:71–73.*

43. W Bowman (1840)—On the minute structure and movements of voluntary muscles, *Phil. Trans. R.S.* *130:457–501.*

44. W Bowman (1842)—On the structure and use of the malpighian bodies of the kidney, *Phil. Trans. R.S.* *132:57–80.*

45. J Quekett (1848)—On the value of the microscope in the determination of minute structures of a doubtful nature, as exemplified in the identification of human skin attached many centuries ago to the doors of churches, *Trans. M.S.L.(O)* *2:151–158.*

46. J Paget—*Report on the chief results obtained by the use of the microscope in the study of human anatomy and physiology.* (London: Churchill, 1842.)

47. J H Bennett (1845)—Introductory address to a course of lectures on histology, and the use of the microscope, *Lancet* *1:517–522.*

48. J Müller—*De glandarum secernentium structura penitiori.* (Lipsiae: Vosii, 1830).

49. F G J Henle—*Symbolae ad anatomiam villorum intestinalium, imprimis eorum epithelii et vasorum lacteorum.* (Berolini: Hirschwald, 1837.)

50. F G J Henle—*Allgemeine Anatomie.* (Leipzig: Voss, 1841.)

51. A H Hassall—*The microscopic anatomy of the human body, in health and disease.* (London: Highley, 2 Vols, 1846, 1849.)

52. J Rosenthal—*De formatione granulosa in nervis aliisque partibus organismi animalis.* (Breslau, 1839.) Original not seen, quoted from Baker, see note 3 above.

53. S Wilks & G T Bettany—*Biographical history of Guy's hospital.* (London: Ward, 1892.) See p. 478.

54. C R B Barrett—*History of the Society of Apothecaries of London.* (London: Elliot Stock, 1905.) See p. 243.

55. J Quekett—*A practical treatise on the use of the microscope.* (London: Bailliere, 3rd edn, 1855.) See the frontispiece for the famous illustration of the histological theatre, itself a rarity at the time, with microscopes arranged for observation, and moving on rails between the students.

56. J Quekett—*Lectures on histology, delivered at the Royal College of Surgeons of England, in the session 1850–51.* (London: Bailliere, 1852.)

57. A Kölliker—*Manual of human histology*. Translated by G Busk and T H Huxley. (London: Sydenham Society, Vol. 1, 1853.) See p. 8.

58. R D Lyons—*A retrospect of the progress of microscopic investigation, and of the more important recent contributions to normal and pathological histology*. (Dublin: Hodges & Smith, 1850.)

59. E H Ackerknecht—*Medicine at the Paris hospital 1794–1848*. (Baltimore: Johns Hopkins Press, 1967.) See pp. 125–126.

60. See the short account in *M.M.J.* 2 :280 (1869) of a brisk discussion in Paris as to merits of the microscope's place in medicine, wherein it was suggested that it was not only valueless but actually detrimental—at so late a date as 1869!

61. R D Lyons—*An apology for the microscope : being the introductory lecture to the first course on microscopic anatomy and physiology, delivered in the theatre of the original school of medicine, during the months of February, March and April, 1851*. (Dublin: Fannin, 1851.)

62. R A von Kölliker—*Handbuch der Gewebelehre des Menschen*. (Leipzig: Engelmann, 1852.)

63. op. cit., note 57 above.

64. T Schwann—*Mikroskopische Untersuchungen über die Uebereinstimmung in der Struktur und dem Wachsthum der Thiere und Pflanzen*. (Berlin: Sander, 1839.)

65. Schleiden had great influence over Schwann, and induced him into a rather blind acceptance of wrong theories of cell formation. A convenient source for both works is: T Schwann—*Microscopical researches into the accordance in the structure and growth of animals and plants*. [with] M J Schleiden—*Contributions to phytogenesis*. Translated by H Smith (London: Sydenham Society, 1847).

66. M Barry (1839)—Researches in embryology: 2nd series, *Phil. Trans. R.S.* *1839 :307–380*.

67. A von Kölliker—*Entwicklungsgeschichte der Cephalopoden*. (Zurich: Meyer & Zeller, 1844.)

68. E H Ackerknecht—*Rudolf Virchow, doctor, statesman, anthropologist*. (Madison: Univ. Wisconsin Press, 1953.)

69. ibid., see pp. 74–76.

70. R Virchow (1855)—Cellular-Pathologie, *Virchow, Archiv* 8 :3–39.

71. Ackerknecht, op. cit. note 68 above. See his p. 83.

72. R L K Virchow—*Die Cellularpathologie in ihrer Begründung auf physiologische und pathologische Gewebelehre*. (Berlin: Hirschwald, 1858.)

73. H Lebert—*Physiologie pathologique, ou recherches cliniques, expérimentales et microscopiques sur l'inflammation, la tuberculisation, les tumeurs, la formation du cal, etc.* (Paris: Baillière, 1845.)

74. J Vogel—*Pathologische Anatomie des menschlichen Körpers*. (Leipzig: Voss, 1845.)

75. C Wedl—*Grundzüge der pathologischen Anatomie*. (Wien: Gerold, 1854.) See also the review of the English translation in *Q.J.M.S.* 4 :225–227.

76. W Dale—*The present state of the medical profession in Great Britain and Ireland, with remarks on the preliminary and moral education of medical and surgical students*. (London: Bennett, 1860.) See p. 42.

77. H Hydén—*The neuron*. (Amsterdam: Elsevier, 1967.)

78. ibid. Chapter 1 by H van der Loos, pp. 1–47.

79. J E Purkinje (1838)—Ueber die gangliösen Körperchen in verschiedenen Theilen des Gehirns, *Ber. Vers. dtsch. Naturf. Aertze* *1838 :179–180*.

80. F Pacini—*Nuovo organi scopteri nel corpo humano*. (Pistoja: Cino, 1840.)

81. W Bowman—*Lectures on the parts concerned in the operations on the eye, and on the structure of the retina*. (London: Longman, 1849.)

82. J A L Clarke (1851)—Researches into the structure of the spinal cord, *Phil. Trans. R.S.* *141 :607–621*.

83. A Corti (1851)—Recherches sur l'organe de l'ouie des mammifères, *Z. wiss. Zool.* *3 :109–169.*

84. H Müller (1851)—Zur Histologie der Netzhaut, *Z. wiss. Zool.* *3 :234–237.*

85. H L F von Helmholtz (1852)—Ueber die Theorie der zusammengesetzten Farben, *Arch. Anat. Physiol. wiss. Med.* *1852 :461–482.*

86. R Wagner & G Meissner (1852)—Ueber das Vorhandensein bisher unbekannter eigenthümlicher Tastkorperchen (Corpuscula tactus) in den Gefühlswärzchen der menschlichen Haut, und über die End-Ausbreitung sensitiver Nerven, *Nachr. Georg-Augusts. Univ. Ges. Wiss. Gottingen* *1852–17–32.*

87. R L K Virchow (1854)—Ueber eine im Gehirn und Rückenmark des Menschen aufgefundene Substanz mit der chemischen Reaction der Cellulose, *Virchow, Archiv* *6 :135–138.*

88. See the revealing review of the book in *Q.J.M.S* *6 :49–58 (1858).*

89. T Nunneley (1858)—On the structure of the retina, *Q.J.M.S.* *6 :217–241.*

90. M J S Schultze (1858)—Ueber die Endingungweise des Hornerven im Labyrinth, *Müller, Archiv* *1858 :343–381.*

91. M J S Schultze (1862)—Untersuchungen über den Bau der Nasenschleimhaut, namentlich die Structur und Endigungsweise der Geruchsnerven bei dem Menschen und den Wirbelthieren, *Abt. naturf. Ges. Halle* *7 :1–100.*

92. M J S Schultze (1866)—Zur Anatomie und Physiologie der Retina, *Arch. mikr. Anat.* *2 :165–174, 175–286.*

93. J J Lister & W Turner (1860)—Some observations on the structure of nerve fibres, *Q.J.M.S.* *8 :29–34.*

94. S Trinchese (1867)—Mémoire sur la terminaison périphérique des nerfs moteurs dans la série animale, *Robin J. Anat.* *4 :485–504.*

95. In *Q.J.M.S.* *16 :44–45 (1868).*

96. J W Hulke (1868)—Notes on the anatomy of the retina of the common porpoise (Phocaena communis), *J. Anat.* *2 :19–25.*

97. J W Hulke (1869)—The histology of the eye, *M.M.J.* *2 :227–252.*

98. S Stricker—*Handbuch der Lehre von den Geweben des Menschen und der Thiere.* (Leipzig: Engelmann, 1869–1872.)

99. In *M.M.J* *4 :286–287 (1870).*

100. S Stricker—*Manual of human and comparative histology.* Translated by H Power. (London: New Sydenham Society, 3 Vols, 1870–73.)

101. L Ranvier (1871)—Recherches sur l'histologie et la physiologie des nerfs, *Compt. Rend. Acad. Sci.* *73 :1168–1171.*

102. L Ranvier—*Leçons sur l'histologie de système nerveux.* (Paris: Savy, 1878.)

103. B Lewis (1876)—A new process of preparing and staining fresh brain for microscopic examination, *M.M.J.* *16 :105–110.*

104. A Kölliker (1890)—Zur feineren Anatomie des centralen Nervensystems, *Zeits. wiss. Zool.* *49 :663–689.*

105. C Golgi—*Sulla fina anatomia degli organi centrali del sistema nervoso.* (Milano: Hoepli, 1886.)

106. S Ramon y Cajal (1892)—Nuevo concepto de la histologia de los centros nerviosos, *Rev. Cienc. méd.* *18 :457–476.*

107. S Ramon y Cajal—*Die Retina der Wirbelthiere.* (Wiesbaden: Bergmann, 1896.)

108. J W Barrett (1886)—The preparation of the eye for histological examination, *Q.J.M.S.* *26 :607–621.*

109. H M Bernard—Studies in the retina: rods and cones in the frog and in some other Amphibia, *Q.J.M.S.* *43 :23–47 (1900), 44 :443–468 (1901), 46 :27–75 (1902), 47 :303–362 (1903).*

110. C Grabower (1902)—Ueber Nervenendigungen im menschlichen Muskel, *Arch. mikr. Anat.* *60 :1–16.*

111. H Joseph (1902)—Untersuchungen über die Stutzsubstanzen des Nervensystems, nebst Erorternungen über deren histogenetische und phylogenetische Deutung, *Arb. Zool. Inst. Wien 13 :335–400.*

112. A W Campbell—*Histological studies on the localisation of cerebral function.* (Cambridge: Univ. Press, 1905.)

113. G V Ciaccio (1863)—On the anatomy of nerve-fibres and cells, and the ultimate distribution of nerve-fibres. Three demonstrations delivered by Professor Lionel S Beale, *Q.J.M.S. 11 :97–105.*

114. L S Beale (1860)—On the distribution of nerves to the elementary fibres of striped muscle, *Phil. Trans. R.S. 1860 :611–618.*

115. L S Beale (1862)—Further observations on the distribution of nerves to the elementary fibres of striped muscles, *Phil. Trans. R.S. 1862 :889–910.*

116. G Thin (1879)—On hyaline cartilage and deceptive appearances produced by reagents, as observed in the examination of a cartilaginous tumour of the lower jaw, *R.S. Proc. 28 :257–260.*

117. J B Haycraft (1881)—Upon the cause of the striation of voluntary muscular tissue, *R.S. Proc. 31 :360–379.*

118. E M Nelson (1891)—Striped muscle fibre of pig, *J.Q.M.C. 11 :1–3.*

119. C Collingwood—*The influence of the microscope upon the progressive advance of medicine.* (Liverpool: Greenwood, 1859.)

120. S T Taylor—*The diary of a medical student during the Victorian period 1860–1864.* (Norwich: Jarrold, 1927.)

121. H H Bellot—*University College London 1826–1926.* (London: Univ. London Press, 1929.) See pp. 166–7.

122. ibid., pp. 314–319.

123. J Burdon-Sanderson (Ed.)—*Handbook for the physiological laboratory.* (London: Churchill, 1873.) Histology was done by Klein, and took about one quarter of the book and $\frac{3}{5}$ of the plates in vol. 2.

124. W Rutherford (1872)—Notes of a course of practical histology for medical students, *Q.J.M.S. 20 :1–21.*

125. H S Atkinson (1873)—The preparation of the brain and spinal cord for microscopical examination, *M.M.J. 10 :27–32.*

126. See, for example, reports of the meetings of the Society in *Q.J.M.S. 21 :303–308 (1873).*

127. A convenient account is in *M.M.J. 10 :266–267 (1873).*

128. See the very brief mention in *M.M.J. 13 :258 (1875).*

129. An amusing account is given in *M.M.J. 16 :206 (1876).*

130. R H Shryock—*The unique influence of the Johns Hopkins University on American medicine.* (Copenhagen: Munksgaard, 1953.)

131. C H Stowell (1883)—Microscopy in the university of Michigan, *Microscope 3 :63–68.*

132. E C Hoyt (1890)—Microscopical education in America, *Microscope 10 :75–81.*

133. J D Cox (1894)—A plea for systematic instruction in the technique of the microscope at the university, *Proc. Amer. Mic. Soc. 15 :1–16.*

134. V A Latham (1896)—What is the best method of teaching microscopical science in medical schools? *Proc. Amer. Mic. Soc. 18 :311–320.*

135. A Flexner—*Medical education in the United States and Canada.* (New York: Carnegie Foundation, 1910.)

136. W Betz (1873)—Die Untersuchungsmethode des Central-nerven-Systems des Menschen, *Arch. mikr. Anat. 9 :101–117.*

137. W G F Waldeyer—*Wie soll man Anatomie lehren und lernen? Rede, gehalten zur Feier des Stiftungstages der militärärzlichen Bildungsanstalten am 2. August 1884.* (Berlin: Hirschwald, 1884.)

138. Flexner, op. cit. note 7 above. See pp. 75–79.

Appendix

First Use of Synthetic Dyes in History: a Chronological Summary

This summary is offered as a quick reference list of first user and date used. The usefulness of this list is restricted because, as explained in the text, first use of any particular stain does not mean that it was thereby established, and also because use in a particular formula, mixture, or procedure is often rather more significant.

Perkin's mauve	1862	Beneke
Paris blue	1863	Waldeyer
Aniline blue	1863	Frey
Basic fuchsin	1863	Waldeyer, Frey, Roberts
Picric acid	1863	Roberts
Indigocarmine	1864	Chrzonszczewsky
Cyanine	1874	Ranvier
Iodine violet	1874	Zuppinger, Huguenin
Alizarin	1874	Lieberkuhn
Methyl violet	1875	Cornil
Eosin	1875	Fischer
Safranin	1877	Ehrlich
Methyl green	1877	Calberla
Bismarck brown	1878	Weigert
Methyl blue	1879	Ehrlich
Nigrosin	1879	Ehrlich
Aurantia	1879	Ehrlich
Tropaeolin	1879	Ehrlich
Acid fuchsin	1879	Ehrlich
Orange G	1879	Ehrlich
Bordeaux	1879	Ehrlich
Methylene blue	1880	Ehrlich
Biebrich scarlet	1880	Schwarze
Iodine green	1881	Stirling, Richardson
Magdala red	1881	Flemming
Chrysoidin	1883	Griesbach

Malachite green	1884	Beneden and Julin
Erythrosin	1885	Gierke
Thionin	1885	Ehrlich
Light green	1886	Griesbach
Congo red	1886	Griesbach
Benzopurpurin	1886	Griesbach
Toluidine blue	1890	Hoyer
Neutral red	1893	Ehrlich
Sudan III	1896	Daddi
Janus green	1898	Ehrlich
Sudan IV	1901	Michaelis
Azocarmine	1905	Heidenhain
Nile blue	1908	Smith
Trypan blue	1909	Goldmann

Index

45 = text-figure   **3 IV** = half-tone plate

Index of Subjects

ACETIC ACID, 60, 61
Acetic-alcohol, 61
Acetic carmine, 68
Acetone, 64
Acid fuchsin, 71
Adams' microtome, 20, 21, **6**
Adhesives for sections, 94
Alcohol, for dehydration, 64
    for fixation, 61
Alcoholic borax carmine, 68
Alizarin, 70
Altmann's fluid, 62, 63
Alum, 69
Ammonia carmine, 68
Amniotic fluid, 59
Aniline blue, 70
Aniline blue/picric acid, 71
Aniline green, 72
Aniline oil, for clearing, 83
    for stains, 71
Army Medical Museum microtome, *213*
Asphalt varnish, 94
Automatic tissue processor, *279, 280*
Automatic advance microtome, first, 263

BACTERIA, techniques for, 38, 39, 44, 71
Bardeen's microtome, *208*
Basic fuchsin, 70
Bausch & Lomb's microtomes, 216, *228*
Beck's microtomes, 50, *128, 224, 230*, **29**
Beck-Becker microtome, *236*
Becker microtome, *244*
Bell's cement, 94
Benzidine dyes, 76
Benzole, 83
Betz microtome, 215
Bismarck brown, 71, 76
Block trimmer, *260*, 270, *271*
Blood, stains for, 72
Boecker's microtome, 203, *222*
Borax carmine, 68
Borrmann's rack, *277*, 279
Botterill's trough, **24**

Bouin's fluid, 62
Bowerbank's microtome, 118
Brandt's microtome, 240
Brasil's fixative, 61
Broek's microtome, *239*
Bruce's microtome, 206, *235*
Brunswick black, 94
Bulloch's microtome, 230

CAJEPUT OIL, 83
Calcium chloride mountant, 92
Caldwell's automatic microtome, 50, 263,
    264, **42**
Cambridge automatic microtome, *264*, **41**
Cambridge flat-cutting microtome, 259,
    261
Cambridge lever-operated microtome,
    240
Cambridge rocking microtome, 250, *251*,
    39
Canada balsam, molten, 90, **II**
    in solution, 90–91
Canada balsam mounting, first use, 88
Canada balsam mounts, 88, **II**
Capanema's microtome, *121*
Capillary-tube mounts, 10, 18
Carbon dioxide, for freezing microtome,
    206
    for narcosis, 59
Carmalum, 68
Carmine, 67
Carnoy's fluid, 61
Cathcart microtome, *202*, **35**
Cathcart–Darlaston microtome, 202, *203*,
    34
Cavity slide, 112, **24**
Cedarwood oil, 83
Cell mounts, 8, 24, 92, **21**
Celloidin infiltration, 81
Celloidin sections, 41
Cellosolve, 64
Cements, 94
Charring, 69

Chlor-zinc-iodide, 61
Chloral hydrate, 59, 93
Chloroform, 59, 80
Chromium trioxide (chromic acid), 59, 60
Chromosomes, techniques for, 60, 61
Clarke's fluid, 61
Clearing agents, 20, 25, 82, **17, 18**
Clearing agents, introduction of, 9, 11, 14, 82
Clove oil, 80, 83
Cocaine, 59
Cochineal, 67
Cole's microtome, *212*
Collodion, *see* celloidin
Compressorium, 114, **25**
Congo red, 71
Coplin jar, 268, *269*
Corrosion methods, 88
Cover glass, 14, 23, 25, 26, 113
Cover glass fixation, 62
Creosote, 80, 83, 93
Currey's microtome, *122*
Curtis microtome, 43, *218*
Custance's microtome, 16, *17*
Cyanin, 70

DALE'S MICROTOME, 206
Dammar mountant, 91
Dark-ground illumination, 8, 10
Dark well, 115, **26**
Decalcification, 80
Deecke's microtome, 216
De Groot microtome, 265, *266*
Dehydration, 64
Delafield's haematoxylin, 69
Delepine's microtome, *205*
Diatoms, 12, 25
  arranged mounts, 294, **48**
Differentiation, 71
Diffusion, 63
Dioxan, 64
Double embedding, 82
Double staining, 68
Dry-cell mounts, 22, 94, **23**

EHRLICH'S HAEMATOXYLIN, 69
Embedding, 41, 76, 134
  in carrot, 41, 77, 330
  in cheese, 134
  in copal varnish, 79
  in cork, 134
  in egg albumen, 78
  in elder pith, 41, 77, 134, 209
  in gelatin, 134

in glue, 78
in glycerine, 78
in gum/sugar, 79
in paraffin wax, 41, 77, 80, 134, 330
in pulped paper, 79
in soap, 78
in sodium stearate, 78
in shellac, 79
in stearine, 78
in tinfoil, 79
in white wax, 78
in waxy liver, 76
in wood, 73
Embedding baths, 271, *272, 273*
  vacuum, *275, 276*
Embedding moulds, *268*
Eosin, 72
Eosin-methylene blue stains, 72
Erlicki's mixture, 61
Erythrocytes, 309–10
Ether, 80, 138
Ethyl chloride, 206
Euparal, 91

FEARNLEY'S MICROTOME, 141, *142*, **32**
Fiori's microtome, 210, *237*
Fish plate, 19, 114, **25**
Fish tube, **26**
Fisher's microtome, 216, *217*
Fixing agents, 39, 63
  action of, 63–4
  introduction of, 13, 59
  Acetic acid, 60, 61
  Alcohol, 61
  Altmann's fluid, 62, 63
  Bouin's fluid, 62
  Carnoy's fluid, 61
  Chromium trioxide (chromic acid), 59, 60
  Clarke's fluid, 61
  Erlicki's mixture, 61
  Flemming's mixtures, 61, 63
  Formaldehyde, 62
  Heat, 59, 62
  Iodine, 62, 69
  Kleinenberg's mixture, 62
  Mercuric chloride, 60, 75
  Muller's mixture, 60
  Nitric acid, 62
  Osmic acid (osmium tetroxide), 61, 62, 321
  Perenyi's fluid, 62
  Picric acid, 62
  Platinum chloride, 62

Potassium dichromate, 61, 75
Regaud's fluid, 61
Susa mixture, 61
Zenker's fixative, 61
Flattening sections, 81
Flatters microtome, 216, *217*
Flatters & Garnett microtome, *217*
Fleming microtome, 137
Flemming's fixing mixtures, 61, 63
Flemming triple stain for cytology, 72
Fluids for cell mounts, 23, 24
Fluid mounts, *see* liquid mounts
Follin microtome, 48, 127, *129*
Forceps, **26**
Formaldehyde, 62
Francotte's microtome, *224*
Freehand sectioning, 116–17
Freeze drying, 63, 64
Freezing microtomes, 41, 79, 135, 137
Fromme's microtomes, *225, 258*

GIBBON'S MICROTOME, 124, *125*
Giemsa stain, 72
Gieson, Van, stain, 72
Globules, 27, **16**
Glycerine, 93
Glycerine jelly, 93
Goadby's fluid, 93
Gold chloride, 74
Gold size, 94
Golgi bodies, 73
Golgi's methods, 75
Gram's stain, 71
Gray's microtome, 203
Grenacher's carmine, 68
Groves microtome, 141, *142*
Gudden's microtome, 215
Gum-glycerine freezing medium, 203
Gum solution for freezing microtome, 138

HAEMATOXYLIN, 67, 69
Haematoxylin, ripening, 70
Haematoxylin/aniline blue, 71
Haematoxylin/eosin, 71
Hailes microtome, 212
Hailes polymicrotome, *214*
Hand microtomes, 209, **33**
Hansen's microtome, 247
Hardening agents, 60
  introduction of, 13
Hart's microtome, 265, *266*
Hawksley's microtome, *133*
Hayes microtome, *204*
Heat, as fixing agent, 59, 62

Heidenhain's iron haematoxylin, 70
Hensen's microtome, 129, *130*
Hett's preparations, 291, 292, **21**
Hewson's preparations, 22–3, **8, 9, 10, 11, 12, 13**
Hildebrand microtome, 246
Hill's microtome, 12, *15*, **2, 3, 4**
His microtome, 129, *131*
Histochemistry, 24, 52, 95
Histology, beginnings of, 309, **14, 15, 16**
  errors in, 326
  first use of word, 308
  German pre-eminence, 312–13, 332
  and medical education, 328
  practical, introduction of, 311, 330
Histology of motor end plates, 321, 326
Histology of nerves and sense organs, 311, 312, 318, 321, 323
Histology of retina, 131, *314*, **316**, *317*, 319, 320, *322*, 323, *324*, 325
Histology of spinal cord, 319, 330
Hoggan's microtome, *219*
Holman's life slide, **24**
Holman's siphon slide, **24**
Hughes' microtome, *138*
Hunt's microtome, 133
Hyrtl's injected preparations, 85, 315, **20**

INDIFFERENT LIQUIDS, 59
Indigo, 68
Indigocarmine, 70
Infiltration, 76, 218
  with paraffin, 80
Injection of fixing agents, 63
Injection media, 10, 84
Injection techniques, 10, 41, 47, 84, **20, 21, 22, II, III, IV**
Insect stage, **26**
Intra-vitam staining, 75
Iodine, 62
Iodine violet, 70
Isinglass, 18

JACKSON'S MICROTOME, 118
Jacob's microtome, 50, 203
Janus green, 71
Jenner's stain, 72
Jung microtomes, 39, 44, *204*, 233, *234*, *247, 248*
Jung lever-operated microtome ('Tetrander'), *239*

KAISER'S JELLY, 79
Katsch microtome, *216*

King's microtome, 221, *223*
Kleinenberg fixing mixture, 62
Koritska microtomes, 235, 237
Korting microtome, *227*, **37, 38**
Krause microtome, *208*
Krefft microtome, *238*

LATHE USED AS MICROTOME, 265, *267*
Leake's microtome, 259, *260*
Leeuwenhoek microscope, 9, **1**
Leishman's stain, 72
Leitz microtomes, 228, *229*, 237
Lelong microtome, *243*
Lewis's microtome, *139*
Leyser–Brandt microtome, *241*
Light green, 71
Liquid droppers, 270, *271*
Liquid mounts, 22, 24, 51, *22*
Lithium carmine, 68
Live box, **24, 25, 26**
Liver, 60
Loeffler's flagellum stains, 71
Loeffler's polychrome methylene blue, 71
Loewe microtome, 44
Logwood, 67, 69
Logwood, with mordant, 69
Luys microtome, 48, 131, *132*

MACERATION, 13, 85, 88
Malachite green, 71
Malassez microtome, *231*, 232
Mallory's phosphotungstic haematoxylin,
    70
Mallory's triple stain, 72
Malpighian bodies of kidney, 11
Marine glue, 94
Mayer's albumen, 95
McCarthy microtome, 137
Medical Microscopical Society, 330
Mercuric chloride, 60, 75
Metachromatic effects in stains, 70
Methyl blue, 71
Methyl salicylate, 83
Methyl violet, 70, 72
Methylal, 64
Methylene blue, 64, 71, 76
Metallic colouring, 74
Mica, 8, 23, 114
Microdissection, 85, **22**
Microdissector, Schmidt's, 85, *86, 87*
Microscopical Society of London, 4, 111,
    311
Microtome, first, 12
    first automatic, 13, 50

first use of word, 47
Adams, 20, *21*, **6**
Army Medical Museum, *213*
Bardeen, *207*
Bausch & Lomb, 216, *228*
Beck, 50, *128*, *224*, *230*, **29**
Beck–Becker, *230*
Becker, *244*
Betz, 215
Boecker, 203, *222*
Bowerbank, 118
Brandt, 240
Broek, *241*
Bruce, 206, *235*
Bulloch, 230
Caldwell automatic, 50, 263, *264*, **42**
Cambridge automatic, *264*, **41**
Cambridge flat-cutting, 259, *261*
Cambridge lever-operated, *240*
Cambridge rocking, 250, *251*, **39**
Capanema, *121*
Cathcart, *202*, **35**
Cathcart–Darlaston, 202, *203*, **34**
Cole, *212*
Currey, *122*
Curtis, 43, *218*
Custance, 16, *17*
Dale, 206
Deecke, 216
De Groot, 265, *266*
Delepine, *205*
Fearnley, 141, *142*, **32**
Fiori, 210, *237*
Fisher, *120*
Flatters, 216, *217*
Flatters & Garnett, *217*
Fleming, 137
Follin, 48, 127, *129*
Francotte, *224*
Fromme, *225, 258*
Gibbon, 124, *125*
Gray, 203
Groves, 141, *142*
Gudden, 215
Hailes, 212
Hailes polymicrotome, *214*
Hansen, 247
Hart, 265, *266*
Hawksley, *133*
Hayes, *204*
Hensen 129, *130*
Hildebrand, 246
Hill, 12, *15*, **2, 3, 4**
His, 129, *131*

Microtome (*cont.*)
Hoggan, *219*
Hughes, *138*
Hunt, 133
Jackson, 118
Jacobs, 50, 203
Jung, 39, 44, *204*, 233, *234*, *247*, *248*
Jung lever-operated ('Tetrander'), *239*
Katsch, *216*
King, 221, *223*
Koritska, 235, 237
Korting, *227*, **37, 38**
Krause, *208*
Krefft, *238*
Leake, 259, *260*
Leitz, 228, *229*, 237
Lelong, *243*
Lewis, *139*
Leyser–Brandt, *241*
Loewe, 44
Luys, 48, 131, *132*
Malassez, *231*, 232
McCarthy, 137
Miehe, *233*
Minot, 252, *254*, **40**
Minot brain, 256, *257*
Minot improved, 254, **40**
Minot sliding, *236*
Minot–Blake, *256*, **40**
Mouchet, 127
Nachet, 48, *246*
Naples, *248*
Notcutt, *126*
Oschatz, 119, *120*
Osterhaut, *207*
Pearson-Teesdale, 214, *215*
Pfeifer, *253*
Pritchard, 117, **28**
Providence, 221, *222*
Quekett, *119*
Radais, 262, *263*
Ranvier, 42, 44, 49, *122*, 123
Reichenbach, 241
Reichert, 39, 211, *229*, 235, 248, *249*, 258, *259*
Reinhold–Giltay, 256, *257*
Rivet, 49, 131, *132*
Rivet–Leiser, 241, *242*
Ross, *122*
Roy, *214*, 221
Rutherford, 50, *136*
Rutherford improved, *137*, *140*, **31**
Ryder, *261*
Ryder improved, *262*

Satterthwaite & Hunt, *201*
Schacht, 42
Schanze, 39, 50
Schiefferdecker, 209, *231*, *232*
Schmidt, 126, *127*, *128*
Schultze, 228
Seiler, 219, *220*
Smith, 124, *125*
Stirling, 133, *134*, *212*
Strasser, 233, *234*
Swift, 219, *220*
Tait, 137, *138*
Taylor, 141, *201*
Thate, *226*
Thoma, 39, 50, 233, *243*
Topping, 34, 48, *118*, **29**
Triepel, 262, *264*
Van der Stad, *252*
Vinassa, *245*
Walmsley, 248, *249*
Welcker, 43, 44, *123*, *124*
Williams, 140, *143*
Windler, *242*
Yankawer, *226*
Zeiss, 39, 215, 223, **36**
Zimmermann–Minot, *255*
Microtomy, first use of word, 117
Miehe microtome, *233*
Minot brain microtome, 256, *257*
Minot microtome, 252, *254*, **40**
Minot improved microtome, *254*, **40**
Minot sliding microtome, *236*
Minot–Blake microtome, *256*, **40**
Mitonchondria, stains for, 73
Mordant, 14, 66, 69
Mouchet microtome, 127
Mounting media, 88
  high refractive index, 92
  low refractive index, 91
  calcium chloride, 92
  canada balsam, introduction of, 88
  canada balsam, molten, 90
  canada balsam, in solution, 90–1
  dammar, 91
  euparal, 91
  glycerine, 93
  glycerine jelly, 93
  indifferent liquids, 59
  Kaiser's jelly, 79
  naphthalene monobromide, 92, **47**
  sandarac, 91
  styrax, 92
  venetian turpentine, 25
Muller's mixture, 60

NACHET MICROTOMES, 48, *246*
Naphthalene monobromide mountant, 92, **47**
Naples microtome, *248*
Narcotising agents, 59
Natural stains, 67
Neumayer's rack, *278*, 279
Neurone, 69, 74, 75, 76
Neutral red, 71, 76
Nigrosin, 71
Nitric acid, 62
Notcutt's microtome, *126*

OPAQUE MOUNTS, 18, **21**
Origanum oil, 83
Oschatz microtome, 119, *120*
Osmic acid (osmium tetroxide), 61, 321
  regeneration, 62
Osterhaut microtome, *207*

PACINIAN CORPUSCLES;, 319
Paper covers, 89, 299
Paraffin wax infiltration, 80
Paris blue, 70
Pearson–Teesdale microtome, 214, *215*
Penetration of fixing agents, 64
Perenyi's fluid, 62
Perkin's mauve, 70
Pfeifer's microtome, *253*
Phagocytic colouring, 75
Phenol, 83
Picric acid, 62, 68, 70
Picro-carmine, 68
Pigeon post, 292
Platinum chloride, 62
Potassium dichromate, 61
Preparations by
  Amadio, J, 295
  Baker, C, 290, 294, 295, 303, **22, 46**
  Beck, R & J, 293, 301, **21, 45**
  Boecker, H, 295, 301, **47**
  Bolton, T, 301, 303
  Bourgogne, C, 294, **45**
  Bourgogne, E, 294, **45**
  Bourgogne, H, 294, **45**
  Bourgogne, J, 294, **21, 45, III**
  Bourgogne Brothers, 294, 298
  Bourgogne & Alliot, 294
  Chevalier, A, 298
  Clark, L C, **22**
  Clarke, H H, **49**
  Clarke & Page, 93, 304, **22, 49**
  Cogit, E, **47**
  Cole, A C, 295, 298, 301, **21, 46**

Cole, M, 298, **46**
Comber, T, 294
Crouch, H, 296
Custance, 16, 21, 22
Dancer, J B, 291, 292, **44**
Darker, W, **II**
Enock, F, 93, 298, 301, **22**
Essex Institute, 297, **47**
Firth, W, 299
Flatters, A, 304, **49**
Flatters & Garnett, 304, **49**
Flatters, Milborne & McKechnie, 304, **49**
Flogel, **49**
Grayson, H J, 297
Hensoldt, M, **47**
Hett, A, 291, 292, **21**
Hewson, W, 22–3, **8, 9, 10, 11, 12, 13**
Hornell, J, 302, **48**
How, J, 301, **47**
Hunter, J J, **45**
Hunter & Sands, 301, **45**
Hyrtl, J, 85, 315, **20**
Jones, 21
Joshua, 301
Klonne & Muller, **47**
Ladd, W, 291, 293
Macartney, J, **III**
Moller, I D, 295, 298, **47**
Murby, T, **47**
Nachet, 298
Naples (Stazione Zoologica), 301, **47**
Nobert, F A, 296
Norman, J T, 291, 295, 298, **44**
Pacini, F, **III**
Pelletan, J, 300
Peragallo, H, **47**
Pillischer, M, 291, 293
Poulton, C, 291
Pritchard, A, 289, **43**
Quekett, J, 77, 115, **22, III, IV**
Rodig, C, 298, **47**
Russell, 301
Sigmund, F, 304, **49**
Sinel, 301, **48**
Sinel & Hornell, 302, **48**
Smith, 301
Smith & Beck, 291, 293
Smith, Beck & Beck, 293
Stevens, S, 291, 293
Suter, R, 292, **21, 44**
Tempere, J, **47**
Thomas, B W, **49**
Thum, E, 295, **47**

Topping, A, 290, 298, **43**
Topping, C M, 77, 112, 119, 290, 295, 299, **21, 43, II**
Van Heurck, H, **47**
Voigtlander, **49**
Walmsley, W H, **47**
Ward, E, 302, **47**
Watson, W, 299, 300, 303, **48**
Webb, W, 298, **48**
West, F L, 291, 293
West, J, 289, **43**
Wheeler, E, 294, 297, 299, 301, **48**
Willats, T, 290
Ypelaar, A, 22, **7, 19**
Pritchard's microtome, 117
Protoplasm, 63
    first use of word, 313
Providence microtome, 221, *222*
Pyrrhol blue, 76
Purkinje cells, 319

QUEKETT'S FLUID, 93
Quekett's microtome, *119*
Quekett's preparations, 77, 115, **22, III, IV**
Quadruple stain, first, 72
Quinoline blue, 76

RADAIS MICROTOME, 262, *263*
Ranvier microtome, 42, 44, 49, *122*, 123
Raspail's freezing method, 135
Regaud's fluid, 61
Reichenbach microtome, 241
Reichert microtomes, 39, 211, *229*, 235, 248, *249*, 258, *259*
Reinhold–Giltay microtome, 256, *257*
Ribbon of sections, first production of, 81
Ringing cements, 94
Ripening of stains, 72
Rivet microtome, 49, 131, *132*
Rivet–Leiser microtome, 241, *242*
Romanovsky stains, 72
Ross microtome, *122*
Roy microtome, *213, 221*
Royal Microscopical Society, 3
Rutherford microtome, 50, *136*
Rutherford improved microtome, *137*, *140*, **31**
Ryder microtome, *261*
Ryder improved microtome, *262*

SAFFRON, 67
Safranin, 71
Sandalwood oil, 83

Sandarac mountant, 91
Satterthwaite & Hunt microtome, *201*
Schacht microtome, 42
Schanze microtomes, 39, 50
Schiefferdecker microtome, *231, 232*
Schiefferdecker hand microtome, 209
Schmidt's microdissector, *86, 87*
Schmidt's microtome, 126, *127, 128*
Schultze microtome, 228
Section stretcher, *269, 270*
Sectioning, 76, **27**
Seiler microtome, 219, *220*
Serial sections, 11, 77, 81
Serial section adhesives, 94
Sieve-dishes, 65
Silver nitrate, 74
Slides, glass, 18, 26, **I**
Slide, sizes of, 111, 112
Sliders, 18, 26, **5, I**
Slider, first illustration of, *19*
Smears, 9, 20
Smith microtome, 124, *125*
Stage forceps, 19, 115, **26**
Stains (a chronological list of date of introduction is given in the appendix)
    Double staining, first record, 68
    Quadruple staining, first record, 72
    Triple staining, first record, 72
    Acetic carmine, 68
    Acid fuchsin, 71
    Alcoholic borax carmine, 68
    Alizarin, 70
    Ammonia carmine, 68
    Aniline blue, 70
    Aniline blue/picric acid, 71
    Aniline green, 72
    Basic fuchsin, 70
    Bismarck brown, 71, 76
    Blood stains, 72
    Borax carmine, 68
    Carmalum, 68
    Carmine, 67
    Cochineal, 67
    Congo red, 71
    Cyanin, 70
    Delafield's haematoxylin, 69
    Ehrlich's haematoxylin, 69
    Eosin-methylene blue stains, 72
    Flemming triple stain for cytology, 72
    Giemsa stain, 72
    Gieson, van, stain, 72
    Gold chloride, 74
    Gram's stain for bacteria, 71

Stains (*cont.*)
Grenacher's carmine, 68
Haematoxylin, 67, 69
Haematoxylin, ripening, 70
Haematoxylin/aniline blue, 71
Haematoxylin/eosin, 71
Heidenhain's iron haematoxylin, 70
Indigocarmine, 70
Iodine violet, 70
Janus green, 71
Jenner's stain, 72
Leishman's stain, 72
Light green, 71
Lithium carmine, 68
Loeffler's flagellum stains, 71
Loeffler's polychrome methylene blue,
  71
Logwood, 67 69
Logwood, with mordant, 69
Malachite green, 71
Mallory's phosphotungstic haematoxy-
  lin, 70
Mallory's triple stain, 72
Methyl blue, 71
Methyl violet, 70, 72
Methylene blue, 64, 71, 76
Metallic staining, 74
Mitochondrion stains, 73
Natural stains, 67
Neutral red, 71, 76
Nigrosin, 71
Paris blue, 70
Perkin's mauve, 70
Picrid acid, 68, 70
Picro-carmine, 68
Pyrrhol blue, 76
Quinoline blue, 76
Romanovsky stains, 72
Saffron stain, 67
Safranin, 71
Silver nitrate, 74
Sudan III, 71
Sudan IV, 71
Ziehl–Neelsen stain, 71
Staining reactions, 66
Staining, introduction of 11, 14, 67
Starch–iodine reaction, 25, 95
Stilling's frozen sections 135

Stirling microtome, 133, *134, 212*
Strasser microtome, 233, *234*
Strauss–Durckheim 'microtome', 116
Styrax mountant, 92
Sudan III, 71
Sudan IV, 71
Susa mixture, 61
Swift microtome, 219, *220*

TAIT MICROTOME, 137, *138*
Talcs, 18
Taylor microtome, 141, *201*
Temporary mounts, 24
Terpineol, 83
Test objects, 25, 296, 297
Thate microtome, *226*
Thoma microtome, 50, 233, *243*
Thwaite's fluid, 93
Timber, structure of, 12
Topping microtome, 34, 48, 118, **29**
Tissue processor, automatic, *279, 280*
Tricpel microtome, 262, *264*
Triple stain, first, 72
Turntable, 115
Turpentine, 80, 82
Turpentine, venetian, 25, 88

VACUUM EMBEDDING, 79, *275, 276*, 277
Valentin knife, 43, 44, 50, 116, 330, **27**
Van der Stad microtome, *252*
Van Gieson's stain, 72
Vinassa microtome, *245*
Vital staining, 68, 75

WALMSLEY MICROTOME, 248, *249*
Watchglass, 19, *26*
Welcker microtome, 43, 44, *123, 124*
Williams microtome, 140, *141*
Windler microtome, *242*

XANTHOPROTEIC REACTION, 95

YANKAWER MICROTOME, *226*
Ypelaar preparations, 22, **7, 9**

ZENKER'S FIXATIVE, 61
Zeiss microtomes, 39, 215, 223, **36**
Ziehl–Neelsen stain, 71
Zimmermann–Minot microtome, *255*

Index of Names

ABBE, E, 92, 327
Ackerknecht, E H, 315, 317
Acland, H, v, 313
Adams, G (elder), 18
Adams, G (younger), 19, **6**
Addison, W, 310
Ady, J E, 36
Alexander, J, 288
Altmann, R, 62
Amadio, F, 295
Andres, A, 109
Apáthy, S, 45, 74
Ardern, L L, 292
Arendt, G, 280
Arnold, A, 69
Arnold, J, 74
Atkinson, H S, 330
Auerbach, L, 74

BAILIFF, A C M le, 25
Baker, C, 294, 305, 306
Baker, H, 9, 18, 29, 61
Baker, J R, 58, 67, 308
Bale, W M, 297
Balfour, F M, 286
Balfour, I B, 13
Barber, M A, 85
Bardeen, C R, 283
Barrett, C R B, 338
Barrett, J W, 325
Barron, A L E, 7, 292
Barron, S L, 250
Barry, M, 317
Bastian, H C, 74, 83, 108
Beale, L S, 32, 33, 68, 84, 85, 96, 326, 327, 328
Bearn, J G, 6, 29
Beck, A, 235
Beck, R, 33, 90, 127
Beer, J J, 66
Behrens, J W, 284
Behrens, W J, 7, 44, 45
Bellot, H H, 341
Benda, C, 73
Beneden, E van, 68, 102
Beneke, F W, 70
Bennett, J H, 311, 312, 330

Berg, W, 64
Bergh, R S, 74
Bernard, H M, 325
Bernhard, W, 271
Bernthsen, A, 72
Berthold, G, 63
Bettany, G T, 338
Betz, W, 215, 334
Bichat, M F X, 309
Billroth, T, 308
Bird, C H G, 209
Blanchard, E, 60, 92
Blum, F, 62
Boecker, W E, 203, 221
Boeker, J, 286
Boerhaave, H, 11
Bohm, A, 45
Bohmer, F, 69
Bolton, T, 37, 303
Bonanni, F, 18
Bond, A, 89
Boneval, R, 56
Born, C, 109
Borrmann, R, 279
Bostock, J, 337
Bottcher, A, 71
Bouffard, G, 76
Bouin, P, 62
Bowerbank, J S, 89, 111, 118
Bowman, W, 60, 311, 319
Boyd, S, 311
Boyle, R, 97
Boys, T, 91
Bracegirdle, B, 2, 103
Bradbury, S, 2, 306
Brandt, A, 240
Brandt, K, 76
Brasil, L, 61
Bresgen, M, 78
Bristol, C L, 98
Broek, A J P van, 238
Brooke, C, 305
Bruce, A, 206, 235
Budge, A, 107
Bulloch, W, 309
Burdon-Sanderson, J, 329
Busk, G, 7
Butschli, O, 64, 80

CAJAL, *see* Ramon y Cajal
Calberla, E, 78
Caldwell, W H, 81
Campbell, A W, 326
Capanema, G F de, 121
Carazzi, D, 52
Carnoy, J B, 61, 107
Caro, H, 66
Carpenter, W R, 33, 80
Cat, C N le, 11
Cathcart, C W, 202
Certes, M A, 76
Chamberlain, C J, 51
Chambers, R, 288
Cheatle, G L, 65
Chevalier, A, 48, 127
Chevalier, C, 25, 47, 117, 298
Chrzonszczewsky, N, 70, 75
Ciaccio, G V, 341
Cittert, P H van, 28
Clark, C H, 51
Clarke, E, 1, 3, 27, 29, 97, 282, 308
Clarke, J A L, 61, 83, 319
Claubry, H G de, 95
Clay, R S, 2
Cobb, N A, 65
Cohn, F, 68
Cohnheim, J, 74
Cole, A C, 36, 40, 79
Cole, F J, 10, 27, 85
Cole, M J, 36, 40
Colin, J J, 95
Collingwood, C, 328
Colman, W S, 40
Conn, H J, 58
Cooke, M C, 34
Cooper, D, 4, 24, 109, 112, 114, 305
Cooper, J T, 89
Cori, C J, 68, 98
Corner, G W, 309
Cornil, M V, 70
Corti, A, 60, 319
Coulier, M P, 48
Court, T H, 2
Cowles, R P, 286
Cox, J D, 341
Craigie, D, 309
Crisp, F, 7
Crookshank, E M, 38
Cross, M I, 40
Currey, F, 122
Curtis, E, 217
Custance, 16, 21

DADDI, L, 103
Dale, H F, 206
Dale, W, 318
Darwin, H, 250
Davies, T, 34
Davis, G E, 35
De Groot, J G, 287
Deane, H, 93
Deby, J, 3
Deecke, W, 216
Delafield, F, 69
Delepine, S, 37
Derham, W, 27
Dippel, L, 44
Disney, A N, 2
Dobell, C, 28
Dobson, J, 23
Dogiel, J, 71
Donne, A, 47, 315
Duesberg, J, 73
Dujardin, F, 47
Duval, M, 81

EHRENBERG, D C G, 67, 75
Ehrlich, P, 46, 63, 69, 71, 72, 76
Enock, F, 93
Erlicki, A N, 61
Evens, E D, 306
Exner, S, 43

FAIRCHILD, D G, 65
Falconar, M, 23
Farrar, C B, 3
Farrar, W V, 66
Fearnley, W, 38, 141
Feltz, V, 74
Field, H H, 82
Fiori, A, 237
Fischer, A, 55, 67
Fisher, G T, 32, 120
Flatters, A, 41, 51
Fleming, W J, 137
Flemming, W, 44, 60, 61, 62, 63, 68, 72, 78
Flesch, M H J, 92
Flexner, A, 308, 334
Flinzer, M C A, 74
Foettinger, A, 97
Fol, H, 81, 97
Folkes, M, 9
Fontana, F, 12
Ford, W W, 309
Foster, M, 286, 329
Foucault, L, 55

Fouilliand, R, 287
Francotte, P, 52, 106
Fredericq, L, 80
Freeman, W H, 103
Frey, H, 43
Friedlander, C, 44
Frison, E, 2
Fuhrmann, F, 277
Furnrohr, A E, 98, 110

Gage, S H, 50
Galeotti, G, 76
Garbini, A, 52
Garnsey, H E F, 13
Garrison, F H, 308
Gaskell, W H, 81
Gehudten, A van, 97
Georgievics, G C T von, 67
Gerard, R, 56
Gerlach, J von, 68, 112
Gibbes, H, 38, 72
Gibbons, W S, 124
Giemsa, G, 72
Gierke, H, 58, 65, 66
Giesbrecht, W, 80, 95, 109
Gieson, J van, 72
Gilson, G, 61, 82, 91
Girod-Chantrans, J, 24, 95
Gleichen, W F F von, 16, 75
Goadby, H, 60, 93
Goldmann, E E, 76
Golgi, C, 63, 75, 325
Goppert, H R, 67
Goring, C R, 25, 26, 31
Gould, C, 26, 31
Grabower, C, 326
Graham, T, 63
Grainger, T H, 309
Gram, C, 71
Gray, F, 2
Gray, M, 203
Gray, P, 2
Grenacher, H, 68
Grew, N, 10
Griesbach, H A, 72, 102, 107
Griffith, J W, 26, 33, 34, 78, 90, 109, 112
Gronland, J, 132
Groult, P, 285
Groves, J W, 141
Grube, A E, 311
Gudden, B A von, 215
Gueguen, F, 107
Gunter, R, 27
Gurlt, E F, 311

Gurr, E, 70
Guyer, M F, 51

Hager, H, 55
Hailes, H F, 212
Hailes, W, 214
Hamilton, D J, 79
Hannover, A, 60
Hardesty, I, 56
Hardy, W B, 64
Harrison, R G, 288
Hartig, T, 67, 74
Harting, P, 10, 26, 92, 281
Hassall, A H, 338
Hawksley, T, 133
Haycraft, J B, 327
Haymaker, W, 103
Heidenhain, M, 61
Heidenhain, R, 96
Henfrey, A, 34, 78, 107
Henle, F G J, 313
Henneberg, A, 285
Henneguy, L F, 3
Hensen, V von, 129
Hermann, E, 71
Hermann, F, 62
Hertwig, C W T R, 79
Hertwig, W A O, 68
Hewson, W, 22, **8, 9, 10, 11, 12, 13**
Heurck, H van, 52, 92
Heys, W H, 91
Hildebrand, H E, 245
Hill, A, 75
Hill, C F, 6
Hill, J, 12, 67, **2, 3, 4**
Himmelweit, F, 100
His, W, 62, 74, 129, 320
Hodgkin, T, 27, 309
Hoff, H E, 23
Hoffman, F W, 277
Hogg, J, 33
Hoggan, G, 218
Holland, J, 24
Holmyard, E J, 66
Holzner, G, 100
Hooke, R, 8
Hornell, J, 302
Hoyer, H F, 107
Hoyt, E C, 341
Hueppe, F, 44
Hughes, A, 3, 308
Hughes, R, 138
Hulke, J W, 323

Hunt, J G, 133
Hunter, R H, 309
Hyatt, J D, 106
Hyden, H, 319
Hyrtl, J, 85, 88, **20**

INGEN-HOUSZ, J, 23
Ison, C H, 306

JACOBSON, J, 59
James, F L, 51, 286
Jenner, L, 72
Joblot, M, 12
Johnston-Davis, H J, 106
Jordan, H, 83
Joseph, H, 326
Julin, C, 102

KADYI, H, 78
Kaiser, H, 79
Kanmacher, F, 29
Keen, W W, 309
Kehrmann, F, 72
Kent, W S, 62
King, J D, 50, 223
Kirkman, H, 104
Klebs, E, 77, 78, 96
Kleinenberg, N, 62
Knecht, E, 67
Koch, G von, 106
Koch, R, 62
Kölliker, A, 43, 314, 316, 317, 325
Kolossow, A, 274
Koritska, F, 235
Kornauth, K, 270
Korotnev, A, 97
Korting, P, 227
Kosobutskii, V I, 28
Kossel, A C L M L, 45, 110
Krause, C, 74
Krause, R, 55, 206
Krefft, P, 237
Krumbhaar, E B, 309
Kruta, V, 337
Kuchin, K Z, 83
Kuhne, W, 88
Kükenthal, W, 44
Kultschizky, N K, 82

LANG, A, 60
Lankester, E, 7
Latham, V A, 73, 334
Latteux, P, 81, 285
Lavdowsky, M, 93

Lawson, H, 7
Lebert, H, 315, 318
Lebrun, H, 107
Le Cat, C N, 11
Ledermueller, M F, 16
Lee, A B, 38, 57, 72, 76, 84, 107, 109, 117
Leeuwenhoek, A van, 8, 9, 10, 67, 1
Leishman, W B, 72
Leitgeb, H, 96
Levere, T H, 29
Levi, G, 52
Lewis, W B, 140, 325
Lieberkühn, J N, 11
Lieberkühn, N, 70
Liebreich, M E O, 106
Lilienfeld, L, 96
Link, D H F, 95
Lister, J J, 27, 111, 309, 321
Loeffler, F, 71
Long, E R, 309
Löwit, M, 74
Lowy, E, 100
Luys, J, 131
Lyons, R D, 315

MADAN, H G, 108
Maddox, R L, 281
Maggi, L, 52
Major, R H, 309
Malassez, L C, 232
Mallory, F B, 70, 72
Malpighi, M, 10, 11, 60
Mann, G, 41, 63, 64, 65, 66, 67
Marsh, S, 35, 80, 94, 284
Martin, J, 82
Martin, J H, 35, 90
Maschke, O, 68
Materna, L, 277
Mathews, A P, 67, 110
Mayer, A F J K, 336
Mayer, P, 68, 69, 107, 109, 285, 287
Mayzel, W, 60
Meissner, G, 319
Meves, F, 73
Meyer, A W, 309
Meyer, E A, 82
Meyer, E von, 66
Michael, A D, 118
Michaelis, L, 103
Michel, M, 48
Microscopical Society of London, 4
Miescher, J F, 67, 96
Milne-Edwards, H, 26
Minot, C S, 237

Möhl, H von, 26, 93, 95, 112
Moldenhawer, J J P, 13
Moleschott, J, 93
Molisch, H, 96
Moll, J W, 55, 286
Monti, A, 96
Moreau, J L, 98
Mosse, M, 55
Mouchet, M, 127
Muijs, W W, 11
Müller, H, 60, 61, 319
Müller, J, 313
Müller, O F, 12
Muys, *see* Muijs

NAKANISHI, K, 76
Needham, J, 77, 79, 137, 276
Needham, J, 309
Neelsen, F, 71, 107
Nelis, C, 97
Nelson, E M, 8, 297, 328
Neumann, E, 78
Neumayer, L, 287
Newman, C, 308
Nietzki, R, 70
Nissl, F, 73
Nobert, F A, 296
Nocht, B A E, 72
Norman, J T, 291
Notcutt, W L, 34, 126
Nowak, J, 286
Nunneley, T, 320

O'MALLEY, C D, 3, 97, 284, 308
Oppel, A, 45
Orth, J, 68
O'Rahilly, R, 308
Oschatz, A, 93, 112, 119
Osterhout, W J V, 206

PACINI, F, 319
Paget, J, 312
Pal, J, 285
Pappenheim, A, 76
Parat, M, 288
Parker, G H, 99
Pearse, A G E, 96
Pelletan, J, 49, 300
Perenyi, J, 62
Perkin, W H, 70
Perls, M, 96
Petersen, H, 96
Petri, R J, 2
Phin, J, 50

Pickstone, J V, 27
Piersol, G A, 283
Pillischer, M, 293
Pollack, B, 55
Polzam, F, 78
Poole, W H, 71
Poulsen, V A, 56
Pratt, J S, 306
Prenant, A, 287
Pringle, A, 277
Pringle, J, 97
Pritchard, A, 25, 26, 31, 32, 75, 88, 92, 290
Pritchard, U, 138, 140
Prudden, J M, 69
Purkinje, J E, 310, 311, 313, 339

QUEKETT, E, 94
Quekett, J, 26, 32, 49, 50, 69, 77, 79, 89, 111, 115, 118, 119, 281, 293, 312, 313

RABL, C, 62
Ramon y Cajal, S, 52, 75, 325
Ranvier, L, 62, 70, 76
Ranvier, L A, 48, 68, 74, 323
Raspail, F V, 25, 30, 95, 135, 318
Rather, L J, 310
Rawitz, B, 67
Reade, J B, 7, 111
Recklinghausen, F von, 74
Reed, F, 297
Regaud, C, 61, 287
Reichel, G C, 67
Reichenbach, H, 241
Reil, J C, 61
Remak, R, 60, 311, 318
Renaut, J L, 71
Reuter, K, 72
Richards, O W, 286
Richardson, B W, 72, 106
Rindfleisch, E, 107
Rio de Lara, D L, 52
Rivet, G, 131
Roberts, W, 62, 70
Robin, C, 47, 48, 84, 85
Romanovsky, D, 72
Rooseboom, M, 22, 88, 92
Roper, F C S, 3
Rosenhof, R von, 12
Rosenthal, J, 338
Rosin, H, 55
Ross, A, 113
Rothig, P, 55
Rousselet, C F, 97, 293

Roy, C S, 214, 219
Royal Microscopical Society, 3, 4, 5
Royal Society of London, 3
Rudneff, M, 98
Rutherford, W, 35, 135, 136, 137, 329, 331
Ruysch, F, 11
Ryder, J A, 273, 281

SALENKA, E, 78
Salensky, W W, 78
Sandritter, W, 96
Sanzo, L, 278
Saurel, L J, 47
Savory, W S, 310
Schacht, H, 42
Schaffer, J, 225, 270, 286
Schallibaum, H, 95
Schering, A, 82
Schiefferdecker, E F P, 45, 72, 82, 107, 209, 232, 255, 285, 286
Schiller, F, 103
Schleiden, M J, 95, 316
Schmidt, H D, 85, 126
Schouten, S L, 85
Schultze, M J S, 61, 88, 320, 324
Schultze, O, 228
Schulze, F E, 61, 95, 287
Schwann, T, 316
Schwarz, E, 68
Schwarz, F, 63
Schweigger-Seidel, F, 71
Seaman, W H, 58
Sehrwald, E, 273
Seiler, C, 50
Severinghaus, A E, 104
Shadbolt, G, 115, 294
Sharpey, W, 329
Shryock, R H, 341
Singer, C, 28, 309
Smith, A H, 56
Smith, G M, 2
Smith, J, 124
Sollas, W J, 79
Solly, R H, 26, 111
Spence, D S, 306
Spengel, J H, 241, 285
Squire, P W, 40
Stadtmüller, F, 104
Starlinger, J, 286
Steinach, E, 65
Stephenson, J W, 91
Stevens, G W W, 292
Stevenson, J, 78

Stieda, L, 83, 94
Stilling, B, 77, 135, 321
Stirling, A B, 133
Stirling, J F, 292
Stirling, W, 40, 88
Stowell, C H, 332
Strasser, H, 233
Stricker, S, 78, 80, 83, 323
Suchannek, H, 107
Suffolk, W T, 35
Suter, R, 292
Swammerdam, J, 11

TAFANI, A, 72
Tait, L, 69, 137
Taylor, S T, 341
Teesdale, W, 284
Teichmann, L, 107
Tellyesniczky, K, 64
Thiersch, H, 68
Thin, G, 327
Thoma, R F K, 233, 243
Thomas, G B, 29
Thornton, R J, 16
Threlfall, R, 81, 95, 264
Todd, R B, 32
Topping, C M, 77
Trembley, A, 12, 75
Triepel, H, 262
Trinchese, S, 321
Trillat, J A, 62
Trzebiński, S, 63
Tuckwell, W, v
Tulk, A, 107
Turner, G L'E, 2, 6, 29, 306
Turner, W, 321

UHMA, A, 76
Underwood, E A, 309
Unna, P G, 72, 101
U.S. War Department, 3

VALENTIN, G G, 116, 310, 311
Van Cittert, P H, 28
Van Marum, M, 22
Van der Pas, P W, 23
Varley, C, 26, 82, 92
Viallanes, H, 82
Vicq-D'Azyr, F, 61
Vinassa, E, 245
Virchow, R L K, 317, 318, 320
Vogel, J, 95, 318
Vosmaer, G C J, 106

Waddington, H J, 109
Wagner, R, 104, 310, 319
Waldeyer, W G F, 69, 70, 335
Walsem, G C von, 286
Ward, E, 302
Warington, R, 109
Watson-Baker, W E, 6, 40
Weber, C O, 67
Weber, E H, 309
Wedl, C, 318
Weigert, C, 55, 73, 102
Welcker, H, 43, 93, 112, 123, 124
Wells, H G, 298
Wendt, A, 310
Wethered, F J, 40
Wheeler, E, 294, 299
Whitman, C O, 56
Wieger, G, 109

Wilks, S, 338
Williams, J, 140
Willkomm, M, 42
Wilson, J, 115
Wissowzky, A, 103
Witham, H T M, 305
Witt, O N, 66
Wolff, M, 286
Wythes, J H, 49

Yankawer, S, 225

Zacharias, E, 67
Zenker, K, 61
Ziehl, F, 71
Zimmermann, A, 55
Zuppinger, H, 70